NeutroGeometry, NeutroAlgebra, and SuperHyperAlgebra in Today's World

Florentin Smarandache
University of New Mexico, USA

Madeline Al Tahan
Abu Dhabi University, UAE

A volume in the Advances in Computer and Electrical Engineering (ACEE) Book Series

IGI Global
PUBLISHER of TIMELY KNOWLEDGE

Published in the United States of America by
IGI Global
Engineering Science Reference (an imprint of IGI Global)
701 E. Chocolate Avenue
Hershey PA, USA 17033
Tel: 717-533-8845
Fax: 717-533-8661
E-mail: cust@igi-global.com
Web site: http://www.igi-global.com

Library of Congress Cataloging-in-Publication Data

Names: Smarandache, Florentin, editor. | Al-Tahan, Madeleine Ali, 1986- editor.
Title: Neutrogeometry, neutroalgebra, and superhyperalgebra in today's world / Florentin Smarandache, Madeleine Al-Tahan, editors.
Description: Hershey, PA : Engineering Science Reference, [2023] | Includes bibliographical references and index. | Summary: "Because we live in a world where the spaces are not homogeneous, but mixed, complex, even ambiguous, this research book shows many applications where the geometric spaces from our real world may be modeled as NeutroGeometric or AntiGeometric ones, where some axioms are partially, partially indeterminate, and partially false, or even totally false"-- Provided by publisher.
Identifiers: LCCN 2022039960 (print) | LCCN 2022039961 (ebook) | ISBN 9781668447406 (hardcover) | ISBN 9781668447413 (paperback) | ISBN 9781668447420 (ebook)
Subjects: LCSH: Geometry, Non-Euclidean.
Classification: LCC QA685 .N477 2023 (print) | LCC QA685 (ebook) | DDC 516.9--dc23/eng20221209
LC record available at https://lccn.loc.gov/2022039960
LC ebook record available at https://lccn.loc.gov/2022039961

This book is published in the IGI Global book series Advances in Computer and Electrical Engineering (ACEE) (ISSN: 2327-039X; eISSN: 2327-0403)

British Cataloguing in Publication Data
A Cataloguing in Publication record for this book is available from the British Library.

All work contributed to this book is new, previously-unpublished material.
The views expressed in this book are those of the authors, but not necessarily of the publisher.

For electronic access to this publication, please contact: eresources@igi-global.com.

Advances in Computer and Electrical Engineering (ACEE) Book Series

Srikanta Patnaik
SOA University, India

ISSN:2327-039X
EISSN:2327-0403

MISSION

The fields of computer engineering and electrical engineering encompass a broad range of interdisciplinary topics allowing for expansive research developments across multiple fields. Research in these areas continues to develop and become increasingly important as computer and electrical systems have become an integral part of everyday life.

The **Advances in Computer and Electrical Engineering (ACEE) Book Series** aims to publish research on diverse topics pertaining to computer engineering and electrical engineering. **ACEE** encourages scholarly discourse on the latest applications, tools, and methodologies being implemented in the field for the design and development of computer and electrical systems.

COVERAGE

- Microprocessor Design
- Computer Hardware
- Algorithms
- Qualitative Methods
- Programming
- Digital Electronics
- VLSI Fabrication
- Chip Design
- Power Electronics
- Analog Electronics

IGI Global is currently accepting manuscripts for publication within this series. To submit a proposal for a volume in this series, please contact our Acquisition Editors at Acquisitions@igi-global.com or visit: http://www.igi-global.com/publish/.

The Advances in Computer and Electrical Engineering (ACEE) Book Series (ISSN 2327-039X) is published by IGI Global, 701 E. Chocolate Avenue, Hershey, PA 17033-1240, USA, www.igi-global.com. This series is composed of titles available for purchase individually; each title is edited to be contextually exclusive from any other title within the series. For pricing and ordering information please visit http://www.igi-global.com/book-series/advances-computer-electrical-engineering/73675. Postmaster: Send all address changes to above address.

Titles in this Series

For a list of additional titles in this series, please visit: http://www.igi-global.com/book-series/

Energy Systems Design for Low-Power Computing
Rathishchandra Ramachandra Gatti (Sahyadri College of Engineering and Management, India) Chandra Singh (Sahyadri College of Engineering and Management, India) Srividya P. (RV College of Engineering, India) and Sandeep Bhat (Sahyadri College of Engineering and Management, India)
Engineering Science Reference • © 2023 • 389pp • H/C (ISBN: 9781668449745) • US $270.00

Artificial Intelligence Applications in Battery Management Systems and Routing Problems in Electric Vehicles
S. Angalaeswari (Vellore Institute of Technology, India) T. Deepa (Vellore Institute of Technology, India) and L. Ashok Kumar (PSG College of Technology, India)
Engineering Science Reference • © 2023 • 342pp • H/C (ISBN: 9781668466315) • US $260.00

Handbook of Research on AI Methods and Applications in Computer Engineering
Sanaa Kaddoura (Zayed University, UAE)
Engineering Science Reference • © 2023 • 632pp • H/C (ISBN: 9781668469378) • US $335.00

Structural and Functional Aspects of Biocomputing Systems for Data Processing
U. Vignesh (Information Technology Department, Vel Tech Rangarajan Dr. Sagunthala R&D Institute of Science and Technology, Chennai, India) R. Parvathi (Vellore Institute of Technology, India) and Ricardo Goncalves (Department of Electrical and Computer Engineering (DEEC), NOVA School of Science and Technology, NOVA University Lisbon, Portugal)
Engineering Science Reference • © 2023 • 246pp • H/C (ISBN: 9781668465233) • US $270.00

5G Internet of Things and Changing Standards for Computing and Electronic Systems
Augustine O. Nwajana (University of Greenwich, UK)
Engineering Science Reference • © 2022 • 308pp • H/C (ISBN: 9781668438558) • US $250.00

IGI Global
PUBLISHER of TIMELY KNOWLEDGE

701 East Chocolate Avenue, Hershey, PA 17033, USA
Tel: 717-533-8845 x100 • Fax: 717-533-8661
E-Mail: cust@igi-global.com • www.igi-global.com

Table of Contents

Detailed Table of Contents

valued neutrosophic hyperring and hyperideals are examined and established. The proposed concept has been applied in debit and credit card number pattern creation. The conclusion of the current effort also includes recommendations for the future.

NeutroGeometries generalize geometries in the same way that NeutroAlgebras generalize universal and partial algebras. NeutroGeometry is not one kind of classical geometry, but it can be a combination of some of them in the same space. For the first time, this chapter introduces notions of NeutroGeometry for finite geometries. Finite geometries consist of incidence structures where the set of points has finite cardinality. Usually, they are either projective or affine geometries. In this chapter, the author is mainly focused on the definition of the mixed projective-affine geometry (MPA geometry), which is a NeutroGeometry that follows the line of Smarandache's ideas of defining mixed geometries. The properties of the MPA spaces are studied, especially the one related to the satisfaction of parallelism for some lines and the no satisfaction for others. Additionally, other approaches to this theory are introduced, where elements of Neutrosophy are combined with the incidence matrices obtained from finite geometries. The chapter explores the advantages to use MPA in cryptography.

Algebraic hyperstructure is a generalized form of the ordinary algebraic structures. Research on algebraic hyperstructures can focus on the theory and its applications. In this chapter, the discussion will focus on the application of algebraic hyperstructures in chemistry. Algebraic hyperstructures play a role in investigating the types of structures found in a chemical reaction. Furthermore, the concept of algebraic hyperstructures is associated with the concept of NeutroAlgebra so that a new type of structure called NeutroHyperstructures is obtained. In this chapter the types of NeutroHyperstructures contained in some chemical reactions will be investigated along with the application of these NeutroHyperstructures.

Soft and neutrosophic sets are used in this chapter as tools for introducing a multi-valued logic for ethics. The introduction of a multi-valued logic in ethics is not a new idea, but there is not any integrated proposal about this reported in the literature until now. The target is not to add another theory about ethics, but instead to create a new basis enabling a modern approach to the subject. The chapter starts by examining the role of human logic and statistical thinking for the creation and evolution of ethics. A brief historical account of the development of ethics follows with emphasis to the moral dilemmas, the existence of which motivates the application of a multi-valued logic. The basic information about fuzzy sets, fuzzy logic, soft sets, and neutrosophic sets, needed for the understanding of the rest of the chapter, are also presented before using soft and neutrosophic sets for a mathematical representation of the ethical rules and in extension of the moral theories.

Neutro geometric topology is obtained by combining the neutro topology and the neutro geometry. Similarly, the hybridization of anti-topology and the anti-geometry produced anti-geometric topology. Considering this knowledge, in this chapter, neutro geometric topology is studied. Undoubtedly the classical geometric topology provides the foundation for introducing the neutro geometric topology. Also, some basic properties and results associated with neutro geometric topology are obtained and proved. Moreover, to justify the proposed study, some examples are given. The future direction of the proposed study is mentioned at the end of the chapter.

Mixed projective-affine-hyperbolic (MPAH) planes belong to both the branches of NeutroGeometries and mixed or Smarandache geometries. This kind of plane is a geometric structure containing a finite number of points. Some lines of the MPAH satisfy the axioms of projective planes, other lines satisfy the axioms of affine planes, and others satisfy the axioms of hyperbolic planes. Therefore, each of the axioms of parallelism is partially satisfied. This chapter describes version 1.0 of a new software called "NeutroGeometry Laboratory" coded in Python by the author that is used for the calculation and visualization related to MPAH planes. This software is easy to use by users once they know the theory of finite mixed projective-affine-hyperbolic planes, and particularly, it supports the study of this topic and finite planes in general.

NeutroGeometry is the axiomatic approach to geometry from the neutrosophic theory. Neutrosophy is the branch of philosophy that studies neutrality. NeutroGeometry is a geometric structure based on at least one axiom, concept, definition, among others, which is only partially satisfied by the elements of the structure, so they are indeterminate since their definition. In AntiGeometry one of these entities is not satisfied by any element. This chapter introduces the theory and concepts of finite NeutroGeometry, which is geometry based on a finite set of points within the subject of NeutroGeometries in the plane. The author calls them finite mixed projective-affine-hyperbolic planes. Here, finite planes are defined so that lines are divided into three subsets; either they satisfy the axioms of projective planes, the axioms of affine planes, or the axioms of Bolyai-Lobachevsky (also called hyperbolic) planes. They demonstrate the main properties of these planes.

Recently, Smarandache introduced NeutroAlgebra and AntiAlgebras. NeutroAlgebras and AntiAlgebras represent a new research subject that is based on real-world examples. He studied the concept of neutro-structures and anti-structures. From the concepts of NeutroAlgebra, he showed that if a statement (theorem, lemma, consequence, property, etc.) is totally true in a classical Algebra, it does not mean that it is also totally true in a NeutroAlgebra or an AntiAlgebra. It depends on the operations and axioms (if they are totally true, partially true, totally false, or partially or totally indeterminate) it is based upon. In this study, the concept of Neutro-Topological neighbourhood and Neutro-Topological base are studied. Some examples of Neutro-Neighbourhood and Neutro-Base are given and compare the results of classical topological neighbourhood and Neutro-Topological Neighbourhood as well as classical topological base and Neutro-Topological Base.

There is a relation between classical topological space and NeutroTopological space. Every classical topological space generates a NeutroTopological space, and every anti-topological space generates a NeutroTopological space. According to NeutroTopological spaces, NeutroTopological spaces have a broader structure. Thus, the neutrosophic theory's (T, I, F) components are added to traditional topological space, yielding a new structure. Topology is a branch of mathematics that works with particular definitions for spatial structure notions, compares them, and analyses the relationships between the structure and the set's qualities. In topology, the initial step is to define a general definition of the fit, followed by an investigation of the connections between topological structures derived from various methodologies. Finding the number of topologies in a set is challenging work. In this chapter, the formulas for finding the number of NeutroTopological spaces having two and three open sets are defined. Further, formulae for non-NeutroTopological space are defined.

Preface

The Classical Algebraic Structures have all operations totally (100%) well-defined, and all axioms (theorems, concepts, ideas, etc.) that are totally (100%) true.

The NeutroAlgebraic Structures have operations or axioms (and in general: theorems, concepts, ideas, etc.) that are not totally (100%) well-defined or respectively totally (100%) true, but only partially well-defined or partially true (and none of them is 0% well-defined or respectively 0% true as in AntiAlgebraic Structures).

And the AntiAlgebraic Structures have at least one operation or one axiom that is 0% well-defined or respectively 0% true.

Similarly, while the Non-Euclidean Geometries [hyperbolic and elliptic geometries] resulted from the total negation of only one specific axiom (Euclid's Fifth Postulate), the AntiGeometry results from the total negation of any axiom (and in general: theorem, concept, idea, etc.) and even of more axioms (theorem, concept, idea, etc.) and in general from any geometric axiomatic system (Euclid's five postulates, Hilbert's 20 axioms, etc.), and the NeutroAxiom results from the partial negation of one or more axioms (and no total negation of no axiom) from any geometric axiomatic system.

Therefore, the NeutroGeometry and AntiGeometry are respectively alternatives and generalizations of the Non-Euclidean Geometries.

As a particular and hybrid case are the Smarandache Geometries (SG) that are mixed Non-Euclidean geometries, and sometimes partially Non-Euclidean and partially Euclidean. Because the real geometric spaces are not pure but hybrid, and the real rules do not uniformly apply to all space's elements, but they have degrees of diversity – applying to some geometrical concepts (point, line, plane, surface, etc.) in a smaller or bigger degree.

From Prof. Dr. Linfan Mao's arXiv.org paper Pseudo-Manifold Geometries with Applications, Cornell University, New York City, USA, 2006, https://arxiv.org/abs/math/0610307:

A Smarandache geometry is a geometry which has at least one Smarandachely denied axiom (1969), i.e., an axiom behaves in at least two different ways within the same space, i.e., validated and invalided, or only invalided but in multiple distinct ways.

The NeutroAlgebra & AntiAlgebra, afterwards the NeutroGeometry and AntiGeometry, and in general the NeutroStructures and AntiStructures in any field of knowledge were introduced by Smarandache in 2019 – 2020, inspired from our real world since the laws do not equally apply to all citizens, that's why they are called NeutroLaws. Several laws apply in some degree to a category of citizens, and in a different degree to another category. Almost always there are exceptions to the law!

The SuperHyperAlgebra is the most general form of algebras, whose operations are defined from $P^m(H)$ to $P^n(H)$, where $P^m(H)$ is the m-th powerset of the set H, or $P^m(H) = P(P\ldots(P(H))\ldots)$, where P is repeated m times. Similarly, for $P^n(H)$.

ORGANIZATION OF THE BOOK

The book is organized into 10 chapters. A brief description of each of the chapters follows:

Chapter 1 defines a new method to deal with NeutroGeometry in true, false, and Refined NeutroGeometry (T,C,I,F). It presents some real-life application examples in 3D computer graphics, Astrophysics, nanostructure, neutrolaw, neutrogender, neutrocitation and quantum space.

Chapter 2 introduces the concepts of interval-valued neutrosophic hyperring and interval-valued neutrosophic hyperideal. It examines the algebraic properties and structural characteristics of the interval-valued neutrosophic hyperring and. Moreover, it applies the proposed concept in debit and credit card number pattern creation. Finally, discusses some recommendations for the future research.

Chapter 3 focuses on the definition of the Mixed Projective-Affine geometry (MPA geometry), which is a NeutroGeometry that follows the line of Smarandache's ideas of defining mixed geometries. It studies the properties of the MPA spaces, especially the one related to the satisfaction of parallelism for some lines and the no satisfaction for others. Additionally, it introduces other approaches to this theory, where elements of Neutrosophy are combined with the incidence matrices obtained from finite geometries. Finally, it explores the advantages to use MPA in cryptography.

Chapter 4 considers applications of neutrohyperstructures in Chemistry. It considers different chemical reactions to verify the connection between the two concepts.

Chapter 5 starts by examining the role of human logic and statistical thinking for the creation and evolution of ethics. A brief historical account of the development of ethics follows with emphasis to the moral dilemmas, the existence of which motivates the application of a multi-valued logic. The basic information about fuzzy sets, fuzzy logic, soft sets and neutrosophic sets, needed for the understanding of the rest of the chapter, is also presented, before using soft and neutrosophic sets for a mathematical representation of the ethical rules, and in extension of the moral theories.

Chapter 6 studies neutro geometric topology. The latter is obtained by combining the neutro topology and the neutro geometry. Similarly, the hybridization of anti topology and the anti geometry produced anti geometric topology. It presents some basic properties and results associated with neutro geometric topology. Moreover, it presents some examples to justify the proposed study. Finally, it discusses some future research directions.

Chapter 7 describes version 1.0 of a new software called "NeutroGeometry Laboratory" coded in python by the author that is used for the calculation and visualization related to MPAH planes. This software is easy to use by users once they know the theory of finite Mixed Projective-Affine-Hyperbolic planes and particularly it supports the study of this topic and finite planes in general.

Chapter 8 introduces and studies the theory and concepts of finite NeutroGeometry, which is geometry based on a finite set of points within the subject of NeutroGeometries in the plane. The latter is called "Finite Mixed Projective-Affine-Hyperbolic planes". Here, finite planes are defined so that lines are divided into three subsets; either they satisfy the axioms of projective planes, the axioms of affine planes, or the axioms of Bolyai-Lobachevsky (also called Hyperbolic) planes.

Chapter 9 studies the concept of Neutro-Topological neighbourhood and Neutro-Topological base. It presents some examples of Neutro-Neighbourhood and Neutro-Base. Moreover, it compares the results of classical topological neighbourhood and Neutro-Topological Neighbourhood as well as classical topological base and Neutro-Topological Base.

Chapter 10 investigates formulas for finding the number of neutroTopological spaces having two and three open sets.

Florentin Smarandache
University of New Mexico, USA

Madeline Al Tahan
Abu Dhabi University, UAE

Acknowledgment

Our gratitude goes first to IGI for giving us the opportunity to publish this project and for their endless support, our authors, and our reviewers. This project would never be completed without the contributions of our authors and the highly appreciated efforts of our reviewers.

Chapter 1
A Review of AntiGeometry and NeutroGeometry and Their Application to Real Life

Carlos Granados
Universidad de Antioquia, Medellín, Colombia

ABSTRACT

Non-Euclidean geometry is part of AntiGeometry, since it has an axiom that is 100% false. Dealing with the NeutroGeometry in true, false, and uncertain regions is of great interest for researchers. Not too many studies have been done on this topic. For that reason, the aim of this work is to define a new method to deal with NeutroGeometry in true, false, and refined NeutroGeometry (T,C,I,F). Furthermore, some real-life application examples in 3D computer graphics, astrophysics, nanostructure, neutro law, neutrogender, neutrocitation, and quantum space are presented.

INTRODUCTION

Neutrosophy is a new branch of philosophy which was introduced by (Smarandache, 2002) which has been of great interesting of researchers who study different topics (applied or pure science) and it studies the origin, nature and scope of neutralities, as well as their interactions with different ideational spectra: (B) is an idea, proposition, theory, event, concept or entity; anti (B) is the opposite of (B); and (neut-B) means neither (B) nor anti (B), i.e. neutrality between the two extremes (Bal eta l., 2018). Its fundamental theory states that every idea <B> tends to be neutralized, diminished, balanced by <nonB> ideas (not only <antiB> as Hegel). <noB> = what is not <B>, <antiB> = the opposite of <B>, and <neutB> = what is neither

DOI: 10.4018/978-1-6684-4740-6.ch001

<B> nor <antiB>. In their classical form <B>, <neutB>, <antiB> are disjointed two by two. Smarandache (2002) defined fundamental notion of neutrosophic sets in the following way: Let R be a universe and N be a subset of R. An element y of R is written with respect to the set N as y (T, I, F) and belongs to N as follows: t% of true, i% of indeterminacy (unknown) and f% of false, where t belongs to T, i belongs to I and f belongs to F. Statically T, I, F are subsets, but dynamically T, I, F are functions or operators that depend on many known and unknown parameters. Following the idea of neutrosophic theory, many topics have been developed such that neutrosophic topology, neutrosophic normed spaces, neutrosophic probability, neutrosophic probability, decision making, neutroAlgebra, neutroGeometry and so on.

In our real world, spaces are not homogeneous, but mixed, complex, even ambiguous. The elements that populate them and the rules that act on them are not perfect, uniform or complete but fragmented and disparate, with unclear and contradictory information, and are not applied in the same degree to each element. The perfect, idealistic, exist only in the theoretical sciences. We live in a multi-space endowed with a multi-structure (Smarandache, 2021). Neither the elements of space nor the rules that govern them are egalitarian, all of them are characterized by degrees of diversity and variation. Indeterminate data and procedures (vague, unclear, incomplete, unknown, contradictory, ignorance, etc.) surround us.

While Non-Euclidean Geometries result from the total negation of a single specific axiom (Euclid's Fifth Postulate), AntiGeometry results from the total negation of any axiom and even more axioms of any geometric axiom system (Euclid's Five Postulates). Therefore, NeutroGeometry and AntiGeometry are respectively alternatives and generalizations of Non-Euclidean Geometries.

Smarandache (2021) proposed: Let's consider a classical geometry concept, it forms the following geometric neutrosophic triplet:

Concept(1, 0, 0), NeutroConcept(T, I, F), AntiConcept (0, 0, 1).

Where $(T, I, F) \notin \{(1, 0, 0), (0, 0, 1)\}$.

Concept(1, 0, 0) means that the degree of truth of the concept is T = 1, I = 0, F = 0, or the Concept is 100% true, 0% indeterminate, and 0% false in the given geometric space.

NeutroConcept (T, I, F) means that the concept is T% true, I% indeterminate, and 0% false in the given geometric space, with $(T, I, F) \in [0, 1]$, and $(T, I, F) \notin \{(1, 0, 0), (0, 0, 1)\}$.

AntiConcept (0, 0, 1) means that T = 0, I = 0, and F = 1, or the Concept is 0% true, 0% indeterminate, and 100% false in the given geometric space.

Smarandache (2021) went from the neutrosophic triplet (Algebra, NeutroAlgebra, AntiAlgebra) to a similar neutrosophic triplet (Geometry, NeutroGeometry,

AntiGeometry), in the same way. Correspondingly from the algebraic structuires, with respect to the geometries, one has in the classical (Euclidean) Geometry, on a given space, all classical geometric Concepts are 100% true (i.e. true for all elements of the space). While in a NeutroGeometry, on a given space, there is at least one NeutroConcept (and no AntiConcept). In the AntiGeometry, on a given space, there is at least one AntiConcept.

With a view to device a practical tool for inference, Belnap (1977) introduced the notion of a four-valued logic. In his work, corresponding to a certain information he considered four possibilities namely T: True, F: False, none: neither true nor false, and both: both true and false. He symbolized these four truth values as {T, F, both, none}, for more notions derived from this paper, we refer the reader to (Das, et al., 2021; Mohanasundari and Mohana, 2020).

Later on (Smarandache, 2013) Smarandache has generalized Belnap's Logic (True, False, Unknown, and Contradiction), Lukasiewicz' Logic (True, False, and Possible), and Kleene's Logic (True, False, Unknown (or Undefined)) to Refined Neutrosophic Set having any $n \geq 2$ components, where the Truth T was split into Sub-Truths $T_1, T_2, \ldots, T_p$, the Indeterminacy I was split into Sub-Indeterminacies, $I_1, I_2, \ldots, I_r$, and the Falsehood F was split into Sub-Falsehoods $F_1, F_2, \ldots, F_s$, where $p, r, s \geq 0$, are positive integers and at least one of them is ≥ 2, with $p+r+s = n$.

Therefore, he also extended the Fuzzy Set to Refined Fuzzy Set, Intuitioinoistic Fuzzy Set to Refined Intuitionoistic Fuzzy Set, and similarly for other fuzzy extension sets.

In the case of Belnap's Logic, the Indeterminacy was split into two sub-indeterminacies: I_1 = Unknown, and I_2 = Contradiction.

In this work, we use the notions presented by Singh (2022) and Smarandache (2021), to carry out an exhaustive analysis of the NeutroGeometry in the cases in which the indeterminacy is divided into two categories, ignorance and contradiction as is was proposed by Smarandache (2013) in n-refined neutrosophic logic; given that these cases can occur in real life and are first line, in this way, several application examples are presented where this sort of act can occur. Additionally, a method for dealing with NeutroGeometry in true, false and neutrality (or indeterminacy) (T,C,I,F) is presented, where T is true, C is contradiction, I is ignorance and F is false. Throughout the development of this work, neutrogeometry (T,C,I,F) will also be written as neutrogeometry.

BACKGROUND

It is well-known that Euclidian-geometry is one of the oldest disciplines of mathematics. Its origin etymological, gives us a clear idea of the activities to which it appears related in its beginnings. This is how many historians find the roots of this science in ancient Egypt, where the foundations geometry solids by simply introducing the measurement of land with the surveyors, who after the annual floods of the Nile, had the task of rebuilding the boundaries of the lands assigned to the settlers, or also linked to the construction of its famous pyramids According to Salazar (1984) The development of modern geometry was carried out by the mathematician German David Hilbert (1862-1943), who made an analysis of Euclidean-geometry in his work Foundation of geometry (1899), reaching the conclusion that only six primitive concepts (point, line, plane, belongs, congruence and between) and 21 postulates, which lay the solid foundations of geometry, thus becoming a science rational and deductive, of which its components are independent, categorical and enough.

Figure 1. The sum of angles A, B, and C in the given triangle is 180° as per Euclidian geometry (i.e., A+B+C=180°)

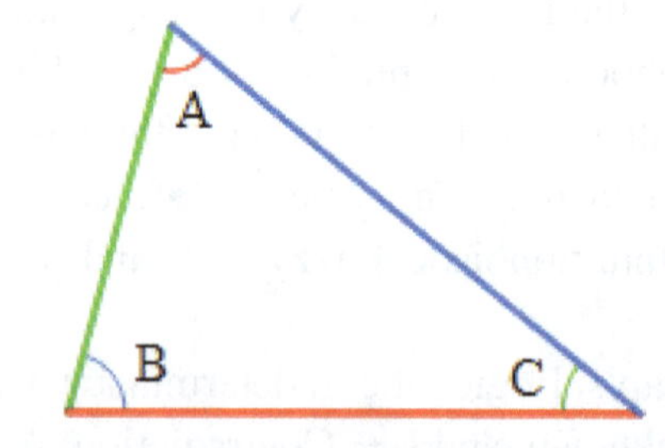

On the other hand, it is called non-Euclidean geometry, to any formal system of geometry were different postulates and propositions in some matter from those established by Euclid in his treatise elements.

Figure 2. An illustrative example of non-Euclidian geometry

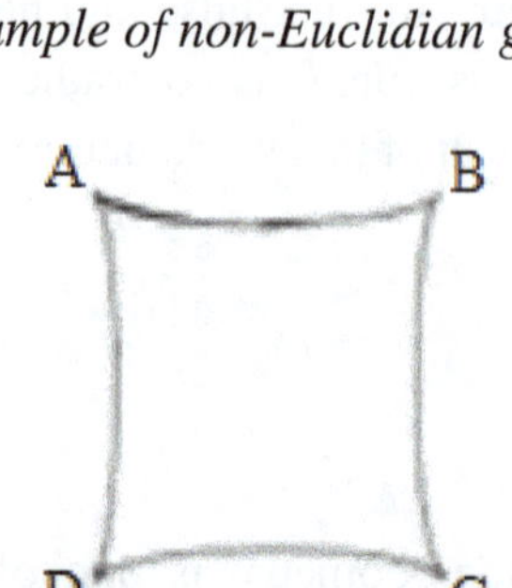

There is not a single system of non-Euclidean geometry, but many, although if the discussion is restricted to homogeneous spaces, in which the curvature of space is the same at each point, in which the points of space are indistinguishable, three formulations of geometries:

1. Euclidean geometry satisfies all five of Euclid's postulates and has zero curvature (i.e., it is assumed to be in flat space so the sum of the three interior angles of a triangle is always 180°).
2. Hyperbolic geometry satisfies only Euclid's first four postulates and has negative curvature (in this geometry, for example, the sum of the three interior angles of a triangle is less than 180°).
3. Elliptic geometry satisfies only Euclid's first four postulates and has positive curvature (in this geometry, for example, the sum of the three interior angles of a triangle is greater than 180°).
4. Spherical geometry is the geometry of the two-dimensional surface of a sphere. It is an example of non-Euclidean geometry. Spherical geometry is the simplest model of elliptical geometry, in which a line has no parallel lines through a given point. In contrast to hyperbolic geometry, in which a line has two parallels, and an infinite number of ultra-parallels, through a given point.

Figure 3. An illustrative example of difference between Euclidian and non-Euclidian geometry

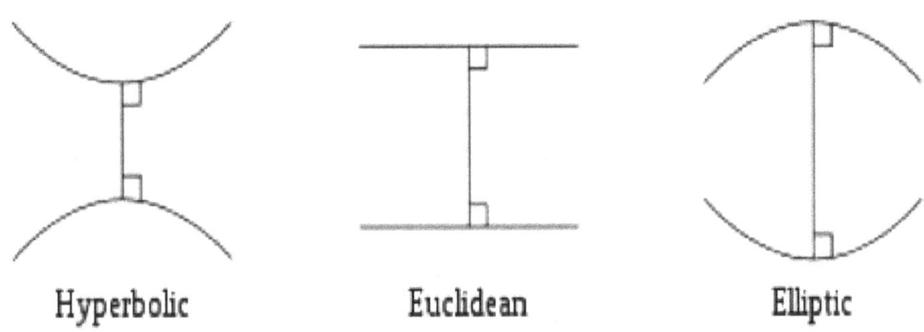

II, III, IV are particular cases of Riemannian geometries, in which the curvature is constant, if the possibility that the intrinsic curvature of the geometry varies from one point to another is allowed, we have a case of general Riemannian geometry, as it happens in the theory of general relativity where gravity causes an inhomogeneous curvature in space-time, the curvature being greater near concentrations of mass, which is perceived as an attractive gravitational field. Smarandache (2021) said that Riemannian geometry, which is called elliptic geometry, is an antigeometry too, since the fifth Euclidean postulate is 100% invalidated in the following antipostulate (second version) place, through a point outside of a line, no parallel can be drawn to that line or (T,I,F)=(0,0,1). Since in this paper indeterminacy factor consists of two divisions namely contradiction (C) and ignorance (I) (Chatterjee et al., 2016), through a point outside of a line, no parallel can be drawn to that line or (T,C,I,F)=(0,0,0,1). This means that for this concept, we form the following geometric neutrosophic triplet:

Concept(1, 0, 0, 0), NeutroConcept(T, C, I, F), AntiConcept (0, 0, 0, 1).

Where (T, C, I, F) $\notin$ {(1, 0, 0, 0), (0, 0, 0, 1)}.

Concept(1, 0, 0, 0) means that the degree of truth of the concept is T = 1, C=0, I = 0, F = 0, or the Concept is 100% true, 0% contradiction, 0% ignorance and 0% false in the given geometric space.

NeutroConcept (T, C, I, F) means that the concept is T% true, C% contradiction, I% ignorance and 0% false in the given geometric space, with (T, C, I, F) $\in$ [0, 1], and (T, C, I, F) $\notin$ {(1, 0, 0,0), (0, 0, 0, 1)}.

AntiConcept (0, 0, 0, 1) means that T = 0, C = 0, I=0 and F = 1, or the Concept is 0% true, 0% contradiction, 0% ignorance and 100% false in the given geometric space.

We go from the algebraic structures, with respect to the geometries, one has in the classical (Euclidean) Geometry, on a given space, and all classical geometric Concepts are 100% true. While in a NeutroGeometry, on a given space, there is at least two NeutroConcept (and no AntiConcept). In the AntiGeometry, on a given space, there is at least one AntiConcept.

How to deal with these sort of phenomenon and characterize them in true, false, or uncertain regions in which these uncertain regions are divided in two parts (contradiction and ignorance) is one of the most crucial tasks. Recently, Singh (2022) presented a method for dealing with one type of indeterminacy, therefore, in the next section, we propose a method to deal with these types of information in neutrogeometry (when indeterminacy is divided in contradiction and ignorance) for multi-decision process and we show some application examples in real life, this method is an extension of the method proposed by (Singh, 2022), but the method proposes in the next section is more general that the method proposed by (Singh, 2022).

METHOD TO DEAL WITH NEUTROGEOMETRY IN TRUE, FALSE, AND INDETERMINACY AND ITS APPLICATION TO REAL LIFE

Step 1. Consider the information with a geometry and its attributes (***B***).

Step 2. Let ***B*** be any non-empty set of a given geometrical information.

Step 3. Define the operator as $\boldsymbol{\varphi}$: B×B→B***m***(B) as (T,C,I,F) $\notin$ {(1,0,0,0),(0,0,0,1)}.

Step 4. In case any mapping is possible, then it can be characterized by:

a. In case for any elements t, u ∈***B*** , the geometry provides a new element in the geometrical space, i.e., t ○ u ⊆ ***B***. It can be considered as true characterization.

b. In case for any elements t, u ∈ ***B***, the geometry provides a new element which does not exist in the geometrical space using the given operator as t ○ u ⊈ ***B***. It can be considered as false regions.

c. In case for any element of t, u ∈ ***B***, the geometry provides a new element which in saddle space and its quantum state is uncertain and it is divided in two unknown parts. This type of element can be considered in neutrogeometry.

Step 5. it defined a function ϑ: T→U, which provides three possibilities:

a. In case a well-defined mapping exists among t and u, then it is called as true regions.

b. In case the mapping is outer-defined mapping between t and u, then it is in false regions.

c. It is unknown whether the mapping exists or not and it thinks but not sure what is the value for the mapping among t and u then the element is in neutrogeometry.

Step 6. In this way, the geometrical space and its characterization can be possible.

Step 7. The similarity among the information sets can be found using the geodesic distance.

Step 8. The geodesic distance provides the shortest path among two neutrogeometric spaces rather than its straight line distance of Euclidean geometry as shown in Figure 4.

Step 9. The information shading to the defined geodesic distance can be considered as cluster for knowledge processing tasks.

Figure 4. The difference between Euclidean and geodesic distance

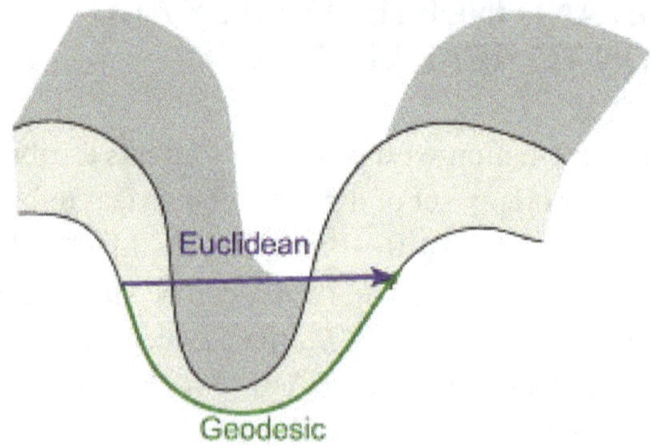

Next, we show some applications with non-Euclidian geometry and NeutroGeometry.

Application 1: Consider there are m non-Euclidean information sets in a given space. Defining the function will take $\boldsymbol{O(m^2)}$ time complexity among them. The characterization of those information sets in neutrogeometry will take maximum $\boldsymbol{O(m^2)}$ time complexity. In this way, the overall time complexity for characterization of non-Euclidean information in true, false, contradiction, ignorance regions may take maximum $\boldsymbol{O(m^3)}$ time complexity.

Figure 5. A NeutroGeometry time complexity information and its visualization

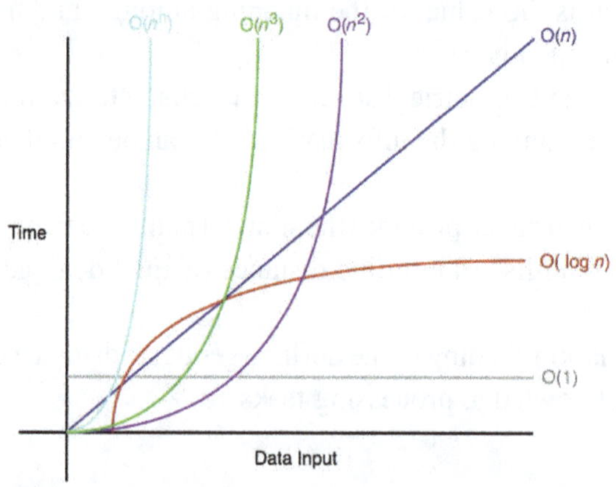

Application 2: Astrophysics is the development and study of physics applied to astronomy. It studies stars, planets, galaxies, black holes and other astronomical objects as physical bodies, including their composition, structure and evolution. Astrophysics uses physics to explain the properties and phenomena of stellar bodies through their laws, formulas and magnitudes. The beginning of astrophysics was possibly in the 19th century when, thanks to the spectra, the physical composition of the stars could be ascertained. Once it was understood that the celestial bodies are composed of the same ones that make up the Earth and that the same laws of physics and chemistry apply to them, astrophysics was born as an application of physics to the phenomena observed by the Earth astronomy. Astrophysics is therefore based on the assumption that the laws of physics and chemistry are universal, that is, that they are the same throughout the universe (Ginzburg, 1979). In this way, Astrophysics is a branch of information with NeutroGeometry. The representation of astronomy and its pattern is based on spherical geometry and its algebra as shown in Figure 6.

Figure 6. A NeutroGeoemtry astrophysics information and its visualization

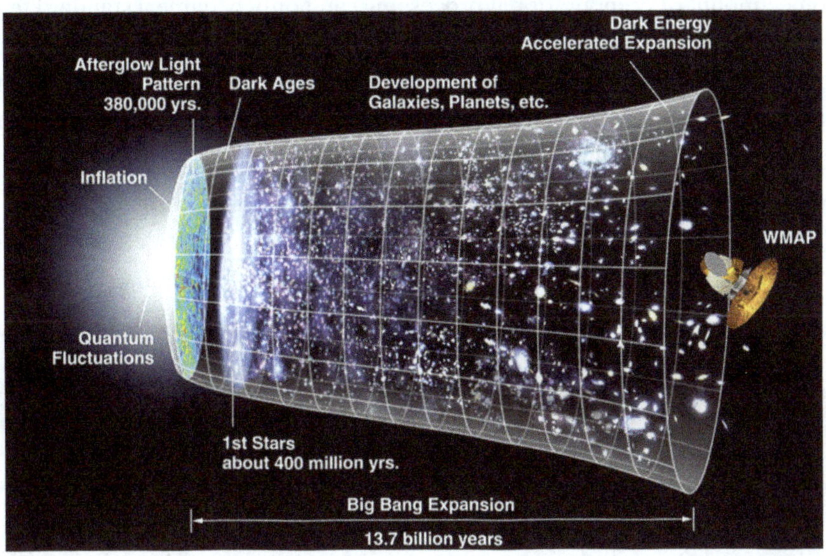

Application 3: A nanostructure is a structure with an intermediate size between molecular and microscopic (micron-sized) structures. Here we are talking about the nanoscale. Generally, these structures experience quantum effects that are not as obvious in larger structures and therefore have special physical properties (Farrow et al., 2007). This case can be represented by Riemannian geometry as can be seen in Figure 7.

Figure 7. A NeutroGeoemtry nanostructure information and its visualization

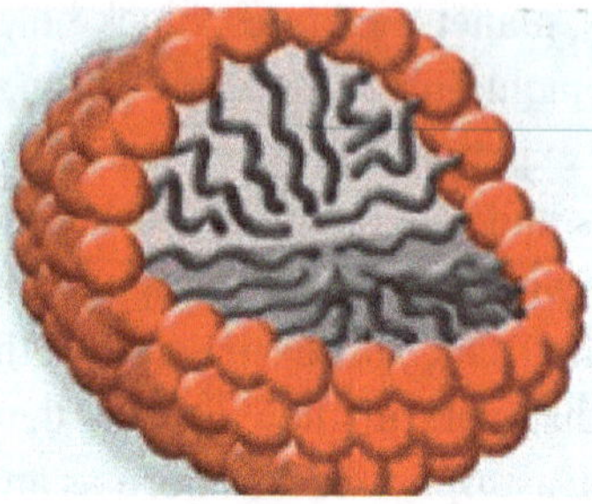

Application 4: In 3D computer graphics (abbreviated CG) is a method of improving the quality of a texture on a surface that is viewed from an oblique angle relative to the projection angle of the image. Texture on a surface, like bilinear filtering and trilinear filtering, anisotropic filtering removes aliasing, but it differs from the previous methods in that it reduces blurring and preserves detail at extreme viewing angles. Anisotropy filtering is relatively heavy (mainly because of memory usage and some amount of computational processing) and only became a standard feature on commercial graphics cards in the late 1990s. Computer graphics is now common in modern boards and can be activated and configured both by the user from the driver configuration, or by graphic applications or video games using programming tools as can be seen in Figure 8.

Figure 8. A NeutroGeoemtry computer graphic information and its visualization

This can be done via steeply angled rather than right angled with respect to the given point which required neutrogeometry. This can be characterized as follows:

1. **True Image (1,0,0,0):** In case the true image is made via enhancing the image which can be represented as (1, 0, 0, 0).
2. **False Image (0,0,0,1):** The image does not provide the true image or provide distinct results can be represented as (0, 0, 0, 1).
3. **NeutroImage (T, C, I, F):** The expert is uncertain about the image and its quality after the enhancement, this means that he/she does not too much about the topic and the does not if the image quality will be as he/she thought.

Application 5: The law in any country is totally uncertain and vague (Singh, 2021; Smarandache, 2021; Kappor and Singh, 2020; Singh, 2022). It depends on hierarchical ordering of citizens and their positional power in the given country which is a neutrogeometric information rather than flat. There are several cases where the same punishment will not be given to each citizen for the same act. This can be defined by:

1. **Law (1, 0, 0, 0):** In case the given law is fully applied on the particular citizen. in this case, the government or court can be considered as unbiased.
2. **AntiLaw (0, 0, 0, 1):** In this case, there is no law defined for the particular act. It used to be observed when a politician or business class people never get punishment under the same law.
3. **NeutroLaw (T,C,I,F):** In this case, the law changes based on person to person, region to region, and, religion to religion. Besides, some laws presented by governments usually contradict what is present in some other decree or do not agree to propose a new law, in turn, there is a high level of ignorance among people, given that they do things out of ignorance and not knowing the law does not exempt them from responsibility. This type of law where partial influence occur by any government or higher authority can be considered as neutro-law. So, the law differs into indeterminacy which is divided in ignorance and contradiction, the hidden pattern in these types of information can be analyzed using neutrogeometry.

Application 6: Neutro-gender law is one of the most suitable examples of neutrogeoemtry information (Singh, 2022) where the law differs based on the gender. This can be characterized as follows:

1. **Women Law (1, 0, 0, 0):** Consider, a woman complains that a man did sexual or mental harassment to her. In this case, the given crime can be accepted immediately without proper proof also.
2. **Men Law (0, 0, 0, 1):** Consider, a man complains because he suffered sexual abuse by a woman or was psychologically violated. In this case, the given crime cannot be accepted right away with providing several proofs also.
3. **NeutroLaw (T,C,I,F):** In case a person who belongs to LGBT community reports about sexual or mental abuse, sometimes nobody body listens, sometimes nobody does not what to do, the laws are not clear for these types of people since there is a contradiction if should be care as a man or as a woman . The law differs for them which shows indeterminacy which is divided in ignorance and contradiction. In this case, the entire information can be considered as uncertain and vague.

Application 7: The characterization of a citation for intellectual measurement cannot be done via flat way like Euclidean geometry (Singh, 2022; Smarandache, 2021). It requires neutrogeometry classification which can be characterized as follows:

1. **Citation (1, 0, 0, 0):** A paper cited by the domain expert, keyword, or methodology matching for the given topic can be considered as relevant citation (1, 0, 0, 0).
2. **Anti-Citation (0, 0, 0, 1):** A paper cited in irrelevant way, a retracted paper citation, a posthumous authors papers citation, same departmental citations beyond the relevant of topic, host conference citation without relevancy, forced citation, and random citation can be considered as Anti-Citation (0, 0, 0, 1).
3. **Neutro-Citation (T,C,I,F):** An article that is self-cited, influenced citations, citations added because peer review was required, articles published in predatory journals, etc. It can be considered a Neutro-Citation (T,C,I,F).

Application 8: Quantum field theory in curved space-time is an extension of standard quantum field theory in which the possibility is contemplated that the space-time through which the field propagates is nevertheless not flat (described by the metric of Minkowski). A generic prediction of this theory is that particles can be generated due to time-dependent gravitational fields, or the presence of horizons Quantum field theory in curved space-time may be required as a first approximation of quantum gravity. The next step consists of a semi classical gravity, in which quantum corrections will be taken into account, due to the presence of matter, on space-time as can be seen in Figure 9. In this way, the traversal criteria do not matter whether you go x steps right and then you go y steps forward and vice versa (Bresar,

Figure 9. A NeutroGeoemtry quantum space information and its visualization

2014). It means the non-commutative geometry cannot be represented precisely which requires fuzzy spherical coordinates.

Application 9: The way in which we consume perishable foods is something that we do not know and we cannot exactly measure whether they are healthy or not, since the companies that produce, mention that they are made under strict health protocols, while some foundations question these methods, since they mention that these are harmful to health. So, it requires non-Euclidean classification which can be characterized as follows:

1. **Health-Food (1, 0, 0, 0):** People who eat healthy without consuming perishable products for the given topic can be considered as relevant Health-food (1, 0, 0, 0).
2. **Anti-Health-Food (0, 0, 0, 1):** People who eat unhealthy food and/or perishable products, knowing how bad these can be for their health. This can be considered as Anti-Health-food (0, 0, 0, 1).
3. **Neutro-Health-Food (T,C,I,F):** People who claim to take care of themselves but eat unhealthy food and people who are unaware of their health status because they don't have the means or because they don't want to. This can be considered a Neutro-Health-food (T,C,I,F).

Application 10: The conservation of the environment is something that has been of great interest and debate of many researchers and non-researchers. In the last decades, the concern for the conservation of the environment has undergone an amazing growth at all levels, and today it must be considered one of the most relevant matters at a scientific, doctrinal and normative level. Indeed, if less than fifty years ago the relationship between human rights and the environment was ignored, today there are numerous binding normative texts that enshrine both the right to a healthy environment and the so-called rights of environmental action, all of which is now preached as necessary to guarantee that present and future generations can develop in a healthy and beneficial environment for human life (García, 2018). But there have always been some entities that have caused a lot of damage to the environment regardless of the consequences. So, it requires neutrogeometric classification which can be characterized as follows:

1. **Environment (1, 0, 0, 0):** People who care for the environment and do not consume the products of companies that affect the well-being of the environment, for the given topic can be considered as relevant Environment (1, 0, 0, 0).
2. **Anti-Health-Food (0, 0, 0, 1):** People who do not care for the environment and consume the products of companies that affect the well-being of the environment, even knowing that this can be harmful to themselves. This can be considered as Anti-environment (0, 0, 0, 1).
3. **Neutro-Environment (T,C,I,F):** People who consume the product of said company and talk about conserving the well-being of the environment and people who are unaware of the reality of the environment because they do not read news about it or are not interested in knowing about the subject, since according to them it is not their convenience. This can be considered a Neutro-environment (T,C,I,F).

CONCLUSION

In this work, we presented a method to deal with NeutroGeometry of type (T,C,I,F). The analysis of this method is showed together with some applications and illustrative examples. For future works, new applications to this method can be introduced and decision-making applications can be presented for a better study and analysis of this topic.

REFERENCES

Bal, M., Shalla, M., & Olgun, N. (2018). Neutrosophic Triplet Cosets and Quotient Groups. *Symmetry, 10*(4), 126. doi:10.3390ym10040126

Belnap, N.D. (1977). A useful four valued logic. *Modern Uses of Multiple Valued Logic*, 9–37.

Boyer, C. (1989). *Historia de la geometría*. Editorial Coimoff.

Bresar, M. (2014). *Introduction to noncommutative algebra*. Springer. doi:10.1007/978-3-319-08693-4

Chatterjee, R., Majumdar, P., & Samanta, S. K. (2016). On some similarity measures and entropy on quadripartitioned single valued neutrosophic sets. *Journal of Intelligent & Fuzzy Systems, 30*(4), 2475–2485. doi:10.3233/IFS-152017

Das, S., Das, R., & Granados, C. (2021). Topology on quadripartitioned neutrosophic sets. *Neutrosophic Sets and Systems, 45*, 54–61.

Farrow, C. L., Juhas, P., Liu, J., Bryndin, D., Božin, E., Bloch, T., Proffen, T., & Billinge, S. (2007). PDFfit2 and PDFgui: Computer programs for studying nanostructure in crystals. *Journal of Physics Condensed Matter, 19*(33), 335219. doi:10.1088/0953-8984/19/33/335219 PMID:21694142

García, E. (2018). El medio ambiente sano: La consolidación de un derecho. *Iuris Tantum Revista Boliviana de Derecho, 25*, 550–569.

Ginzburg, V. L. (1979). *Theoretical Physics and Astrophysics*. Pergamon Press.

Kapoor, P., & Singh, P. (2020). IT act crime pattern analysis using regression and correlation matrix. *Proc. 8th Int. Conf. Reliabil., Infocom Technologies and Optimization*, 1102–1106.

Mohanasundari, M., & Mohana, K. (2020). Quadripartitioned Single valued Neutrosophic Dombi Weighted Aggregation Operators for Multiple Attribute Decision Making. *Neutrosophic Sets and Systems, 32*, 107–122.

Salazar, L. (1984). *Fundamentos de la geometría euclidiana*. Seccional Manizales.

Singh, P. (2021). AntiGeometry and NeutroGeometry characterization of Non-Euclidean data set. *J. Neutrosophic Fuzzy Syst, 1*(2), 24–33. doi:10.54216/JNFS.010102

Singh, P. K. (2022). Data with Non-Euclidean Geometry and Its Characterization. *Journal of Artificial Intelligence and Technology, 2*(1), 3–8.

Smarandache, F. (2002). *Neutrosophy, a new Branch of Philosophy*. Infinite Study.

Smarandache, F. (2013). N-valued refined neutrosophic logic and its applications to physics. *Progress in Physics, 4*, 143-146. http://fs.unm.edu/RefinedNeutrosophicSet.pdf

Smarandache, F. (2021). NeutroGeometry & AntiGeometry are alternatives and generalizations of the Non-Euclidian geometries. *Neutrosophic Sets and Systems, 46*, 456–477.

Chapter 2
New Aspects of Neutrosuperhyper Algebra With Its Application

M. Lathamaheswari
Hindustan Institute of Technology and Science, India

S. Sudha
Hindustan Institute of Technology and Science, India

Said Broumi
https://orcid.org/0000-0002-1334-5759
University Hassan II, Morocco

Florentin Smarandache
https://orcid.org/0000-0002-5560-5926
University of New Mexico, USA

ABSTRACT

Security system is the process of developing, adding, and testing security features within applications to prevent security vulnerabilities against threats such as unauthorized access and modification. The implementation of such secure message passing protocols and the best, most dependable computing algorithms are made possible by hyperideal architecture. In this chapter, the authors introduced the concepts of interval-valued neutrosophic hyperring and interval-valued neutrosophic hyperideal. The algebraic properties and structural characteristics of the interval-valued neutrosophic hyperring and hyperideals are examined and established. The proposed concept has been applied in debit and credit card number pattern creation. The conclusion of the current effort also includes recommendations for the future.

DOI: 10.4018/978-1-6684-4740-6.ch002

INTRODUCTION

The grade of membership in fuzzy sets is often represented as a single real integer in the interval (0,1). The interval-valued fuzzy set model developed by Truksen was used to reduce the uncertainty in the fuzzy set model's grade of membership. Atanassov first developed intuitionistic fuzzy sets, a generalization of fuzzy sets, in 1986. In Bordbar, Cristea, and Novak (2017), this model was the same as interval valued fuzzy sets. The type of information that frequently exists in real-world situations but cannot be handled by intuitionistic fuzzy sets is indeterminate information. To overcome these issue Smarandache created the concept of neutrosophic model. Each premise in neutrosophic logic has a degree of truth (T), an indeterminacy (I), and a falsity (F), where T, I, and F are standard or non-standard subsets of $]^{-}0,1^{+}[$ as might be seen in (Smarandache, F., 2003,2015). Numerous neutrosophic structures, such as neutrosophic groups, neutrosophic rings, neutrosophic modules, neutrosophichypergoups, neutrosophichyperrings, neutrosophic loops, and many more, have been developed since the advent of this theory. (Marty, 1934) first proposed the hyperstructure idea in 1934. (Corsini, 1993) developed the notion of hyperring homomorphism as well as the concept of hyperring and its generic form. The term H_υ-ring, H_υ-subring, and left and right H_υ-ideal of a H_υ-ring were all created by (Vougiousklis, 1995) as expansions of the notions linked to hyperings that Corsini had previously introduced. The triple (R,+,·) is a hyperring in general if and only if the hyperoperations "+" and "•" are such that (R,+) is a hypergroup, (R,•) is a semihypergroup, and "•" is distributive with respect to "+". In essence, these structures are rings with roughly modified axioms. Researchers in the field of algebraic hyperstructures have studied many conception of hyperrings. Krasner, for instance, invented a type of hyperring in (Krasner, 1983) in which "•"is a binary operation and "+" is a hyperoperation.

Many different mathematical theories make extensive use of theory of hypersturctures. In (Rosenfeld, 1971) Rosenfeld initiated the study of fuzzy algebra, and this was later extended to various fuzzy-based models such intuitionistic fuzzy sets, fuzzy soft sets, and ambiguous soft sets. Recent research on fuzzy soft rings and ideals, vague soft groups, and fuzzy soft ring may be found in (Selvachandran & Salleh, 2015, 2016). Fuzzy hyperalgebraic theory was created as a result of research on fuzzy algebra. The idea of a ring's fuzzy ideals, which was first proposed by Liu (1982). The broadening of the fuzzy hyperideal that Davvaz introduced in 1998. Bharathi and Vimala (2016) introduced the idea of the gamma-ideal. The fuzzy gamma-ideal was subsequently expanded by Vimala and Arokiareeta in 2018. Davvaz (2008, 2016) developed the ideology of hyperrings and hyperideals. Al-Tahan et al. (2021) introduced the structure of hyperrings and hyperideals. The hypergroup and hyperring theory for vague soft sets were developed in Wang and

Inthumathi (2021). Abirami and Murugesan introduced intuitionistic fuzzy gamma hypergroup and intuitionistic fuzzy gamma hypermodule as a generalization of the fuzzy hypergroup and the usual fuzzy gamma hypermodule in the year of 2016. Hila and Naka (2017) investigated fuzzy hyperideals via T-fuzzy hyperideals under triangular norm T. Hameed et al. (2022) developed the characterizations of single valued neutrosophic hyperring and hyperideals in terms of algebraic features and structural characteristics. Bordbar et al. (2017) proposed the notion of height of a prime hyperideal in a Krasner hyperring with its desirable properties. Selvachandran and Salleh (2015) extended the concept of algrbraic hyperstructures of vague soft sets associated with hyperrings and hyperideals. As of above research finding there is less contribution papers under neutrosophic hyperring and neutrosophic hyperideal this motivated us to develop the same under interval valued neutrosophic hyperrings and interval-valued neutrosophic hyperideals to fill the research gap. In this paper we develop the theory of interval-valued neutrosophic hyperrings and interval-valued neutrosophic hyperideals. In second section, basic concepts have been given. In third section, interval-valued neutrosophic hyperring has been defined and proven some related theorems. In fourth section, interval-valued neutrosophic hyperideal has been developed with desirable theorems. In fifth section proposed concept has been applied in number creation of debit and credit cards and in last section conclusion of the present work is given and future directions were also discussed.

PRELIMINARIES

In this section, we provided some basic definition related to interval valued neutrosophic sets, neutrosophichyperring, neutrosophichyperideal.

Definition 2.1. (Sahni et al., 2021) Let ‘X’ be any universal set and let ‘A’ be the subset of ‘X’ defined as:

$$A=\left\{\left\langle x:\left[Tr_A^L,Tr_A^U\right],\left[Ind_A^L,Ind_A^U\right],\left[Fal_A^L,Fal_A^U\right]\right\rangle x\in X\right\},$$

where $Tr_A(x)$ denotes truth membership function, $Ind_A(x)$ is the indeterminacy function and $Fal_A(x)$ is the falsity membership function and mapping of the function $Tr_A(x)$, $Ind_A(x), Fal_A(x): X\rightarrow[0,1]$ satisfying the condition $0\leq Tr_A(x)+Ind_A(x)+Fal_A(x)\leq 3$ where

$$Tr_A(x_j) = [Tr_A^L(x_j), Tr_A^U(x_j)],$$
$$Ind_A(x_j) = [Ind_A^L(x_j), Ind_A^U(x_j)],$$
$$Fal_A(x_j) = [Fal_A^L(x_j), Fal_A^U(x_j)] \subseteq [0,1],$$

then it is called an interval valued neutrosophic set (IVNS).

Definition 2.2. (Sahni et al., 2021) An Interval-valued neutrosophic empty set 'A' is empty if and only if inf $Tr_A(x)$=sup $Tr_A(x)$, inf $Ind_A(x)$=sup $Ind_A(x)$=1 and inf $Fal_A(x)$=sup $Fal_A(x)$=0, for all $x \in X$.

Definition 2.3 (Sahni et al., 2021) An IVNS 'A' is contained in the other IVNS 'B', $A \subseteq B$, if and only if

inf $Tr_A(x) \leq$ inf $Tr_B(x)$, sup $Tr_A(x) \leq$ sup $Tr_B(x)$

inf $Ind_A(x) \leq$ inf $Ind_B(x)$, sup $Ind_A(x) \leq$ sup $Ind_B(x)$

inf $Fal_A(x) \leq$ inf $Fal_B(x)$, sup $Fal_A(x) \leq$ sup $Fal_B(x)$ for all $x \in X$.

Definition 2.4. (Sahni et al., 2021) The complement of an IVNS 'A' is denoted by A^c and is defined by

$$Tr_{A^c} = Fal_A(x)$$

$$\inf Ind_{A^c}(x) = 1 - \sup Ind_A(x)$$

$$\sup Ind_{A^c}(x) = 1 - \inf Ind_A(x)$$

$$Fal_{A^c}(x) = Tr_A(x)$$

Definition 2.5. (Sahni et al., 2021) The union of two IVNSs 'A' and 'B' is an IVNS, 'C' can be written as $C = A \cup B$ is defined as follow

inf $Tr_c(x)$ = max{inf $Tr_A(x)$, inf $Tr_B(x)$}

sup $Tr_c(x)$ = max{sup $Tr_A(x)$, sup $Tr_B(x)$}

inf $Ind_c(x)$ = min{inf $Ind_A(x)$, inf $Ind_B(x)$}

sup $Ind_c(x)$ = min{sup $Ind_A(x)$, sup $Ind_B(x)$}

inf $Fal_c(x)$ = min{inf $Fal_A(x)$, inf $Fal_B(x)$}

sup $Fal_c(x)$ = min{sup $Fal_A(x)$, sup $Fal_B(x)$} for all $x \in X$

Definition 2.6. (Sahni et al., 2021) The intersection of two IVNSs 'A' and 'B' is an IVNS, 'C' can be written as $C=A \cap B$ is defined as follow

inf $Tr_c(x)$ = min{inf $Tr_A(x)$, inf $Tr_B(x)$}

sup $Tr_c(x)$ = min{sup $Tr_A(x)$, sup $Tr_B(x)$}

inf $Ind_c(x)$ = max{inf $Ind_A(x)$, inf $Ind_B(x)$}

sup $Ind_c(x)$ = max{sup $Ind_A(x)$, sup $Ind_B(x)$}

inf $Fal_c(x)$ = max{inf $Fal_A(x)$, inf $Fal_B(x)$}

sup $Fal_c(x)$ = max{sup $Fal_A(x)$, sup $Fal_B(x)$}for all $x \in X$.

Definition 2.7. (Sahni et al.,) Let (R, +, •) be any hypering and let I be an indeterminate. The hyperstructure ($R(I)$,+,•) generated by 'R' and 'I', that is, $R(I)$= <$R \cup I$>, For all (a,bI), $(c,dI) \in R(I)$ with $b \neq 0$ or $d \neq 0$, we define

$$(a,bI)\cdot(c,dI)=\{(x,yI):x\in a\cdot c, y\in a\cdot d\cup b\cdot c\cup b\cdot d\}.$$

Definition 2.8. (Sahni et al., 2021) Let ($R(I)$, +, •) be a neutrosophichyperring and let $A(I)$ be a neutrosophicsubhyperring of $R(I)$.

(i) $A(I)$ is called a left neutrosophichyperideal if $(r,sI)\cdot(a,bI)\subseteq A(I)$ for all $(r,sI) \in R(I)$ and $(a,bI) \in A(I)$.
(ii) $A(I)$ is called a right neutrosophichyperideal if $(a,bI)\cdot(r,sI)\subseteq A(I)$ for all $(r,sI) \in R(I)$ and $(a,bI) \in A(I)$.
(iii) $A(I)$ is called a neutrosophichyperideal if $A(I)$ is both left and right neutrosophic hyperideal.

Let 'A' be a IVNS over R. 'A' is called IVNHR over R, if the following conditions are satisfied.

Definition 2.9. (Preethi et al., 2019) Let 'A' be a single-valued neutrosophic set (SVNS) over R. 'A' is called Single-Valued Neutrosophic Hyperring (SVNHR) over R, if the following conditions are satisfied.

(i) $\forall\, x,m \in R$,

$$\min\{Tr_A(m),Tr_A(n)\} \le \inf\{Tr_A(l): l \in m+n\}, \max\{Ind_A(m),Ind_A(n)\} \ge \sup\{Ind_A(l): l \in m+n\}$$

and $\max\{Fal_A(m), Fal_A(n)\} \ge \sup\{Fal_A(l): l \in m+n\}$.

(ii) $\forall\, x,m \in R$, there exists $n \in R$, such that $m \in x+n$ and

$$\min\{Tr_A(x),Tr_A(m)\} \le Tr_A(n), \max\{Ind_A(x),Ind_A(m)\} \ge Ind_A(n)$$

and $\max\{Fal_A(x), Fal_A(m)\} \ge Fal_A(n)$

(iii) $\forall\, x,m \in R$ there exists $l \in R$, such that $m \in l+x$ and

$$\min\{Tr_A(x),Tr_A(m)\} \le Tr_A(l), \max\{Ind_A(x),Ind_A(m)\} \ge Ind_A(l)$$

and $\max\{Fal_A(x), Fal_A(m)\} \ge Fal_A(l)$

(iv) $\forall\, x,m \in R$,

$$\min\{Tr_A(m),Tr_A(n)\} \le \inf\{Tr_A(l): l \in m \circ n\}, \max\{Ind_A(m),Ind_A(n)\} \ge \sup\{Ind_A(l): l \in m \circ n\}$$

and $\max\{Fal_A(m), Fal_A(n)\} \ge \sup Fal_A(l);\ l \in m \circ n$.

Definition 2.10. (Preethi,D., Rajareega, S., Vimala, J.,& Smarandache, F.,2019)

Let 'A' be a SVNS over R. Then 'A' is single-valued neutrosophic left (resp. right) hyperideal over R, if,

(i) $\forall\, x,m \in R$,

$$\min\{Tr_A(m),Tr_A(n)\} \le \inf\{Tr_A(l): l \in m+n\}, \max\{Ind_A(m),Ind_A(n)\} \ge \sup\{Ind_A(l): l \in m+n\}$$

and $\max\{Fal_A(m), Fal_A(n)\} \geq \sup\{Fal_A(l): l \in m+n\}$

(ii) $\forall\, x,m \in R,$ there exists $n \in R$, such that $m \in x+n$ and

$$\min\{Tr_A(x), Tr_A(m)\} \leq Tr_A(n), \max\{Ind_A(x), Ind_A(m)\} \geq Ind_A(n)$$

and $\max\{Fal_A(x), Fal_A(m)\} \geq \sup Fal_A(n)$

(iii) $\forall\, x,m \in R,$ there exists $l \in R$, such that $m \in l+x$ and

$$\min\{Tr_A(x), Tr_A(m)\} \leq Tr_A(l), \max\{Ind_A(x), Ind_A(m)\} \geq Ind_A(l)$$

and $\max\{Fal_A(x), Fal_A(m)\} \geq Fal_A(l)$;

(iv) $\forall\, x,m \in R,$

$$Tr_A(n) \leq \inf\{Tr_A(l): l \in m \circ n\} \quad (resp.\ Tr_A(m) \leq \inf\{Tr_A(l): l \in m \circ n\}),$$

$$Ind_A(n) \geq \sup\{Ind_A(l): l \in m \circ n\} \quad (resp.\ Ind_A(m) \geq \sup\{Ind_A(l): l \in m \circ n\})$$

and

$$Fal_A(n) \geq \sup\{Fal_A(l): l \in m \circ n\} \quad (resp.\ Fal_A(m) \geq \sup\{Tr_A(l): l \in m \circ n\}).$$

'A' is a single-valued neutrosophic left (resp. right) hyperideals of R. From conditions (i),(ii) and (iii) 'A' is a single-valued neutrosophicsubhypergroup of $(R,+)$.

Definition 2.11. (Krishna Praba et al., 2021)

In 2014, Ridvan used the scoring function approach to compare two IVNNs. The IVNS grades are compared using the scoring formula. This function illustrates the idea that IVNS increase in size with increasing value, using this idea, paths can be rated.

Let $IV^U = (Tr^L, Ind^L, Fal^L)$ and $IV^L = (Tr^U, Ind^U, Fal^U)$ be an interval valued neutrosophic number, then the score function

$$s\left(IV^L, IV^U\right) = \left(\frac{1}{4}\right) \times \left[2 + Tr^L + Tr^L - 2Ind^L - 2Ind^U - Fal^L - Fal^U\right]$$

INTERVAL-VALUED NEUTROSOPHIC HYPERRINGS (IVNHR)

In this section we define IVNHR and demonstrated a few related theorems.

Definition 3.1. IVNHR

Let 'A' be a IVNS over R. 'A' is called IVNHR over R, if the following conditions are satisfied.

(i) $\forall\, x,m \in R,$

$$\min\left\{Tr_A^{\ L}(m), Tr_A^{\ L}(n)\right\} \le \inf\left\{Tr_A^{\ L}(l) : l \in m+n\right\}$$
$$\min\left\{Tr_A^{\ U}(m), Tr_A^{\ U}(n)\right\} \le \inf\left\{Tr_A^{\ U}(l) : l \in m+n\right\}$$

$$\max\left\{Ind_A^{\ L}(m), Ind_A^{\ L}(n)\right\} \ge \sup\left\{Ind_A^{\ L}(l) : l \in m+n\right\}$$
$$\max\left\{Ind_A^{\ U}(m), Ind_A^{\ U}(n)\right\} \ge \sup\left\{Ind_A^{\ U}(l) : l \in m+n\right\}$$

$$\max\left\{Fal_A^{\ L}(m), Fal_A^{\ L}(n)\right\} \ge \sup\left\{Fal_A^{\ L}(l) : l \in m+n\right\}$$

and

$$\max\left\{Fal_A^{\ U}(m), Fal_A^{\ U}(n)\right\} \ge \sup\left\{Fal_A^{\ U}(l) : l \in m+n\right\}$$

(ii) $\forall\, x,m \in R$ there exists $n \in R$, such that $m \in x+n$ and

$$\min\left\{Tr_A^{\ L}(x), Tr_A^{\ L}(m)\right\} \le Tr_A^{\ L}(n)$$
$$\min\left\{Tr_A^{\ U}(x), Tr_A^{\ U}(m)\right\} \le Tr_A^{\ U}(n)$$

$$\max\left\{Ind_A^L(x), Ind_A^L(m)\right\} \geq Ind_A^L(n)$$
$$\max\left\{Ind_A^U(x), Ind_A^U(m)\right\} \geq Ind_A^U(n)$$

$$\max\left\{Fal_A^L(x), Fal_A^L(m)\right\} \geq Fal_A^L(n)$$

and

$$\max\left\{Fal_A^U(x), Fal_A^U(m)\right\} \geq Fal_A^U(n)$$

(iii) ∀ *x,m*∈*R* there exists *n*∈*R*, such that *m*∈*l*+*x* and

$$\min\left\{Tr_A^L(x), Tr_A^L(m)\right\} \leq Tr_A^L(n)$$
$$\min\left\{Tr_A^U(x), Tr_A^U(m)\right\} \leq Tr_A^U(n)$$

$$\max\left\{Ind_A^L(x), Ind_A^L(m)\right\} \geq Ind_A^L(n)$$
$$\max\left\{Ind_A^U(x), Ind_A^U(m)\right\} \geq Ind_A^U(n)$$

$$\max\left\{Fal_A^L(x), Fal_A^L(m)\right\} \geq Fal_A^L(n)$$

and

$$\max\left\{Fal_A^U(x), Fal_A^U(m)\right\} \geq Fal_A^U(n)$$

(iv) ∀ *x,m*∈*R*,

$$Tr_A^L(n) \leq \inf\left\{Tr_A^L(l) : l \in m \circ n\right\} \quad \left(resp.\ Tr_A^L(m) \leq \inf\left\{Tr_A^L(l) : l \in m \circ n\right\}\right),$$

$$Tr_A^U(n) \leq \inf\left\{Tr_A^U(l) : l \in m \circ n\right\} \quad \left(resp.\ Tr_A^U(m) \leq \inf\left\{Tr_A^U(l) : l \in m \circ n\right\}\right)$$

$$Ind_A^L(n) \geq \sup\left\{Ind_A^L(l) : l \in m \circ n\right\} \quad \left(resp.\ Ind_A^L(m) \geq \sup\left\{Ind_A^L(l) : l \in m \circ n\right\}\right)$$

$$Ind_A^U(n) \geq \sup\{Ind_A^U(l) : l \in m \circ n\} \quad \left(resp.\ Ind_A^U(m) \geq \sup\{Ind_A^U(l) : l \in m \circ n\}\right)$$

and

$$Fal_A(n) \geq \sup\{Fal_A(l) : l \in m \circ n\} \quad \left(resp.\ Fal_A(m) \geq \sup\{Tr_A(l) : l \in m \circ n\}\right).$$

Example 3.2. IVNS-Subhyperring

A family of t-level sets of IVNSs over R is a subhyperring of R is given below for both upper and lower bound:

$$A_t = \left\{ \begin{array}{l} m \in R : Tr_A^L \geq t, Tr_A^L \geq t \\ \qquad Tr_A^U \geq t, Tr_A^U \geq t \\ \qquad Ind_A^L \geq t, Ind_A^L \geq t \\ \qquad Ind_A^U \geq t, Ind_A^U \geq t \\ \qquad Fal_A^L \geq t, Fal_A^L \geq t \\ \qquad Fal_A^U \geq t, Fal_A^U \geq t \end{array} \right\}, \text{ for all } t \in [0,1].$$

Then 'A' is a single-valued neutrosophic hyperring over R.

Theorem 3.3. 'A' is a IVNS over R. Then 'A' is interval-valued neutrosophic hypering over 'R' if and only if 'A' is interval-valued neutrosophic semi hyper group over $(R,\circ)$ and also a interval-valued neutrosophichypergroupover $(R,+)$.

Proof. From the Definition 3.1 the proof of this theorem is obviously true.

Theorem 3.4. Let 'A' and 'B' be interval-valued neutrosophic hyperrings over R. Then $A \cap B$ is an interval-valued neutrosophichyperring over R if it is non-null.

Proof. Let 'A' and 'B' are interval-valued neutrosophic hyperrings over R, by definition 3.1,

$$A \cap B = \{m, Tr_{A\cap B}^L(m), Tr_{A\cap B}^U(m), Ind_{A\cap B}^L(m), Ind_{A\cap B}^U(m), Fal_{A\cap B}^L(m), Fal_{A\cap B}^U(m), m \in R\},$$

where

$$Tr_{A\cap B}{}^{L}(m)=\min\left(Tr_{A}{}^{L}(m),Tr_{B}{}^{U}(m)\right),$$
$$Ind_{A\cap B}{}^{L}(m)=\max\left(Ind_{A}{}^{L}(m),Ind_{B}{}^{U}(m)\right),.$$
$$Fal_{A\cap B}{}^{L}(m)=\max\left(Fal_{A}{}^{L}(m),Fal_{B}{}^{U}(m)\right)$$

Then for all $m,n\in R$, we have the following observations. We prove all the four conditions for the truth membership terms.

(i)
$$\begin{aligned}
&\min\left(Tr_{A\cap B}{}^{L}(m),Tr_{A\cap B}{}^{L}(n)\right)\\
&=\min\left\{\min\left(Tr_{A}{}^{L}(m),Tr_{B}{}^{L}(m),\min\left(Tr_{A}{}^{L}(n),Tr_{B}{}^{L}(n)\right)\right)\right\}\\
&\leq\min\left\{\min\left(Tr_{A}{}^{L}(m),Tr_{B}{}^{L}(n),\min\left(Tr_{A}{}^{L}(m),Tr_{B}{}^{L}(n)\right)\right)\right\}\\
&\leq\min\left\{\inf\left\{Tr_{A}{}^{L}(l):l\in m+n\right\},\inf\left\{Tr_{B}{}^{L}(l):l\in m+n\right\}\right\}\\
&\leq\inf\left\{\min\left\{Tr_{A}{}^{L}(l),Tr_{B}{}^{L}(l)\right\}:l\in m+n\right\}
\end{aligned}$$

$$\min\left(Tr_{A\cap B}{}^{L}(m),Tr_{A\cap B}{}^{L}(n)\right)=\inf\left\{T_{A\cap B}{}^{L}(l);l\in m+n\right\}$$

$$\begin{aligned}
&\min\left(Tr_{A\cap B}{}^{U}(m),Tr_{A\cap B}{}^{U}(n)\right)\\
&=\min\left\{\min\left(Tr_{A}{}^{U}(m),Tr_{B}{}^{U}(m),\min\left(Tr_{A}{}^{U}(n),Tr_{B}{}^{U}(n)\right)\right)\right\}\\
&\leq\min\left\{\min\left(Tr_{A}{}^{U}(m),Tr_{B}{}^{U}(n),\min\left(Tr_{A}{}^{U}(m),Tr_{B}{}^{U}(n)\right)\right)\right\}\\
&\leq\min\left\{\inf\left\{Tr_{A}{}^{U}(l):l\in m+n\right\},\inf\left\{Tr_{B}{}^{U}(l):l\in m+n\right\}\right\}\\
&\leq\inf\left\{\min\left\{Tr_{A}{}^{U}(l),Tr_{B}{}^{U}(l)\right\}:l\in m+n\right\}
\end{aligned}$$

$$\min\left(Tr_{A\cap B}{}^{U}(m),Tr_{A\cap B}{}^{U}(n)\right)=\inf\left\{Tr_{A\cap B}{}^{U}(l);l\in m+n\right\}$$

Similarly, for indeterminacy and falsity the results as follows:

$$\max\left(Ind_{A\cap B}{}^{L}(m),Ind_{A\cap B}{}^{L}(n)\right)=\sup\left\{Ind_{A\cap B}{}^{L}(l);l\in m+n\right\}$$
$$\max\left(Ind_{A\cap B}{}^{U}(m),Ind_{A\cap B}{}^{U}(n)\right)=\sup\left\{Ind_{A\cap B}{}^{U}(l);l\in m+n\right\}$$

and

$$\max\left(Fal_{A\cap B}{}^{L}(m), Fal_{A\cap B}{}^{L}(n)\right) = \sup\left\{Fal_{A\cap B}{}^{L}(l); l \in m+n\right\}$$
$$\max\left(Fal_{A\cap B}{}^{U}(m), Fal_{A\cap B}{}^{U}(n)\right) = \sup\left\{Fal_{A\cap B}{}^{U}(l); l \in m+n\right\}$$

(ii) $\forall\, x,m \in R$ there exists $n \in R$, such that $m \in x+n$, then it follows that:

$$\min\left(Tr_{A\cap B}{}^{L}(m), Tr_{A\cap B}{}^{L}(n)\right)$$
$$= \min\left\{\min\left(Tr_{A}{}^{L}(m), Tr_{B}{}^{L}(m), \min\left(Tr_{A}{}^{L}(n), Tr_{B}{}^{L}(n)\right)\right)\right\}$$
$$\leq \min\left\{\min\left(Tr_{A}{}^{L}(m), Tr_{B}{}^{L}(n), \min\left(Tr_{A}{}^{L}(m), Tr_{B}{}^{L}(n)\right)\right)\right\}$$
$$\leq \min\left(Tr_{A}{}^{L}(l), Tr_{B}{}^{L}(l)\right)$$

$$\min\left(Tr_{A\cap B}{}^{L}(m), Tr_{A\cap B}{}^{L}(n)\right) = Tr_{A\cap B}{}^{L}(l)$$

$$\min\left(Tr_{A\cap B}{}^{U}(m), Tr_{A\cap B}{}^{U}(n)\right)$$
$$= \min\left\{\min\left(Tr_{A}{}^{U}(m), Tr_{B}{}^{U}(m), \min\left(Tr_{A}{}^{U}(n), Tr_{B}{}^{U}(n)\right)\right)\right\}$$
$$\leq \min\left\{\min\left(Tr_{A}{}^{U}(m), Tr_{B}{}^{U}(n), \min\left(Tr_{A}{}^{U}(m), Tr_{B}{}^{U}(n)\right)\right)\right\}$$
$$\leq \min\left(Tr_{A}{}^{U}(l), Tr_{B}{}^{U}(l)\right)$$

$$\min\left(Tr_{A\cap B}{}^{U}(m), Tr_{A\cap B}{}^{U}(n)\right) = Tr_{A\cap B}{}^{U}(l)$$

Similarly, for indeterminacy and falsity the results as follows:

$$\max\left(Ind_{A\cap B}{}^{L}(m), Ind_{A\cap B}{}^{L}(n)\right) = Ind_{A\cap B}{}^{L}(l)$$
$$\max\left(Ind_{A\cap B}{}^{U}(m), Ind_{A\cap B}{}^{U}(n)\right) = Ind_{A\cap B}{}^{U}(l)$$

$$\max\left(Fal_{A\cap B}{}^{L}(m), Fal_{A\cap B}{}^{L}(n)\right) = Fal_{A\cap B}{}^{L}(l)$$
$$\max\left(Fal_{A\cap B}{}^{U}(m), Fal_{A\cap B}{}^{U}(n)\right) = Fal_{A\cap B}{}^{U}(l)$$

(iii) $\forall\, x,m \in R$, there exists $l \in R$, such that $m \in l+x$ and

$$\min\left\{Tr_{A\cap B}{}^{L}(x), Tr_{A\cap B}{}^{L}(a)\right\} \leq Tr_{A\cap B}{}^{L}(l)$$
$$\min\left\{Tr_{A\cap B}{}^{U}(x), Tr_{A\cap B}{}^{U}(a)\right\} \leq Tr_{A\cap B}{}^{U}(l)$$

$$\max\left\{Ind_{A\cap B}{}^{L}(x), Ind_{A\cap B}{}^{L}(a)\right\} \geq Tr_{A\cap B}{}^{L}(l)$$
$$\max\left\{Ind_{A\cap B}{}^{U}(x), Ind_{A\cap B}{}^{U}(a)\right\} \geq Tr_{A\cap B}{}^{U}(l)$$

$$\max\left\{Fal_{A\cap B}{}^{L}(x), Fal_{A\cap B}{}^{L}(a)\right\} \geq Fal_{A\cap B}{}^{L}(l)$$
$$\max\left\{Fal_{A\cap B}{}^{U}(x), Fal_{A\cap B}{}^{U}(a)\right\} \geq Fal_{A\cap B}{}^{U}(l).$$

(iv) $\forall\, x,m \in R$,

$$\min\left\{Tr_{A\cap B}{}^{L}(m), Tr_{A\cap B}{}^{L}(n)\right\} \leq \inf\left\{Tr_{A\cap B}{}^{L}(l) : l \in m \circ n\right\}$$
$$\min\left\{Tr_{A\cap B}{}^{U}(m), Tr_{A\cap B}{}^{U}(n)\right\} \leq \inf\left\{Tr_{A\cap B}{}^{U}(l) : l \in m \circ n\right\}$$

$$\max\left\{Fal_{A\cap B}{}^{L}(m), Fal_{A\cap B}{}^{L}(n)\right\} \geq \sup\left\{Ind_{A\cap B}{}^{L}(l) : l \in m \circ n\right\}$$
$$\max\left\{Ind_{A\cap B}{}^{U}(m), Ind_{A\cap B}{}^{U}(n)\right\} \geq \sup\left\{Ind_{A\cap B}{}^{U}(l) : l \in m \circ n\right\}$$

$$\max\left\{Fal_{A\cap B}{}^{L}(m), Fal_{A\cap B}{}^{L}(n)\right\} \geq \sup\left\{Fal_{A\cap B}{}^{L}(l) : l \in m \circ n\right\}$$

and

$$\max\left\{Fal_{A\cap B}{}^{U}(m), Fal_{A\cap B}{}^{U}(n)\right\} \geq \sup\left\{Fal_{A\cap B}{}^{U}(l) : l \in m \circ n\right\}$$

Hence, from (i)-(iv) $A \cap B$ is an interval-valued neutrosophichyperring over R.

Theorem 3.5. Let 'A' be an interval-valued neutrosophichyperring over R. Then for every $t \in [0,1]$, $A_t \neq \varnothing$ is a subhyperring over R.

Proof. Let 'A' be a interval-valued neutrosophichyperring over R, $\forall t \in [0,1]$ and $m,n \in A_t$. Then

$$Tr_A^L(m), Tr_B^L(n) \geq t, Tr_A^U(m), Tr_B^U(n) \geq t,$$
$$Ind_A^L(m), Ind_B^L(n) \geq t, Ind_A^U(m), Ind_B^U(n) \geq t,$$
$$Fal_A^L(m), Fal_B^L(n) \geq t, Fal_A^U(m), Fal_B^U(n) \geq t.$$

Since 'A' is a interval-valued neutrosophic sub hypergroup of $(R,+)$, we have the following:

$$\inf\{Tr_A^L(l) : l \in m+n\} \geq \min\{Tr_A^L(m), Tr_A^L(n)\} \geq \min\{t,t\} = t$$
$$\inf\{Tr_A^U(l) : l \in m+n\} \geq \min\{Tr_A^U(m), Tr_A^U(n)\} \geq \min\{t,t\} = t$$

$$\sup\{Ind_A^L(l) : l \in m+n\} \leq t$$
$$\sup\{Ind_A^U(l) : l \in m+n\} \leq t$$

and

$$\sup\{Fal_A^L(l) : l \in m+n\} \leq t$$
$$\sup\{Fal_A^U(l) : l \in m+n\} \leq t$$

This implies that $l \in A_t$ and then for every $l \in m+n$, we obtain $m+n \subseteq A_t$. As such, for every $l \in A_t$, we obtain $l + A_t \subseteq A_t$. Now let $m.l \in A_t$. Then

$$Tr_A^L(m), Tr_A^L(l) \geq t, Tr_A^U(m), Tr_A^U(l) \geq t,$$
$$Ind_A^L(m), Tr_A^L(l) \geq t, Ind_A^U(m), Ind_A^U(l) \geq t,$$
$$Fal_A^L(m), Fal_A^L(l) \geq t, Fal_A^U(m), Fal_A^U(l) \geq t.$$

'A' is a interval-valued neutrosophic subhypergroup of $(R,+)$, there exists $n \in R$ such that $m \in l+n$ and

$$Tr_A^L(n) \geq \min\left(Tr_A^L(m), Tr_A^L(l)\right) \geq t,$$
$$Tr_A^U(n) \geq \min\left(Tr_A^U(m), Tr_A^U(l)\right) \geq t,$$

$$Ind_A^L(n) \geq \max\left(Ind_A^L(m), Ind_A^L(l)\right) \leq t,$$
$$Ind_A^U(n) \geq \max\left(Ind_A^U(m), Ind_A^U(l)\right) \leq t,$$

$$Fal_A^L(n) \geq \max\left(Fal_A^L(m), Fal_A^L(l)\right) \leq t,$$
$$Fal_A^U(n) \geq \max\left(Fal_A^U(m), Fal_A^U(l)\right) \leq t,$$

and this implies $n \in A_t$. Therefore, we obtain $l + A_t \subseteq A_t$. As a result, A_t is superhypergroup of $(R,+)$.

Let $m,n \in A_t$, then

$$Tr_A^L(m), Tr_A^L(n) \geq t, Tr_A^U(m), Tr_A^U(n) \geq t,$$
$$Ind_A^L(m), Ind_A^L(n) \geq t, Ind_A^U(m), Ind_A^U(n) \geq t,$$

and

$$Fal_A^L(m), Fal_A^L(n) \geq t, Fal_A^U(m), Fal_A^U(n) \geq t.$$

Since 'A' is interval-valued neutrosophic subsemihypergroup of $(R,\circ)$,then for all $m,n \in R$, we have the following:

$$\inf\left\{Tr_A^L(l) : l \in m \circ n\right\} \geq \min\left\{Tr_A^L(m), Tr_A^L(n)\right\} = t,$$
$$\inf\left\{Tr_A^U(l) : l \in m \circ n\right\} \geq \min\left\{Tr_A^U(m), Tr_A^U(n)\right\} = t,$$

$$\sup\left\{Ind_A^L(l) : l \in m \circ n\right\} \leq \min\left\{Ind_A^L(m), Ind_A^L(n)\right\} = t,$$
$$\sup\left\{Ind_A^U(l) : l \in m \circ n\right\} \leq \min\left\{Ind_A^U(m), Ind_A^U(n)\right\} = t,$$

$$\sup\left\{Fal_A^L(l) : l \in m \circ n\right\} \leq \min\left\{Fal_A^L(m), Fal_A^L(n)\right\} = t,$$
$$\sup\left\{Fal_A^U(l) : l \in m \circ n\right\} \leq \min\left\{Fal_A^U(m), Fal_A^U(n)\right\} = t.$$

This implies that $l \in A_t$. Therefore, for every $m,n \in A_t$ we obtain $l \in m \circ n \in P^*(R)$. Hence A_t is a subhyperring over R.

Theorem 3.6. Let 'A' be a interval-valued neutrosophic over R. Then the following statements are equivalent:

(i) 'A' is a interval-valued neutrosophic hyperring over R.
(ii) $\forall t \in [0,1]$, a non-empty A_t is sub hyperring over R.

Proof. (i)⇒(ii) $\forall t \in [0,1]$ by Theorem 3.5, A_t is sub hyperring over R.

(ii)⇒(i) Assume that A_t is a subhyperring over R, Let $m,n \in A_t$ and therefore $m+n \subseteq A_{t_0}$. Then for every $l \in m+n$ we have

$$Tr_A^{\ L}(l) \geq t_0, Tr_A^{\ U}(l) \geq t_0, Ind_A^{\ L}(l) \leq t_0, Ind_A^{\ U}(l) \leq t_0, Fal_A^{\ L}(l) \leq t_0, Fal_A^{\ U}(l) \leq t_0$$

which implies that:

$$\min\left(Tr_A^{\ L}(m), Tr_A^{\ L}(n)\right) \leq \inf\left\{Tr_A^{\ L}(l) : l \in m+n\right\},$$
$$\min\left(Tr_A^{\ U}(m), Tr_A^{\ U}(n)\right) \leq \inf\left\{Tr_A^{\ U}(l) : l \in m+n\right\},$$

$$\max\left(Ind_A^{\ L}(m), Tr_A^{\ L}(n)\right) \geq \sup\left\{Ind_A^{\ L}(l) : l \in m+n\right\},$$
$$\max\left(Ind_A^{\ U}(m), Ind_A^{\ U}(n)\right) \leq \sup\left\{Ind_A^{\ U}(l) : l \in m+n\right\},$$

$$\max\left(Fal_A^{\ L}(m), Fal_A^{\ L}(n)\right) \geq \sup\left\{Fal_A^{\ L}(l) : l \in m+n\right\},$$
$$\max\left(Fal_A^{\ U}(m), Fal_A^{\ U}(n)\right) \leq \sup\left\{Fal_A^{\ U}(l) : l \in m+n\right\}.$$

Therefore, condition (i) of definition 3.1 has been verified.

Next, let $\forall\ x,m \in R$, for every $t_1 \in [0,1]$ which means that there exists $n \in A_t$, such that $m \in x \circ n$. Since $n \in A_t$ we have

$$Tr_A^{\ L}(n) \geq t_1 = \min\left(Tr_A^{\ L}(m), Tr_A^{\ L}(l)\right),$$
$$Tr_A^{\ U}(n) \geq t_1 = \min\left(Tr_A^{\ U}(m), Tr_A^{\ U}(l)\right),$$

$$Ind_A^{\ L}(n) \leq t_1 = \max\left(Ind_A^{\ L}(m), Ind_A^{\ L}(l)\right),$$
$$Ind_A^{\ U}(n) \leq t_1 = \max\left(Ind_A^{\ U}(m), Ind_A^{\ U}(l)\right),$$

and

$$Fal_A^{\ L}(n) \leq t_1 = \max\left(Fal_A^{\ L}(m), Fal_A^{\ L}(l)\right),$$
$$Fal_A^{\ U}(n) \leq t_1 = \max\left(Fal_A^{\ U}(m), Fal_A^{\ U}(l)\right).$$

Therefore condition (ii) of definition 3.1 has been verified. Compliance to condition (iii) of definition 3.1 can be proven in a similar manner.

Thus 'A' is a interval-valued neutrosophic subhypergroup of (R,+). Now since A_t is a subsemihypergroup of the semihypergroup (R,∘), we have the following.

Let $m, n \in A_{t_2}$ and therefore we have $m \circ n \in A_{t_2}$. Thus, for every $l \in m \circ n$, we obtain

$$Tr_A^{\ L}(l) \geq t_2, Tr_A^{\ U}(l) \geq t_2, Ind_A^{\ L}(l) \leq t_2, Ind_A^{\ U}(l) \leq t_2, Fal_A^{\ L}(l) \leq t_2, Fal_A^{\ U}(l) \leq t_2,$$

and therefore it follows that:

$$\min\left(Tr_A^{\ L}(m), Tr_A^{\ L}(n)\right) \leq \inf\left\{Tr_A^{\ L}(l) : l \in m \circ n\right\},$$
$$\min\left(Tr_A^{\ U}(m), Tr_A^{\ U}(n)\right) \leq \inf\left\{Tr_A^{\ U}(l) : l \in m \circ n\right\},$$

$$\max\left(Ind_A^{\ L}(m), Ind_A^{\ L}(n)\right) \geq \sup\left\{Ind_A^{\ L}(l) : l \in m \circ n\right\},$$
$$\max\left(Ind_A^{\ U}(m), Ind_A^{\ U}(n)\right) \geq \sup\left\{Ind_A^{\ U}(l) : l \in m \circ n\right\},$$

and

$$\max\left(Fal_A^{\ L}(m), Fal_A^{\ L}(n)\right) \geq \sup\left\{Fal_A^{\ L}(l) : l \in m \circ n\right\},$$
$$\max\left(Fal_A^{\ U}(m), Fal_A^{\ U}(n)\right) \geq \sup\left\{Fal_A^{\ U}(l) : l \in m \circ n\right\},$$

which proves the condition (iv) of definition 3.1 has been verified. Hence 'A' is a interval-valued neutrosophic hyperring over R.

INTERVAL-VALUED NEUTROSOPHIC HYPERIDEALS (IVNHI)

We define IVNHI and demonstrate a few related theorems in this section.

Definition 4.1. Interval-Valued Neutrosophic Left (Resp. Right) hyperideal

Let 'A' be a IVNHI over R. Then 'A' is interval-valued neutrosophic left (resp. right) hyperideal over R, if,

(i) $\forall\, m,n \in R,$

$$\min\left\{Tr_A^{\ L}(m), Tr_A^{\ L}(n)\right\} \le \inf\left\{Tr_A^{\ L}(l) : l \in m+n\right\}$$
$$\min\left\{Tr_A^{\ U}(m), Tr_A^{\ U}(n)\right\} \le \inf\left\{Tr_A^{\ U}(l) : l \in m+n\right\}$$

$$\max\left\{Ind_A^{\ L}(m), Ind_A^{\ L}(n)\right\} \ge \sup\left\{Ind_A^{\ L}(l) : l \in m+n\right\}$$
$$\max\left\{Ind_A^{\ U}(m), Ind_A^{\ U}(n)\right\} \ge \sup\left\{Ind_A^{\ U}(l) : l \in m+n\right\}$$

$$\max\left\{Fal_A^{\ L}(m), Fal_A^{\ L}(n)\right\} \ge \sup\left\{Fal_A^{\ L}(l) : l \in m+n\right\}$$

and

$$\max\left\{Fal_A^{\ U}(m), Fal_A^{\ U}(n)\right\} \ge \sup\left\{Fal_A^{\ U}(l) : l \in m+n\right\}$$

(ii) $\forall\, x,m \in R$, there exists $n \in R$, such that $m \in x+n$ and

$$\min\left\{Tr_A^{\ L}(x), Tr_A^{\ L}(m)\right\} \le Tr_A^{\ L}(n)$$
$$\min\left\{Tr_A^{\ U}(x), Tr_A^{\ U}(m)\right\} \le Tr_A^{\ U}(n)$$

$$\max\left\{Ind_A^{\ L}(x), Ind_A^{\ L}(m)\right\} \ge Ind_A^{\ L}(n)$$
$$\max\left\{Ind_A^{\ U}(x), Ind_A^{\ U}(m)\right\} \ge Ind_A^{\ U}(n)$$

$$\max\left\{Fal_A^{\ L}(x), Fal_A^{\ L}(m)\right\} \ge Fal_A^{\ L}(n)$$

and

$$\max\left\{Fal_A^{U}(x), Fal_A^{U}(m)\right\} \geq Fal_A^{U}(n)$$

(iii) $\forall\, x,m \in R$ there exists $n \in R$, such that $m \in l+x$ and

$$\min\left\{Tr_A^{L}(x), Tr_A^{L}(m)\right\} \leq Tr_A^{L}(n)$$
$$\min\left\{Tr_A^{U}(x), Tr_A^{U}(m)\right\} \leq Tr_A^{U}(n)$$

$$\max\left\{Ind_A^{L}(x), Ind_A^{L}(m)\right\} \geq Ind_A^{L}(n)$$
$$\max\left\{Ind_A^{U}(x), Ind_A^{U}(m)\right\} \geq Ind_A^{U}(n)$$

$$\max\left\{Fal_A^{L}(x), Fal_A^{L}(m)\right\} \geq Fal_A^{L}(n)$$

and

$$\max\left\{Fal_A^{U}(x), Fal_A^{U}(m)\right\} \geq Fal_A^{U}(n)$$

(iv) $\forall\, x,m \in R$,

$$Tr_A^{L}(n) \leq \inf\left\{Tr_A^{L}(l) : l \in m \circ n\right\} \quad \left(resp.\ Tr_A^{L}(m) \leq \inf\left\{Tr_A^{L}(l) : l \in m \circ n\right\}\right),$$

$$Tr_A^{U}(n) \leq \inf\left\{Tr_A^{U}(l) : l \in m \circ n\right\} \quad \left(resp.\ Tr_A^{U}(m) \leq \inf\left\{Tr_A^{U}(l) : l \in m \circ n\right\}\right)$$

$$Ind_A^{L}(n) \geq \sup\left\{Ind_A^{L}(l) : l \in m \circ n\right\} \quad \left(resp.\ Ind_A^{L}(m) \geq \sup\left\{Ind_A^{L}(l) : l \in m \circ n\right\}\right)$$

$$Ind_A^{U}(n) \geq \sup\left\{Ind_A^{U}(l) : l \in m \circ n\right\} \quad \left(resp.\ Ind_A^{U}(m) \geq \sup\left\{Ind_A^{U}(l) : l \in m \circ n\right\}\right)$$

and

$$Fal_A^L(n) \geq \sup\{Fal_A^L(l) : l \in m \circ n\} \quad \left(resp.\ Fal_A^L(m) \geq \sup\{Tr_A^L(l) : l \in m \circ n\}\right)$$

$$Fal_A^U(n) \geq \sup\{Fal_A^U(l) : l \in m \circ n\} \quad \left(resp.\ Fal_A^U(m) \geq \sup\{Tr_A^U(l) : l \in m \circ n\}\right).$$

'A' is a interval-valued neutrosophic left (resp. right) hyperideal of R. From conditions (i),(ii) and (iii) 'A' is interval-valued neutrosophicsubhypergroup of $(R,+)$.

Definition 4.2. Inter-Valued Neutrosophic Hyperideals (IVNHI)

Let 'A' be a IVNHI over R. Then 'A' is IVNHI over R, if the following conditions are satisfied:

(i) $\forall\, x,m \in R$,

$$\min\{Tr_A^L(m), Tr_A^L(n)\} \leq \inf\{Tr_A^L(l) : l \in m+n\}$$
$$\min\{Tr_A^U(m), Tr_A^U(n)\} \leq \inf\{Tr_A^U(l) : l \in m+n\}$$

$$\max\{Ind_A^L(m), Ind_A^L(n)\} \geq \sup\{Ind_A^L(l) : l \in m+n\}$$
$$\max\{Ind_A^U(m), Ind_A^U(n)\} \geq \sup\{Ind_A^U(l) : l \in m+n\}$$

$$\max\{Fal_A^L(m), Fal_A^L(n)\} \geq \sup\{Fal_A^L(l) : l \in m+n\}$$

and

$$\max\{Fal_A^U(m), Fal_A^U(n)\} \geq \sup\{Fal_A^U(l) : l \in m+n\}$$

(ii) $\forall\, x,m \in R$ there exists $n \in R$, such that $m \in x+n$ and

$$\min\{Tr_A^L(x), Tr_A^L(m)\} \leq Tr_A^L(n)$$
$$\min\{Tr_A^U(x), Tr_A^U(m)\} \leq Tr_A^U(n)$$

$$\max\left\{Ind_A^{\ L}(x), Ind_A^{\ L}(m)\right\} \geq Ind_A^{\ L}(n)$$
$$\max\left\{Ind_A^{\ U}(x), Ind_A^{\ U}(m)\right\} \geq Ind_A^{\ U}(n)$$

$$\max\left\{Fal_A^{\ L}(x), Fal_A^{\ L}(m)\right\} \geq Fal_A^{\ L}(n)$$

and

$$\max\left\{Fal_A^{\ U}(x), Fal_A^{\ U}(m)\right\} \geq Fal_A^{\ U}(n)$$

(iii) $\forall\, x,m \in R$ there exists $n \in R$, such that $m \in l + x$ and

$$\min\left\{Tr_A^{\ L}(x), Tr_A^{\ L}(m)\right\} \leq Tr_A^{\ L}(n)$$
$$\min\left\{Tr_A^{\ U}(x), Tr_A^{\ U}(m)\right\} \leq Tr_A^{\ U}(n)$$

$$\max\left\{Ind_A^{\ L}(x), Ind_A^{\ L}(m)\right\} \geq Ind_A^{\ L}(n)$$
$$\max\left\{Ind_A^{\ U}(x), Ind_A^{\ U}(m)\right\} \geq Ind_A^{\ U}(n)$$

$$\max\left\{Fal_A^{\ L}(x), Fal_A^{\ L}(m)\right\} \geq Fal_A^{\ L}(n)$$

and

$$\max\left\{Fal_A^{\ U}(x), Fal_A^{\ U}(m)\right\} \geq Fal_A^{\ U}(n)$$

(iv) $\forall\, x,m \in R$,

$$\max\left\{Tr_A^{\ L}(m), Tr_A^{\ L}(n)\right\} \leq \inf\left\{Tr_A^{\ L}(l) : l \in m \circ n\right\},$$
$$\max\left\{Tr_A^{\ U}(m), Tr_A^{\ U}(n)\right\} \leq \inf\left\{Tr_A^{\ U}(l) : l \in m \circ n\right\},$$

$$\max\left\{Ind_A^{\ L}(m), Ind_A^{\ L}(n)\right\} \geq \sup\left\{Ind_A^{\ L}(l) : l \in m \circ n\right\},$$
$$\max\left\{Ind_A^{\ U}(m), Ind_A^{\ U}(n)\right\} \geq \sup\left\{Ind_A^{\ U}(l) : l \in m \circ n\right\},$$

$$\max\left\{Fal_A^{\ L}(m), Fal_A^{\ L}(n)\right\} \geq \sup\left\{Fal_A^{\ L}(l) : l \in m \circ n\right\}$$

and

$$\max\left\{Fal_A^{\ U}(m), Fal_A^{\ U}(n)\right\} \geq \sup\left\{Fal_A^{\ U}(l) : l \in m \circ n\right\}$$

From condition (i), (ii) and (iii) 'A' is an interval-valued neutrosophic subhyperigroup of $(R,+)$. Conditions (iv) indicate both interval-valued neutrosophic left hyperideal and inter-valued neutrosophic right hyperideal. Hence 'A' is a inter-valued neutrosophic hyper ideal of R.

Example 4.2a. A family of t-level sets of IVNSs over R is a subhyperring of R is given below for both upper and lower bound:

$$A_t = \left\{\begin{aligned} m \in R : & Tr_A^{\ L} \geq t, Tr_A^{\ L} \geq t \\ & Tr_A^{\ U} \geq t, Tr_A^{\ U} \geq t \\ & Ind_A^{\ L} \geq t, Ind_A^{\ L} \geq t \\ & Ind_A^{\ U} \geq t, Ind_A^{\ U} \geq t \\ & Fal_A^{\ L} \geq t, Fal_A^{\ L} \geq t \\ & Fal_A^{\ U} \geq t, Fal_A^{\ U} \geq t \end{aligned}\right\}, \text{ for all } t \in [0,1].$$

Then 'A' is a single-valued neutrosophic hyperideal over R.

Theorem 4.3. Let 'A' be a non-null IVNS over R. 'A' is a IVNS over R. A is a interval-valued neutrosophic hyperideal over 'R' if and only if 'A' is a interval-valued neutrosophic hyper group over $(R,+)$ and also 'A' is both interval-valued neutrosophic left hyper ideal and a interval-valued neutrosophic right hyper ideal of R.

Proof. This is straight forward by definitions 4.1 and 4.2.

Theorem 4.4. Let 'A' and 'B' be two interval-valued neutrosophic hyper ideals over 'R'. Then $A \cap B$ is a interval-valued neutrosophichyperideal over R if it is non-null

Proof. Let 'A' and 'B' be two interval-valued neutrosophic hyper ideals over R. By definition 4.2,

$$A \cap B = \left\{m, Tr_{A\cap B}^{\ L}(a), Ind_{A\cap B}^{\ L}(a), Fal_{A\cap B}^{\ L}(a) : a \in R\right\},$$

Where

$$Tr_{A\cap B}{}^{L}(m)=\min\left\{Tr_{A}{}^{L}(m),Tr_{B}{}^{L}(m)\right\},$$
$$Ind_{A\cap B}{}^{L}(m)=\max\left\{Tr_{A}{}^{L}(m),Tr_{B}{}^{L}(m)\right\}$$

and

$$Fal_{A\cap B}{}^{L}(m)=\max\left\{Fal_{A}{}^{L}(m),Fal_{B}{}^{L}(m)\right\}.$$

Similarly

$$A\cap B=\left\{\left\langle m,Tr_{A\cap B}{}^{U}(a),Ind_{A\cap B}{}^{U}(a),Fal_{A\cap B}{}^{U}(a)\right\rangle:a\in R\right\},$$

where

$$Tr_{A\cap B}{}^{U}(m)=\min\left\{Tr_{A}{}^{U}(m),Tr_{B}{}^{U}(m)\right\},$$
$$Ind_{A\cap B}{}^{U}(m)=\max\left\{Tr_{A}{}^{U}(m),Tr_{B}{}^{U}(m)\right\}$$

And

$$Fal_{A\cap B}{}^{U}(m)=\max\left\{Fal_{A}{}^{U}(m),Fal_{B}{}^{U}(m)\right\}.$$

Then ∀ $x,m\in R$, we have the following. We only prove all the four conditions for the truth membership term $Tr_{A}{}^{L},Tr_{B}{}^{U},Fal_{A}{}^{L},Fal_{B}{}^{U}$.

(i) $\min\left(Tr_{A\cap B}{}^{L}(m),Tr_{A\cap B}{}^{L}(n)\right)$

$$=\min\left\{\min\left(Tr_{A}{}^{L}(m),Tr_{B}{}^{L}(m),\min\left(Tr_{A}{}^{L}(n),Tr_{B}{}^{L}(n)\right)\right)\right\}$$
$$\le\min\left\{\min\left(Tr_{A}{}^{L}(m),Tr_{B}{}^{L}(n),\min\left(Tr_{A}{}^{L}(m),Tr_{B}{}^{L}(n)\right)\right)\right\}$$
$$\le\min\left\{\inf\left\{Tr_{A}{}^{L}(l):l\in m+n\right\},\inf\left\{Tr_{B}{}^{L}(l):l\in m+n\right\}\right\}$$
$$\le\inf\left\{\min\left\{Tr_{A}{}^{L}(l),Tr_{B}{}^{L}(l)\right\}:l\in m+n\right\}$$

$$\min\left(Tr_{A\cap B}{}^{L}(m), Tr_{A\cap B}{}^{L}(n)\right) = \inf\left\{T_{A\cap B}{}^{L}(l); l \in m+n\right\}$$

$$\begin{aligned}
&\min\left(Tr_{A\cap B}{}^{U}(m), Tr_{A\cap B}{}^{U}(n)\right) \\
&= \min\left\{\min\left(Tr_{A}{}^{U}(m), Tr_{B}{}^{U}(m), \min\left(Tr_{A}{}^{U}(n), Tr_{B}{}^{U}(n)\right)\right)\right\} \\
&\leq \min\left\{\min\left(Tr_{A}{}^{U}(m), Tr_{B}{}^{U}(n), \min\left(Tr_{A}{}^{U}(m), Tr_{B}{}^{U}(n)\right)\right)\right\} \\
&\leq \min\left\{\inf\left\{Tr_{A}{}^{U}(l) : l \in m+n\right\}, \inf\left\{Tr_{B}{}^{U}(l) : l \in m+n\right\}\right\} \\
&\leq \inf\left\{\min\left\{Tr_{A}{}^{U}(l), Tr_{B}{}^{U}(l)\right\} : l \in m+n\right\}
\end{aligned}$$

$$\min\left(Tr_{A\cap B}{}^{U}(m), Tr_{A\cap B}{}^{U}(n)\right) = \inf\left\{Tr_{A\cap B}{}^{U}(l); l \in m+n\right\}$$

Similarly, for indeterminacy and falsity the results as follows:

$$\max\left(Ind_{A\cap B}{}^{L}(m), Ind_{A\cap B}{}^{L}(n)\right) = \sup\left\{Ind_{A\cap B}{}^{L}(l); l \in m+n\right\}$$
$$\max\left(Ind_{A\cap B}{}^{U}(m), Ind_{A\cap B}{}^{U}(n)\right) = \sup\left\{Ind_{A\cap B}{}^{U}(l); l \in m+n\right\}$$

and

$$\max\left(Fal_{A\cap B}{}^{L}(m), Fal_{A\cap B}{}^{L}(n)\right) = \sup\left\{Fal_{A\cap B}{}^{L}(l); l \in m+n\right\}$$
$$\max\left(Fal_{A\cap B}{}^{U}(m), Fal_{A\cap B}{}^{U}(n)\right) = \sup\left\{Fal_{A\cap B}{}^{U}(l); l \in m+n\right\}$$

(ii) $\forall\, x, m \in R$, there exists $n \in R$, such that $m \in x+n$, then it follows that:

$$\begin{aligned}
&\min\left(Tr_{A\cap B}{}^{L}(m), Tr_{A\cap B}{}^{L}(n)\right) \\
&= \min\left\{\min\left(Tr_{A}{}^{L}(m), Tr_{B}{}^{L}(m), \min\left(Tr_{A}{}^{L}(n), Tr_{B}{}^{L}(n)\right)\right)\right\} \\
&\leq \min\left\{\min\left(Tr_{A}{}^{L}(m), Tr_{B}{}^{L}(n), \min\left(Tr_{A}{}^{L}(m), Tr_{B}{}^{L}(n)\right)\right)\right\} \\
&\leq \min\left(Tr_{A}{}^{L}(l), Tr_{B}{}^{L}(l)\right)
\end{aligned}$$

$$\min\left(Tr_{A\cap B}{}^{L}(m), Tr_{A\cap B}{}^{L}(n)\right) = Tr_{A\cap B}{}^{L}(l)$$

$$\min\left(Tr_{A\cap B}{}^{U}(m), Tr_{A\cap B}{}^{U}(n)\right)$$
$$= \min\left\{\min\left(Tr_{A}{}^{U}(m), Tr_{B}{}^{U}(m), \min\left(Tr_{A}{}^{U}(n), Tr_{B}{}^{U}(n)\right)\right)\right\}$$
$$\leq \min\left\{\min\left(Tr_{A}{}^{U}(m), Tr_{B}{}^{U}(n), \min\left(Tr_{A}{}^{U}(m), Tr_{B}{}^{U}(n)\right)\right)\right\}$$
$$\leq \min\left(Tr_{A}{}^{U}(l), Tr_{B}{}^{U}(l)\right)$$

$$\min\left(Tr_{A\cap B}{}^{U}(m), Tr_{A\cap B}{}^{U}(n)\right) = Tr_{A\cap B}{}^{U}(l)$$

Similarly, for indeterminacy and falsity the results as follows:

$$\max\left(Ind_{A\cap B}{}^{L}(m), Ind_{A\cap B}{}^{L}(n)\right) = Ind_{A\cap B}{}^{L}(l)$$
$$\max\left(Ind_{A\cap B}{}^{U}(m), Ind_{A\cap B}{}^{U}(n)\right) = Ind_{A\cap B}{}^{U}(l)$$

$$\max\left(Fal_{A\cap B}{}^{L}(m), Fal_{A\cap B}{}^{L}(n)\right) = Fal_{A\cap B}{}^{L}(l)$$
$$\max\left(Fal_{A\cap B}{}^{U}(m), Fal_{A\cap B}{}^{U}(n)\right) = Fal_{A\cap B}{}^{U}(l).$$

(iii) $\forall\, x,m \in R$, there exists $l \in R$, such that $m \in l+x$ and

$$\min\left\{Tr_{A\cap B}{}^{L}(x), Tr_{A\cap B}{}^{L}(a)\right\} \leq Tr_{A\cap B}{}^{L}(l)$$
$$\min\left\{Tr_{A\cap B}{}^{U}(x), Tr_{A\cap B}{}^{U}(a)\right\} \leq Tr_{A\cap B}{}^{U}(l)$$

$$\max\left\{Ind_{A\cap B}{}^{L}(x), Ind_{A\cap B}{}^{L}(a)\right\} \geq Tr_{A\cap B}{}^{L}(l)$$
$$\max\left\{Ind_{A\cap B}{}^{U}(x), Ind_{A\cap B}{}^{U}(a)\right\} \geq Tr_{A\cap B}{}^{U}(l)$$

$$\max\left\{Fal_{A\cap B}{}^{L}(x), Fal_{A\cap B}{}^{L}(a)\right\} \geq Fal_{A\cap B}{}^{L}(l)$$
$$\max\left\{Fal_{A\cap B}{}^{U}(x), Fal_{A\cap B}{}^{U}(a)\right\} \geq Fal_{A\cap B}{}^{U}(l).$$

(iv) $\forall\, x,m \in R$,

$$\max\left\{Tr_{A\cap B}{}^{L}(m), Tr_{A\cap B}{}^{L}(n)\right\} \leq \inf Tr_{A\cap B}{}^{L}(c) : c \in m \circ n$$
$$\max\left\{Ind_{A\cap B}{}^{L}(m), Ind_{A\cap B}{}^{L}(n)\right\} \geq \sup\left\{Ind_{A\cap B}{}^{L}(l) : l \in m \circ n\right\}$$

$$\max\left\{Ind_{A\cap B}{}^{U}(m), Ind_{A\cap B}{}^{U}(n)\right\} \geq \sup\left\{Ind_{A\cap B}{}^{U}(l) : l \in m \circ n\right\}$$
$$\max\left\{Fal_{A\cap B}{}^{L}(m), Fal_{A\cap B}{}^{L}(n)\right\} \geq \sup\left\{Fal_{A\cap B}{}^{L}(l) : l \in m \circ n\right\}$$

and

$$\max\left\{Fal_{A\cap B}{}^{U}(m), Fal_{A\cap B}{}^{U}(n)\right\} \geq \sup\left\{Fal_{A\cap B}{}^{U}(l) : l \in m \circ n\right\}$$

Hence, it is verified that $A\cap B$ is a IVNHI over R.

Theorem 4.5. Let 'A' be a interval-valued neutrosophic hyper ideals over R. Then for every $t\in[0,1]$, $A_t\neq\varnothing$ is a subhyperideal over R.

Proof. Let 'A' be a interval-valued neutrosophichyperideal over R, $\forall\, t\in[0,1]$ and $m,n\in A_t$. Then

$$Tr_A{}^{L}(m), Tr_B{}^{L}(n) \geq t, Tr_A{}^{U}(m), Tr_B{}^{U}(n) \geq t,$$
$$Ind_A{}^{L}(m), Ind_B{}^{L}(n) \geq t, Ind_A{}^{U}(m), Ind_B{}^{U}(n) \geq t,$$
$$Fal_A{}^{L}(m), Fal_B{}^{L}(n) \geq t, Fal_A{}^{U}(m), Fal_B{}^{U}(n) \geq t.$$

Since 'A' is an interval-valued neutrosophic sub hypergroup of $(R,+)$, we have the following:

$$\inf\left\{Tr_A{}^{L}(l) : l \in m+n\right\} \geq \min\left\{Tr_A{}^{L}(m), Tr_A{}^{L}(n)\right\} \geq \min\{t,t\} = t$$
$$\inf\left\{Tr_A{}^{U}(l) : l \in m+n\right\} \geq \min\left\{Tr_A{}^{U}(m), Tr_A{}^{U}(n)\right\} \geq \min\{t,t\} = t$$

$$\sup\left\{Ind_A{}^{L}(l) : l \in m+n\right\} \leq t$$
$$\sup\left\{Ind_A{}^{U}(l) : l \in m+n\right\} \leq t$$

and

$$\sup\left\{Fal_A^{\,L}(l): l \in m+n\right\} \leq t$$
$$\sup\left\{Fal_A^{\,U}(l): l \in m+n\right\} \leq t$$

This implies that $l \in A_t$ and then for every $l \in m+n$, we obtain $m+n \subseteq A_t$. As such, for every $l \in A_t$, we obtain $l + A_t \subseteq A_t$. Now let $m, l \in A_t$. Then

$$Tr_A^{\,L}(m), Tr_A^{\,L}(l) \geq t, Tr_A^{\,U}(m), Tr_A^{\,U}(l) \geq t,$$
$$Ind_A^{\,L}(m), Tr_A^{\,L}(l) \geq t, Ind_A^{\,U}(m), Ind_A^{\,U}(l) \geq t,$$
$$Fal_A^{\,L}(m), Fal_A^{\,L}(l) \geq t, Fal_A^{\,U}(m), Fal_A^{\,U}(l) \geq t.$$

'A' is an interval-valued neutrosophic subhyperideal group of $(R,+)$, there exists $n \in R$ such that $m \in l+n$ and

$$Tr_A^{\,L}(n) \geq \min\left(Tr_A^{\,L}(m), Tr_A^{\,L}(l)\right) \geq t,$$
$$Tr_A^{\,U}(n) \geq \min\left(Tr_A^{\,U}(m), Tr_A^{\,U}(l)\right) \geq t,$$

$$Ind_A^{\,L}(n) \geq \max\left(Ind_A^{\,L}(m), Ind_A^{\,L}(l)\right) \leq t,$$
$$Ind_A^{\,U}(n) \geq \max\left(Ind_A^{\,U}(m), Ind_A^{\,U}(l)\right) \leq t,$$

$$Fal_A^{\,L}(n) \geq \max\left(Fal_A^{\,L}(m), Fal_A^{\,L}(l)\right) \leq t,$$
$$Fal_A^{\,U}(n) \geq \max\left(Fal_A^{\,U}(m), Fal_A^{\,U}(l)\right) \leq t,$$

and this implies $n \in A_t$. Therefore, we obtain $l + A_t \subseteq A_t$. As a result, A_t is superhyperideal group of $(R,+)$.

Let $m, n \in A_t$, then

$$Tr_A^{\,L}(m), Tr_A^{\,L}(n) \geq t, Tr_A^{\,U}(m), Tr_A^{\,U}(n) \geq t,$$
$$Ind_A^{\,L}(m), Ind_A^{\,L}(n) \geq t, Ind_A^{\,U}(m), Ind_A^{\,U}(n) \geq t$$

and

$$Fal_A^{\ L}(m), Fal_A^{\ L}(n) \geq t, Fal_A^{\ U}(m), Fal_A^{\ U}(n) \geq t.$$

Since ‘A’ is interval-valued neutrosophic subsemihyperideal group of $(R,\circ)$,then for all $m,n\in R$, we have the following:

$$\inf\left\{Tr_A^{\ L}(l): l\in m\circ n\right\} \geq \min\left\{Tr_A^{\ L}(m), Tr_A^{\ L}(n)\right\} = t,$$
$$\inf\left\{Tr_A^{\ U}(l): l\in m\circ n\right\} \geq \min\left\{Tr_A^{\ U}(m), Tr_A^{\ U}(n)\right\} = t,$$

$$\sup\left\{Ind_A^{\ L}(l): l\in m\circ n\right\} \leq \min\left\{Ind_A^{\ L}(m), Ind_A^{\ L}(n)\right\} = t,$$
$$\sup\left\{Ind_A^{\ U}(l): l\in m\circ n\right\} \leq \min\left\{Ind_A^{\ U}(m), Ind_A^{\ U}(n)\right\} = t,$$

$$\sup\left\{Fal_A^{\ L}(l): l\in m\circ n\right\} \leq \min\left\{Fal_A^{\ L}(m), Fal_A^{\ L}(n)\right\} = t,$$
$$\sup\left\{Fal_A^{\ U}(l): l\in m\circ n\right\} \leq \min\left\{Fal_A^{\ U}(m), Fal_A^{\ U}(n)\right\} = t.$$

This implies that $l\in A_t$. Therefore, for every $m,n\in A_t$ we obtain $l\in m\circ n\in P^*(R)$. Hence A_t is a subhyperideal over R.

APPLICATION OF NEUTROSOPHICYYPERRING

Neutrosophichyperring (NHR) is frequently utilized as a system architecture for communication. Due to its low cost, a set of uniform node degree from its symmetric architecture and flexibility of use for any routing techniques, the NHR is ideal for mid-sized to large scale multicomputer systems and networks of workstations. The implementation of secure message passing protocols and the most dependable computing algorithms are made possible by NHR architecture.

The Government of India initiated Digital India aimed to convert the country into a knowledge-based society and economy. As the country moves towards a cashless environment after demonetization the initial awe and confusion have given way to a flurry of concerns. The issue is solved by following the digital concept, which motivates and drive simplicity to execute financial transactions. Large amount of cash is not required to carry, use of plastic cards or even stand in line to use an ATM. Additionally, the simpler and safer way to spend money while traveling. In this circumstance, card usage has increased. Actually, all credit and debit cards are printed with numbers (generally 16 digits). One of the difficult tasks in creating cards is coming up with the card number without using it twice. This denotes a special account number for a card and provide certain details about the card issuer and the account it is linked to. For obvious reasons, using random numbers won't work because they follow a pattern. Therefore, NHR plays a vital role in this sector of pattern creation.

The following Figures indicates some NHR:

Figure 1. 8-HR

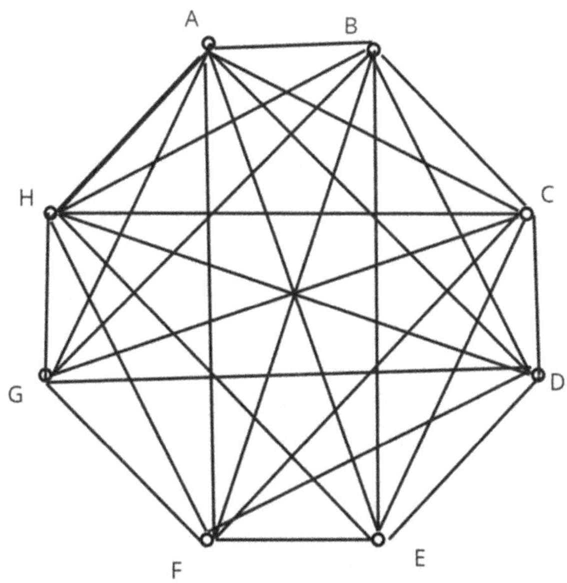

In Figure 1, Figure 2, and Figure 3, we represent A, B, C…U. If we fix Interval-valued neutrosophic number, in each node and combine other node's numbers for generation of credit card number without repetition. Similarly we can fix other nodes and combine the same node with other nodes we can generate multiple set of credit, debit or any other card numbers without any repetitions.

For example: Fix-A and combine A with all other nodes.
Fix-B and combine B with other nodes.

Figure 2. 12-HR

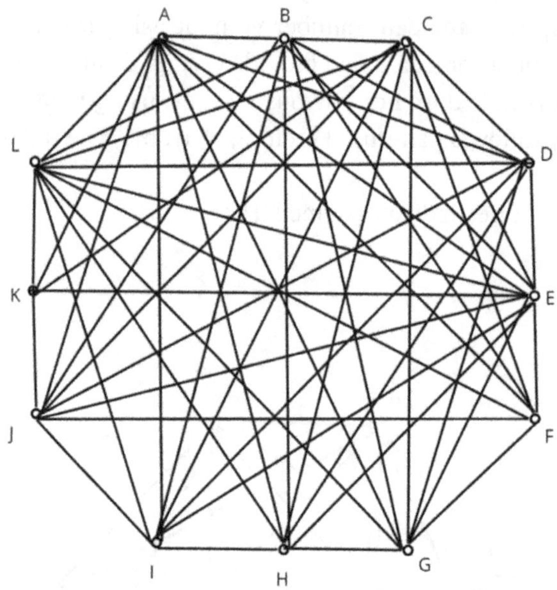

Figure 3. 20-HR

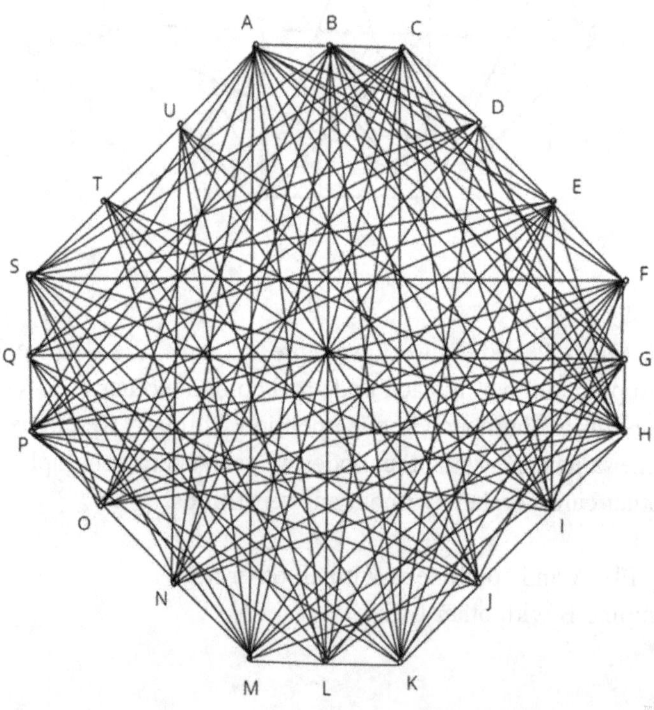

Similarly we can combine all the other nodes as follows

Table 1. Interval-valued neutrosophic numbers for 8-HR

Edges	IVNNs	Score Values (Def:2.11)
A-B	⟨[0.5,0.2], [0.4,0.1], [0.3,01]⟩	0.325
A-C	⟨[0.6,03], [0.5,0.1], [0.4,02]⟩	0.275
A-D	⟨[0.4,0.3], [0.6,0.1], [0.2,0.2]⟩	0.225
A-E	⟨[0.8,0.3], [0.4,0.1], [0.1,0.2]⟩	0.45
A-F	⟨[0.8,0.3], [0.7,0.1], [0.6,0.2]⟩	0.175
A-G	⟨[0.7,0.3], [0.6,0.1], [0.2,0.2]⟩	0.3
A-H	⟨[0.9,0.3], [0.4,0.5], [0.3,0.4]⟩	0.175
B-C	⟨[0.6,0.3], [0.2,0.5], [0.3,0.4]⟩	0.2
B-D	⟨[0.5,0.2], [0.3,0.1], [0.3,0.3]⟩	0.325
B-E	⟨[0.7,0.2], [0.7,0.1], [0.4,0.2]⟩	0.175
B-F	⟨[0.3,0.7], [0.1,0.2], [0.4,02]⟩	0.45
B-G	⟨[0.9,0.3], [0.3,0.1], [0.1,0.2]⟩	0.525
B-H	⟨[0.6,0.3], [0.3,0.1], [0.2,0.2]⟩	0.425
C-D	⟨[0.6,0.2], [0.5,0.4], [0.2,0.1]⟩	0.2
C-E	⟨[0.7,0.2], [0.6,0.3], [0.2,0.1]⟩	0.16
C-F	⟨[0.8,0.5], [0.7,0.4], [0.6,0.2]⟩	0.075
C-G	⟨[0.5,0.4], [0.7,0.1], [0.3,0.2]⟩	0.2
C-H	⟨[0.9,0.1], [0.8,0.2], [0.7,0.2]⟩	0.03
D-E	⟨[0.7,0.2], [0.3,0.4], [0.8,0.2]⟩	0.125
D-F	⟨[0.5,0.4], [0.8,0.3], [0.1,0.2]⟩	0.1
D-G	⟨[0.5,0.3], [0.3,0.4], [0.3,0.1]⟩	0.2575
D-H	⟨[0.8,0.5], [0.9,0.3], [0.7,0.2]⟩	0
E-F	⟨[0.6,0.8], [0.4,0.7], [0.3,0.1]⟩	0.2
E-G	⟨[,0.6,0.4], [0.4,0.3], [0.3,01]⟩	0.3
E-H	⟨[0.2,0.9], [0.1,0.8], [0.3,0.7]⟩	0.075
F-G	⟨[0.3,0.6], [0.5,0.2], [0.4,0.3]⟩	0.2
F-H	⟨[0.1,0.9], [0.2,0.8], [0.2,0.8]⟩	0
G-H	⟨[0.3,0.7], [0.1,0.8], [0.5,0.2]⟩	0.125

Because the score value for the appropriate edge will be high, it can be seen from the preceding Table 1 that the B-G edge combination will be the best. Next high score value between A-E and B-F paths. The edges D-H, F-H are considered as invalid edges because the score values for those edges are tends to 0. So we are not able to combine those edges to generate the card numbers. The same type of IVNNs may be fixed for 12-HR and 20-HR (Figure 2 and Figure 3), and utilizing score values, we can find the optimum way to produce the credit and debit card numbers.

Actually fuzzy is only based on membership function, where an intuitionistic number depends on both the membership and non-membership function. Whereas neutrosphic contains truth, indeterminacy and falsity membership function, which fuzzy and intuitionistic number tend to fail. Hence if we are using interval valued neutrosophic numbers, we are able to generate more numbers comparatively to fuzzy and intuitionistic number. This is the main advantage of using NHR.

FUTURE RESEARCH DIRECTIONS

The results in this chapter may be considered as a base for any possible study in the field of interval-valued neutrosophic hyperring and hyperideals. The future direction is to extend this paper into different types of neutrosophic environments and specially under neutrosophic superhypergraph.

CONCLUSION

This chapter introduced in order to protect security applications from dangers like illegal access and alteration, security systems are developed, added to, and tested within the applications themselves. To protect this the NHR is the best choice for medium-sized to large-scale multicomputer systems and networks of workstations because of its low cost, fixed uniform node degree resulting from its symmetric architecture, and flexibility in usage of any routing protocols. Hence we established hyperstructure for IVNS model through hyperalgebraic structures such as hyperrings and hyperideals. The properties to these structure were defined and verified. The proposed concept has been applied in generation of credit and debit card numbers. The future direction is to establish the hyperalgebraic theory in hyperrings and hyperideals under neutrosophic superhypergraph environment.

Funding

This research received no specific grant from any funding agency in the public, commercial, or not-for-profit sectors

ACKNOWLEDGMENT

The authors are grateful to the referees for their valuable comments and suggestions.

REFERENCES

Al-Tahan, M., Davvaz, B., Smarandache, F., & Anis, O. (2021). On some NeutroHyperstructures. *Symmetry*, *13*(4), 535. doi:10.3390ym13040535

Al-Tahan, M., Smarandache, F., & Davvaz, B. (2021). NeutroOrderedAlgebra: Applications to semigroups. *Neutrosophic Sets and System*, *39*, 133–147.

Atanassov, K. (1986). Intuitionistic. *Fuzzy Sets and Systems*, *20*(1), 87–96. doi:10.1016/S0165-0114(86)80034-3

Corsini, P. (1993). Prolegomena of Hypergroup Theory. Udine Aviani Editore.

Davvaz, B. (1998). On Hυ–rings and fuzzy Hυ–ideal. J*ournal of Fuzzy Mathematics*, 6(*1*), 33–42.

Marty, F. (1934). Sur une generalization de la notion de groupe. *Proceedings of the 8th Congress on Mathmatics, Scandinaves*, 45–49.

Smarandache, F. (1998). Neutrosophy. Neutrosophic Probability, Set, and Logic. ProQuest Information & Learning.

Smarandache, F. (2020). NeutroAlgebra is a generalization of partial algebra. *International Journal of Neutrosophic Science*, *2*, 8–17. doi:10.54216/IJNS.020103

Smarandache, F. (2020). Introduction to NeutroAlgebraic structures and AntiAlgebraic structures (revisited). *Neutrosophic Sets and System*, *31*, 1–16.

Vougiouklis, T. (1994). *Hyperstructures and Their Representations*. Hadronic Press, Inc.

ADDITIONAL READING

Abirami, S., & Murugesan, S. (2016). Intuitionistic Fuzzy Gamma Hypergroup and Intuitionistic Fuzzy Gamma Hypermodule. *Advances in Fuzzy Mathematics*, *11*, 1–13.

Bharathi, P., & Vimala, J. (2016). The role of fuzzyγ–ideals in a commutative γ–group. Gl*obal Journal of Pure and Applied Mathematics,* 12*(3)*, 2067–2074.

Bordbar, H., Cristea, I., & Novak, M. (2017). Height of hyperideals in noetherian Krasner hyperrings. Applied. *Mathematics and Physics*, *79*(2), 31–42.

Davvaz, B. (2013). Polygroup Theory and Related Systems. World Scientific Publishing Co., Pte. Ltd.

Davvaz, B., & Leoreanu-Fotea, V. (2008). *Hyperring Theory and Applications*. International Academic Press.

Davvaz, B., & Omidi, S. (2016). Basic notions and properties of ordered semihyperrings. *Categ. Gen. Algebr. Struct. Appl.*, *4*(1), 43–62.

Hameed, M., Ahamad, Z., Broumi, S., & Ali, S. (2022). On Characterizations of single-valued neutrosophic hyperrings and hyperideals. *Neutrosophic Sets and Systems, 50*, 212-234.

Hila, K., & Naka, K. (2017). Characterizations of Hyperrings by Fuzzy Hyperideals with respect to A t-norm. *Fuzzy Information and Engineering*, *9*(4), 437–454. doi:10.1016/j.fiae.2017.12.003

Krasner, M. (1983). A class of hyperrings and hyperfields. *International Journal of Mathematics and Mathematical Sciences*, *6*(2), 307–311. doi:10.1155/S0161171283000265

Krishna Praba, S., Broumi, S., & Smarandache, F. (2020). Interval valued Neutrosophic Shortest path problem by A* Algorithm. *International Journal of Neutrosophic Science, 11*(1), 53-61.

Liu, W. J. (1982). Fuzzy invariant subgroups and fuzzy ideals. *Fuzzy Sets and Systems*, *8*(2), 133–139. doi:10.1016/0165-0114(82)90003-3

Preethi, D., Rajareega, S., Vimala, J., & Smarandache, F. (2019). Single-valued neutrosophic hyperrings and single-valued neutrosophic hyperideals. *Neutrosophic Sets and System, 29*, 131-149.

Rosenfeld, A. (1971). Fuzzy groups. *Journal of Mathematical Analysis and Applications*, *35*(3), 512–517. doi:10.1016/0022-247X(71)90199-5

Sahni, M., Merigó, J. M., Jha, B. K., & Verma, R. (Eds.). (2021). Mathematical Modeling, Computational Intelligence Techniques and Renewable Energy. Advances in Intelligent Systems and Computing. doi:10.1007/978-981-15-9953-8

Selvachandran, G., & Salleh, A. R. (2015). Algebraic hyperstructure of vague soft sets associated with hyperrings and hyperideals. *TheScientificWorldJournal, 2015*, 1–12. doi:10.1155/2015/780121 PMID:26247058

Selvachandran, G., & Salleh, A. R. (2015). Hypergroup theory applied to fuzzy soft sets. *Global Journal of Pure and Applied Mathematics*, *11*(2), 825–835.

Selvachandran, G., & Salleh, A. R. (2016). Fuzzy soft ideals based on fuzzy soft spaces. *Far East Journal of Mathematical Sciences*, *99*(3), 429–438.

Turksen, I. (1986). Interval valued fuzzy sets based on normal forms. *Fuzzy Sets and Systems*, *20*(2), 191–210. doi:10.1016/0165-0114(86)90077-1

Vimala, J., & Arokiareeta, J. (2018). Ideal approach on lattice ordered fuzzy soft group and its application in selecting best mobile network coverage among travelling paths. *Journal of Engineering and Applied Sciences (Asian Research Publishing Network)*, *13*(7), 2414–2421.

Wang, C., Inthumathi, V., & Pavithra, M. (2021). Vague soft separation axioms and vague soft continuous functions in vague soft topological spaces. *Italian Journal of Pure and Applied Mathematics, 46*, 1047-1067.

Wang, H., Smarandache, F., Zhang, Y. Q., & Sunderraman, R. (2010). Single-valued neutrosophic sets. *Multispace Multistructure*, *4*, 410–413.

Chapter 3
Introduction to the Finite NeutroGeometries:
The Mixed Projective-Affine Geometry

Erick González Caballero
Technological University of Havana, Cuba

ABSTRACT

NeutroGeometries generalize geometries in the same way that NeutroAlgebras generalize universal and partial algebras. NeutroGeometry is not one kind of classical geometry, but it can be a combination of some of them in the same space. For the first time, this chapter introduces notions of NeutroGeometry for finite geometries. Finite geometries consist of incidence structures where the set of points has finite cardinality. Usually, they are either projective or affine geometries. In this chapter, the author is mainly focused on the definition of the mixed projective-affine geometry (MPA geometry), which is a NeutroGeometry that follows the line of Smarandache's ideas of defining mixed geometries. The properties of the MPA spaces are studied, especially the one related to the satisfaction of parallelism for some lines and the no satisfaction for others. Additionally, other approaches to this theory are introduced, where elements of Neutrosophy are combined with the incidence matrices obtained from finite geometries. The chapter explores the advantages to use MPA in cryptography.

DOI: 10.4018/978-1-6684-4740-6.ch003

INTRODUCTION

One of the most recent concepts in neutrosophic theory is NeutroAlgebra. Classical algebras consist of a set of elements, one or more internal operations over these elements, and a set of axioms that must be fulfilled for 100% of them. Meanwhile, NeutroAlgebras change this last condition to the one that establishes that at least one axiom or internal operation is satisfied by a percentage other than 0% and 100% of the elements, (Smarandache, 2020a, 2020b).

This theory includes a more general concept called Neutrosophication such that a concept, theory, structure, among others, denoted by <A> can be divided into a triad <A>, <NeutA>, and <AntiA>; where <AntiA> is the opposite of <A>, while <NeutA> is neither <A> nor <AntiA>. Considering the algebras, if <A> represents an algebra (group, ring, skew-field, field) then <AntiA> represents the structure where one of the axioms or an internal law is not satisfied for 100% of the elements, while <NeutA> is not an <A> nor an <AntiA> because in at least one axiom or an internal operation there are strictly less than 100% of elements that satisfy this or those conditions.

Later, F. Smarandache defined NeutroGeometries from Neutroalgebras, (Smarandache, 2021). This is because Geometries can be seen as a special type of algebras, due to they consist of elements (points, lines, planes, hyperplanes) and axioms that must be fulfilled in 100% of the cases. The most classical example is the Euclidean geometry, where a series of axioms hold. The most controversial of them is the fifth Euclid's postulate wherein its most popular version says that given a straight line l and a point P outside it, one and only one line parallel to l passes through P. However, in NeutroGeometry at least one axiom is fulfilled by less than 100% of the elements and in more than 0% at the same time. Likewise, Neutrosophication is the method to obtain NeutroGeometry from Geometry.

These ideas, although very recent, have a more distant historical antecedent in time with the definition of Smarandache Geometries that consist of geometries that in the same space contain diverse types of classical Geometries, (Iseri, 2002). In the Smarandache Geometry the model of Euclid's fifth postulate is satisfied for some elements of the space and others do not satisfy it in two different ways, either no parallel passes through the point P, which are the so-called Riemannian Geometries, or there are an infinite number of parallels passing through P, which are called hyperbolic geometries. These models have been of interest to physics because they are a new approach to geometry, where mixed geometries generalize and at the same time model in a more truthful way the behavior of physical spaces that is the basis of Albert Einstein's Theory of Relativity and the theory of parallel universes, (Rabounski, 2010).

However, NeutroGeometries can be extended to other branches of geometry that are of interest to mathematics and its applications. This chapter deals with finite geometries, which consist of two types of elements, the points and the lines, whose sets are of finite cardinality, (Dembowski, 1997; Kiss & Szonyi, 2020). Within finite geometries, two types of them are studied: projective geometries and affine geometries. None of them is Euclidean because they are not infinite.

Projective spaces and especially finite projective planes do not contain parallels, since one of its axioms specifies that any two lines intersect at a point. Whereas finite affine geometries contain bundles of lines that form a parallelism equivalence class. Both geometries are related to each other since an affine space can be obtained from a projective space and vice versa, (Dissett, 2000; Hirschfeld, 1979).

These geometries can be represented by combinatorial mathematics or linear algebra. Both, projective and affine geometries are based on incidence structures, which are relations where if the point P passes through the line l, it is said that P is incident with l and also l is incident with P. As mentioned above, although projective and affine spaces can be obtained each one from the other, the axiom related to the parallelism of projective spaces is contradictory with its similar axiom in affine spaces.

From the algebraic point of view, a projective geometry of dimension n can be associated with a vector space of dimension n+1, where each element of the geometry such as points, lines, planes, and so on, is associated with a subspace of the vector space, for example, the points are subspaces of dimension 1, the lines are subspaces of dimension 2, the planes of dimension 3 and so on. Whereas, affine spaces are associated with the cosets of translations of vector subspaces of an n-dimensional vector space.

However, projective and affine spaces are not vector spaces. Projective spaces are defined from vector spaces without the null element with the quotient over an equivalence relation, where two vectors are equivalent if one of them is equal to the product of the other by a non-zero scalar. On the other hand, an affine space is obtained from the translation of the vector subspaces, where the origin of coordinates is not necessarily taken into account. Finite geometries have applications in code theory, cryptography, and Latin squares, (Beutelspacher, 1990; Kärteszi, 1976; Kou, Lin, & Fossorier, 2001; O'Keefe, 1993; Zumbragel, 2016).

This chapter aims to extend for the first time finite geometry theory to the framework of NeutroGeometries. We mainly utilize the axiomatic and algebraic tools of this theory and we obtain mixed spaces with projective properties for some elements of the space and with affine properties for others. An algorithm is provided to obtain these mixed spaces and some of their properties are studied. Likewise, the quantification of the degrees in which the mixed space is projective, affine, and indeterminate is included.

Additionally, more tools are developed that allow the conjunction of Neutrosophy with finite geometries. An interesting idea is the use of incidence matrices that allow the matrix representation of the incidence between points and lines, which can be generalized when the symbol “i” is included to represent indeterminacy. In this way, we work with geometries that can be both projective and affine, and these ideas will be linked to the theory of neutrosophic graphs. The study of the application of these new structures to problems in other branches of knowledge is explored, especially the convenience to use them in cryptography, e.g., in “secret sharing” and “authentication code”.

The chapter is divided into sections, the first of which contains the state of the art in NeutroGeometries, finite vector spaces, and finite geometries. Later, mixed projective-affine geometries are defined and their properties are established and demonstrated. Then, in another section, other notions are introduced that link NeutroGeometries with neutrosophic incidence matrices and their properties. One section is dedicated to analyzing the applications of this new theory in cryptography. The chapter ends with discussions and conclusions.

BACKGROUND

Definition 1 (Smarandache, 2020a): Let S be a classical geometric space, and <A> be a geometric concept (such as: postulate, axiom, theorem, property, function, transformation, operator, theory, etc.). The <AntiA> is the opposite of <A>, while <NeutA> (also called <NeutroA>) is the neutral (or indeterminate) part between <A> and <AntiA>. That is to say, the <NeutroA> contains axioms or geometric structures which are simultaneously satisfied by part of the elements of S and are not satisfied by another part of elements of S.

This process is known as neutrosophication. So, the neutrosophication tri-sections S into three subspaces, where <A> contains only totally true geometric concepts, <AntiA> contains at least one totally false geometric concept, whereas, <NeutA> is not <A> and neither <AntiA>, i.e., there is at least one geometric concept satisfied by some elements and also it is not satisfied by other elements.

Finite geometries are defined as geometric structures with a finite number of points (Dembowski, 1997). That is, they are structures that contain a finite set of elements called points, a finite set of lines defined from the points that pass through them, a finite set of planes, each one formed by points and lines, and so on. A hyperplane is defined as a geometric subspace with the dimension of the space minus one. The finite geometric spaces satisfy certain axioms, thus it is possible to identify these spaces with classical algebraic structures, especially vector spaces defined on finite-dimensional fields, they are the so-called Galois geometries.

Finite NeutroGeometries can also be defined as Neutrogeometries constructed in the framework of finite geometries.

Two fundamental types of finite geometric spaces are projective spaces and affine spaces. Projective spaces are elliptical by definition, which means that there are no parallel lines, every pair of lines intersect at a point, every pair of planes intersect on a line, and so on. On the other hand, affine spaces maintain equivalence relations of parallelism between lines. These spaces are directly related to other branches of mathematics such as orthogonal Latin squares or the combinatory theory. They have applications in code theory and cryptography. This section is devoted to exposing the details of the theory of finite geometries.

Definition 2 (Dembowski, 1997): An *incidence structure* is a triple $\boldsymbol{S}=(\mathcal{P},\mathcal{L},\mathcal{I})$, where $\mathcal{P},\mathcal{L},\mathcal{I}$ are sets satisfying: $\mathcal{P}\cap\mathcal{L}=\varnothing$ and $\mathcal{I}\subseteq\mathcal{P}\times\mathcal{L}$. The elements of $\mathcal{P}$ are called *points*, those of $\mathcal{L}$ *lines*, and $\mathcal{I}$ is a symmetric relation called *incidence*.

Usually, a line $l\in\mathcal{L}$ is defined as a subset of $\mathcal{P}$ having two or more points. Also, generally $P\mathcal{I}l$ and $l\mathcal{I}P$ denote that *P* is contained in the line *l* and the line *l* contains the point *P*, respectively, instead of the notations $(P,l)\in\mathcal{I}$ or $(l,P)\in\mathcal{I}$. We say that "*P* is incident with *l*" or "*l* is incident with *P*", or "*P* and *l* are incident".

Definition 3 (Dembowski, 1997): The *dual structure* $\bar{\boldsymbol{S}}=(\bar{\mathcal{P}},\bar{\mathcal{L}},\bar{\mathcal{I}})$ of the incidence structure $\boldsymbol{S}=(\mathcal{P},\mathcal{L},\mathcal{I})$ is defined by $\bar{\mathcal{P}}=\mathcal{L}$, $\bar{\mathcal{L}}=\mathcal{P}$, and for $P\in\mathcal{P}$ and $l\in\mathcal{L}$; $l\bar{\mathcal{I}}P$ if and only if $P\mathcal{I}l$, and $P\bar{\mathcal{I}}l$ if and only if $l\mathcal{I}P$. I.e., the points are interchanged with lines.

Definition 4 (Dembowski, 1997): If $\boldsymbol{S}=(\mathcal{P},\mathcal{L},\mathcal{I})$ is an incidence structure, $\mathcal{Q}\subset\mathcal{P}$ and $\mathcal{R}\subset\mathcal{L}$, then a *substructure* of $\boldsymbol{S}$ is the incidence structure $\boldsymbol{T}=(\mathcal{Q},\mathcal{R},\mathcal{J})$ with $\mathcal{J}=\mathcal{I}\cap(\mathcal{Q}\times\mathcal{R})$.

Definition 5 (Dembowski, 1997): Let $\boldsymbol{S}=(\mathcal{P},\mathcal{L},\mathcal{I})$ be an incidence structure, where points and lines are numbered in an arbitrary and fixed way: $P_1,P_2,\ldots,P_m$ and $l_1,l_2,\ldots,l_n$. Then the following two matrices are called *incidence matrices* $C=(c_{ij})$ and $D=(d_{ij})$, where

$$c_{ij}=\begin{cases}1, & \text{if } P_i\mathcal{I}l_j\\ 0, & \text{otherwise}\end{cases} \text{ and } d_{ij}=\begin{cases}1, & \text{if } l_i\mathcal{I}P_j\\ 0, & \text{otherwise}\end{cases}.$$

Definition 6 (Cameron, 2000; Zumbragel, 2016): A *projective space* is an incidence structure $\boldsymbol{P}=(\mathcal{P},\mathcal{L},\mathcal{I})$ with the following axioms:

(i) For each pair of two distinct points $P,\ Q\in\mathcal{P}$ there is a unique line that is incident with *P* and *Q*.

(ii) If A, B, C, D are four points such that the lines AB and CD meet, then also the lines AC and BD meet (*Veblen-Young axiom*), see Figure 1.
(iii) Each line is incident with at least three points.

The graphical illustration of the Veblen-Young axiom is shown in Figure 1.

Figure 1. Pictorial representation of the Veblen-Young axiom

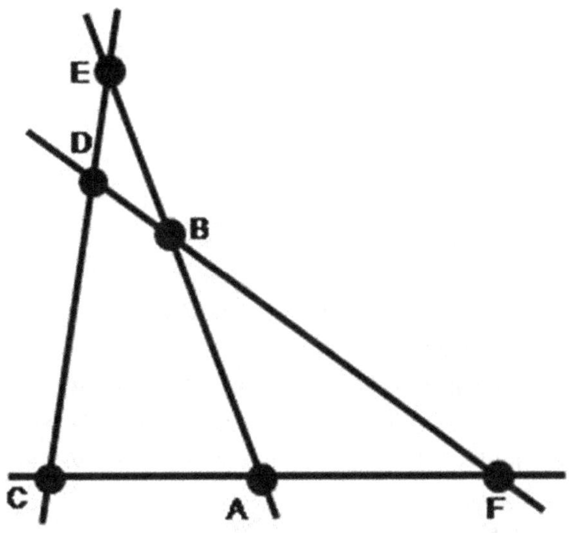

A *projective plane* is a non-degenerate projective space with the Veblen-Young axiom replaced by the stronger statement: Any two lines have at least one point in common.

Definition 7 (Zumbragel, 2016): An *affine space* is an incidence structure $A = (\mathcal{P}, \mathcal{L}, \mathcal{I})$ with the following axioms:

1. For each pair of two distinct points P, $Q \in \mathcal{P}$ there is a unique line that is incident with P and Q.
2. For each point $P \in \mathcal{P}$ and each line $l \in \mathcal{L}$, there exists a unique line $r \in \mathcal{L}$ such that $P\mathcal{I}r$, and l and r do not meet. This is called *parallelism* in **A** and is denoted by $l \| r$.

3. If A, B, C are non-collinear points, A' and B' are points such that $AB \parallel A'B'$ then two lines, one parallel to AC containing A' and the other one parallel to BC passing by B' intersect in a point C' (*triangle axiom*).
4. Each line is incident with at least two points.

The *order* of the projective space and the affine space is related to the number of points contained in one line.

Definition 8 (Dembowski, 1997; Hirschfeld, 1979): A *finite field* is a set K closed under two operations +, × such that:

- (K,+) is an Abelian group with identity 0, i.e., the operation + is associative, there is a neutral element 0, every element has an opposite, and it is commutative.
- (K\{0},×) is an abelian group with identity 1.
- For all $x,y,z \in K$, $x(y+z)=xy+xz$, $(x+y)z=xz+yz$ (Distributivity).
- K has a finite number of elements.

Theorem 1 (Dembowski, 1997; Hirschfeld, 1979): A finite field has q elements, where q is the power of a prime number.

A *finite vector space* is a vector space over a finite field.

Algebraically, a projective space is represented by a finite (n+1)-dimensional vector space over a field K with origin 0. It is denoted by V=V(n+1,K), (Dembowski, 1997; Hirschfeld, 1979).

Given $X,Y \in V \setminus \{0\}$ and for some basis $X=(x_0,x_1,\ldots,x_n)$, $Y=(y_0,y_1,\ldots,y_n)$, then X is equivalent to Y if there is some $\lambda \in K \setminus \{0\}$ such that $y_i=\lambda x_i$ for all i. This is an equivalence relation.

Thus, the set of equivalence classes is the *n-dimensional projective space over K* and it is denoted by PG(n,K). The points of PG(n,K) are the 1-dimensional subspaces of V, the "lines" are the 2-dimensional subspaces of V, the "planes" are the 3-dimensional subspaces of V, and the "hyperplanes" are the (n-1)-dimensional subspaces of V.

Moreover, a point of PG(n,K) is incident with the line in PG(n,K) if the corresponding 1-dimensional subspace is contained in the corresponding 2-dimensional subspace, and also, the intersection of two 3-dimensional spaces (two planes) is a 2-dimensional space (one line).

A useful formula from the classical linear algebra that associates dimensions of vector subspaces is the following:

$$dim(U+W) = dim(U) + dim(W) - dim(U \cap W) \tag{1}$$

Where, supposing U and V are vector subspaces of a vector space, then U∩W is the intersection of them, and U+W= {*x*+*y*: *x*∈U and *y*∈V}.

On the other hand, the affine spaces can be represented by vector spaces of dimension n over the field K, denoted by AG(n,K), where a k-dimensional subspace of AG(n,K) is an additive coset (or translate) of a k-dimensional vector subspace of the vector space (Cameron, 2000; Zumbragel, 2016). In AG(n,K) the points, lines, planes, hyperplanes are the cosets of the subspaces of rank 0, 1, 2, and n-1, respectively. That is to say, if x is a vector of AG(n,K), it is a point if w is a subspace of AG(n,K) of rank 1 generated by one single vector, then x+w is a line, and so on.

There is a relation between projective spaces and affine spaces, an affine space can be defined as a projective space minus one hyperplane and its subspaces (Cameron, 2000; Dembowski, 1997). Given an affine space, we obtain a projective space adding one point for connecting the set of parallel lines and finally defining a hyperplane that contains every of these added points.

It is well-known that parallelism in affine spaces is an equivalence relation.

Given **P** is a finite projective space with dimension d and order n. Let **U** be a t-dimensional subspace of **P**, then we have the following properties (Zumbragel, 2016):

1. The number of points of **U** is $\frac{n^{t+1}-1}{n-1}$.
2. The number of lines of **U** incident with a fixed point of **U** is $n^{t-1}+\ldots+n+1$.
3. The total number of lines of **U** is $\frac{\left(n^{t}+n^{t-1}+\cdots+n+1\right)\left(n^{t-1}+\cdots+n+1\right)}{n+1}$.
4. The number of hyperplanes of **P** is $n^{d}+\ldots+n+1$.
5. The number of hyperplanes of **P** incident with a fixed point is $n^{d-1}+\ldots+n+1$.

Given ***S***= PG(n,K) a projective space, there is a dual space $\bar{S}$ whose points are the hyperplanes of ***S*** and its hyperplanes are the points of ***S***. In general, the k-dimensional subspaces of $\bar{S}$ are the (d-k-1)-dimensional subspaces of ***S*** (Hirschfeld, 1979). Therefore, any theorem T involving the concepts of points, hyperplanes, and incidence, has its dual theorem $\bar{\mathrm{T}}$ interchanging points by hyperplanes and hyperplanes by points. Thus, in PG(2, K) point and line are dual, whereas in PG(3, K) point and plane are dual, and the line is dual with itself. This is the so-called *Principle of Duality*, which is not true in affine spaces.

Given the affine space AG(n, K) of dimension d (Zumbragel, 2016):

1. The number of points of AG(n, K) is n^{d}.
2. The number of points of any t-dimensional subspace is n^{t}.
3. The number of hyperplanes of AG(n, K) is $n^{d}+\ldots+n$.

Especially the theoretical results of the existence of orthogonal Latin squares are related to the existence of projective planes of certain orders. This topic has fascinated multiple mathematicians throughout history. Due to the combinatorial nature of finite geometry we can find problems with millions of possible combinations to find a projective plane of a certain order as small as n = 10, which finally was shown that it does not exist, (Lam, 1991; Lam, Thiel, & Swiercz, 1989). It had previously been shown that there is no such case as a projective plane of order 6. Some theorems seek the necessary or sufficient conditions for the existence of projective planes. An important result is the Bruck-Ryser Theorem which says the following:

Theorem 2 (Lam, 1991): If $n \equiv 1,2 \pmod 4$, then a necessary condition for the existence of a finite projective plane of order n is that integers x, y exist satisfying $x^2+y^2=n$.

However, there are computational approaches to the subject, where the mathematical demonstrations obtained from the implementation of an algorithm and the performance of combinatorial calculations are carried out (Lam et al., 1989).

MIXED PROJECTIVE-AFFINE SPACES

This section is dedicated to introducing the concept of Mixed Projective-Affine spaces, which is the main contribution of this chapter.

Definition 9: A *Mixed Projective-Affine space* (MPA space) $\mathcal{MPA} = (\mathcal{P}, \mathcal{L}, \mathcal{I})$ is an incidence structure such that there is a subset of $\mathcal{S}_p \subsetneq \mathcal{P} \cup \mathcal{L}$ which satisfies the axiomatic of the projective spaces and another subset $\mathcal{S}_a \subsetneq \mathcal{P} \cup \mathcal{L}$ which satisfies the axiomatic of the affine spaces, where $\mathcal{S}_p \cup \mathcal{S}_a = \mathcal{P} \cup \mathcal{L}$.

Definition 10: A *Mixed Projective-Affine subspace* of the MPA $\mathcal{MPA} = (\mathcal{P}, \mathcal{L}, \mathcal{I})$ is an MPA which is a subset of $\mathcal{MPA} = (\mathcal{P}, \mathcal{L}, \mathcal{I})$.

Proposition 1: Let $\mathcal{MPA} = (\mathcal{P}, \mathcal{L}, \mathcal{I})$ be an MPA then it is a NeutroGeometry.

Proof. There is an axiom that is satisfied by affine spaces and not satisfied by projective spaces, it is that for each line *l* and each point *P* non-incident with *l*, there is a unique line *r* which is incident with *P* and is parallel to *l*. Thus, the elements of $\mathcal{P} \cup \mathcal{L}$ with $\mathcal{I}$ restricted to $\mathcal{S}_p$ satisfy that every two lines are not parallel, and the elements of $\mathcal{P} \cup \mathcal{L}$ with $\mathcal{I}$ restricted to $\mathcal{S}_a$ satisfy that every line has a parallel not equal to itself. On the other hand, the axiom (iii) of affine spaces in Definition 7 is satisfied by the lines of $\mathcal{S}_a$, but makes nonsense in $\mathcal{S}_p$, because this axiom is valid only if there is parallelism, and says nothing where parallelism does not occur. Additionally, the Veblen-Young axiom is satisfied by $\mathcal{S}_p$ and not necessarily by

$\mathcal{S}_a$. So, the elements of $\mathcal{S}_p$ are exclusively projective and those of $\mathcal{S}_a$ are exclusively affine, and $\mathcal{MPA}$ is a NeutroGeometry.□

Proposition 2: Given a projective space *p* and an affine space *a*, there is an MPA that can be formed from them.

Proof. Let $p=(\mathcal{P},\mathcal{L},\mathcal{I})$ be a projective space and $P\in\mathcal{P}$, then $\mathcal{MPA}_p=(\mathcal{P}',\mathcal{L}',\mathcal{I}')$ such that $\mathcal{P}'=\mathcal{P}\setminus\{P\}$, $\mathcal{L}'=\{l':l'=l\setminus\{P\}$ for all $l\in\mathcal{L}\}$ and $\mathcal{I}'=\mathcal{I}\cap(\mathcal{P}'\times\mathcal{L}')$, such that it is an incidence structure. In other words, we eliminate only one point of the projective space. Thus, the lines incident with the point from the projective space become parallels when *P* is removed and remain non-parallels for the non-removed points. So, the space is partially projective and partially affine.

Now, let $a=(\mathcal{P},\mathcal{L},\mathcal{I})$ be an affine space, then let us define $\mathcal{H}\subsetneq\{[l]:l\in\mathcal{L}\}$ and $|\mathcal{H}|=n$, where [*l*] is the parallel class of *l*. Let us define $l'=l\cup(\{[l]\}\cap\mathcal{H})$, and for each plane *α* in a let $l_\alpha =$ {[l]: *l* is a line in $\alpha\cap\mathcal{H}$ }. Then, we define $\mathcal{P}'=\mathcal{P}\cup\mathcal{H}$ and $\mathcal{L}'=\{l':l\in\mathcal{L}\}\cup\{l_\alpha$: *α* is a plane in a}. $\mathcal{MPA}_a=(\mathcal{P}',\mathcal{L}',\mathcal{I}')$ is an MPA space. In other words, for each of some parallel classes we add a point incident with them, and also we add a hyperplane containing those new points. Thus, the newly added points and the hyperplane containing them become in part of a projective axiomatic, and the other ones satisfy the affine axiomatic.□

Remark 2: These procedures generalize the procedure to convert a projective space into an affine space when $\mathcal{P}'=\mathcal{P}\setminus\mathcal{H}$ and also the procedure to convert an affine space into a projective space when $\mathcal{H}=\{[l]:l\in\mathcal{L}\}$, respectively.

Example 1: Let us perform this procedure on the Fano plane (Kärteszi, 1976), which is the smallest projective plane, it has order n = 2, so each line has n+1 = 3 points and each point is incident with exactly n+1 = 3 lines. It has a total of 7 points and 7 lines. Figure 2 represents the transformation performed on the Fano plane to obtain $\mathcal{MPA}_p=(\mathcal{P}',\mathcal{L}',\mathcal{I}')$.

The procedure consisted in removing the point *F* in the Fano plane and we obtained three lines *BD*, *CG*, and *AE* satisfying the axiom of parallelism of affine planes; whereas lines *AGD*, *BGE*, *CDE*, and *ABC* have a unique point in common with the others, like in projective planes. So, this is partially a projective plane and partially an affine plane.

Figure 3 represents the second procedure, starting from an affine plane and ending with an $\mathcal{MPA}_a$ plane.

This procedure consisted in connecting the parallel lines *EG* and *AC* in point *B* and the parallels *AG* and *CE* in point *D* and later *B* and *D* are connected by line *BD*.

Figure 2. Fano plane on the left and $\mathcal{MPA}_p$ on the right. Parallels are represented in red lines.

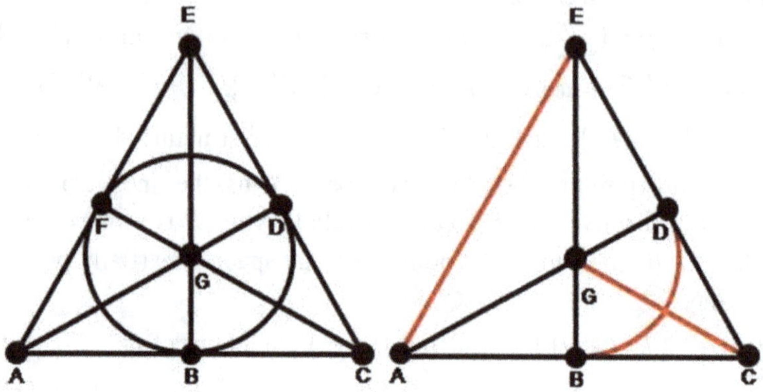

Figure 3. Affine plane on the left and $\mathcal{MPA}_a$ on the right. Parallels are represented in the same color lines in red, green, and blue.

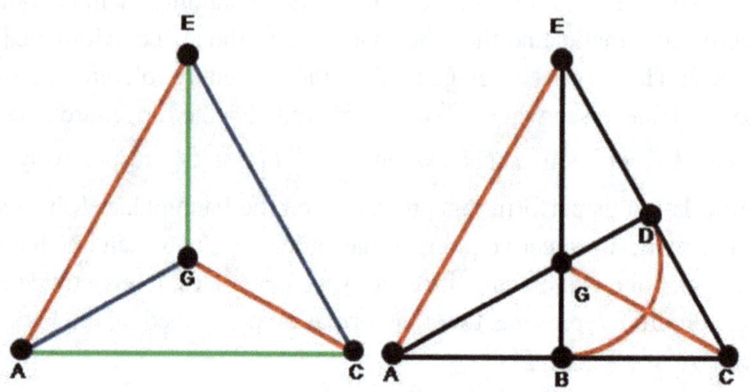

Remark 3: Let us note that in both procedures we left two points of the line *BDF* (Figure 2) and we added two points in Figure 3, otherwise the first axiom of incidence structures is violated. Also, even in the affine plane of Figure 3, the triangular axiom (Definition 7 (iii)) cannot be applied because it needs at least five different points, so it is technically not violated.

See that according to Figures 2 and 3 every point in the MPA is incident with three lines, but the line *BD* is only incident with two points, so the new plane does not satisfy the Principle of Duality. Moreover, the plane contains six points and seven lines.

It is well known that every projective space is an affine space recovered by a hyperplane. Here, we obtained that between the projective space and its corresponding affine space there is at least an MPA space, which recovers the affine space and is recovered by the projective space.

The procedures to obtain $\mathcal{MPA}_p$ or $\mathcal{MPA}_a$ can be extended in interesting ways, the following example illustrates the result of $\mathcal{MPA}_p$ when it is performed once over one hyperplane and later over another hyperplane removing a point not contained in the first line, however we should pay attention, because this algorithm not always yield to an MPA plane.

Example 2: Now, let us consider the projective plane of order n = 3, see Figure 4 (Zangalho-Raposo, 2014).

Figure 4. Projective plane of order 3. Every line is represented by one single color.

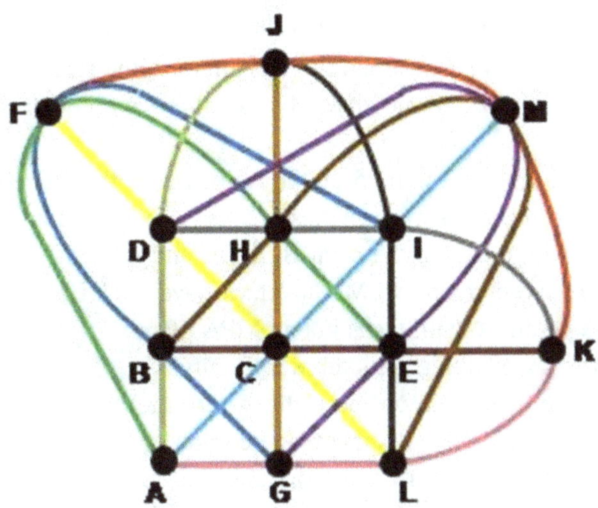

See that every line has 4 points and every point is incident with 4 lines. The number of points and lines is 13.

Figure 5. On the left, $\mathcal{MPA}_p$ obtained from the projective plane of order 3 removing point F. On the right, $\mathcal{MPA}_p$ obtained after also removing point C.

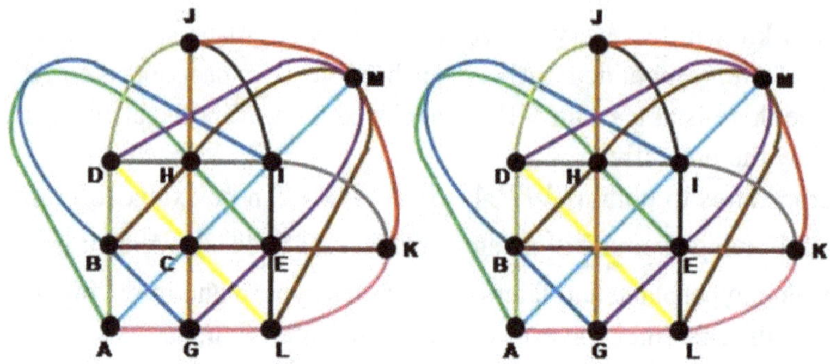

Let us make the procedure $\mathcal{MPA}_p$ removing the point F from the line FJKM, see Figure 5 on the left, and later over that result let us remove the point C, see Figure 5 on the right. When the procedure is made once (Figures 2 and 3 on the right and Figure 5 on the left) we obtain that the relation r=s or r||s, where r and s are two lines of the $\mathcal{MPA}_p$ or they are parallels in $\mathcal{MPA}_a$. This is easy to prove, because if s has affine properties then there is another line r such that r||s. On the other hand, if s is part of the projective subspace there is no other line s such that r||s, but r = s.

However, the example in Figure 5 on the right, where the procedure is applied simultaneously to two different hyperplanes, not necessarily the equivalence relation is satisfied, e.g., in Figure 5 on the right BGI||DL and DL||BEK, but BGI∦BEK, therefore the relation is not transitive. Moreover, we have that DL||AIM and DL||AHE, so for A pass two parallels, and this is not an MPA plane.

Another interesting example is pictured in Figure 6.

In Figure 6 there is a plane on the right, where the point *F* is removed only from the line *BDF*. Thus, this is a new plane contained in the Fano plane, which contains the MPA plane in Figure 1 on the right. So, between the projective plane and its equivalent affine plane, there are other non-MPA planes. Additionally, in the example we have *CFG*||*BD*, *BD*||*AEF*, and *CFG*∦*AEF*, therefore these lines do not form an affine subplane and this is not an MPA plane.

Figure 6. Fano plane on the left and non-projective, non-affine, and non-MPA plane on the right

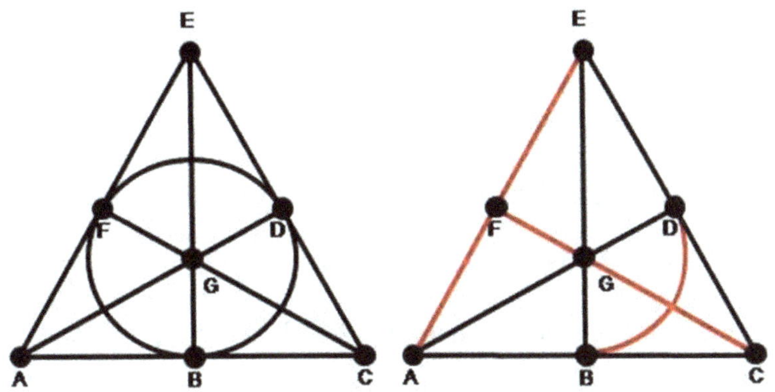

Proposition 3: Let $\mathcal{MPA} = (\mathcal{P}, \mathcal{L}, \mathcal{I})$ be an MPA space. Let us define two sets of lines, they are the following:

1. $\mathcal{PL} = \{l \in \mathcal{L} : \forall s \in \mathcal{L}$ in the same plane of $l, \exists P \in \mathcal{P}$ s.t. $l \cap s = P\}$, i.e., this is the set of lines in $\mathcal{MPA}$ which satisfy they have a common point with the other lines in the same plane.
2. $\mathcal{AL} = \{l \in \mathcal{L} : \forall P \in \mathcal{P}$ and not $P\mathcal{I}l \exists! s \in \mathcal{L}$ in the same plane of l, s.t. $l \| s$ and $P\mathcal{I}s\}$, i.e., $\mathcal{AL}$ is the set of lines in $\mathcal{MPA}$ which satisfy that for every point *P* no incident with *l* there is a unique line in the plane of *l* such that *l*||*s* and *s* is incident with *P*.

Then, $\mathcal{PL} \cap \mathcal{AL} = \varnothing$.

Proof. Let us suppose that there is a line $l \in \mathcal{L}$ such that $l \in \mathcal{PL} \cap \mathcal{AL}$. Let *P* be a point no incident with *l*. $l \in \mathcal{AL}$ implies that there is another line $s \in \mathcal{L}$, both contained in the same plane such that $P\mathcal{I}s$ and *l*||*s*. On the other hand, $l \in \mathcal{PL}$ implies that *l* and *s* are both incident with a common point in the plane, which contradicts that *l*||*s*, so the line *l* does not exist and $\mathcal{PL} \cap \mathcal{AL} = \varnothing$.□

Remark 4: See that the MPA cannot be divided into two disjoint sets of points.

The consequence of this proposition shall be applied in further results throughout this chapter. The first one is the following corollary:

Corollary 1: Every MPA space has MPA subspaces that are either projective spaces, affine spaces, MPA spaces, or a spaces that is none of them.

Proof. Let us take a subset of lines in the MPA and span algebraically the space from this subset. If this is a subset of $\mathcal{PL}$ we span a projective subspace, if this is a subset of $\mathcal{AL}$ we span an affine space, and if we take lines from both $\mathcal{PL}$ and $\mathcal{AL}$ we obtain an MPA space or other. □

The previous proposition allows us to define a quantification of the degree to which one MPA is projective and affine. This measure is defined in the following:

Definition 11: Let $\mathcal{MPA}$ be a Mixed Projective-Affine space, the *Degree of Type of Geometry* (DTG) is a triad (*p,i,a*), where $p = \frac{|\mathcal{PL}|}{|\mathcal{L}|}$, $a = \frac{|\mathcal{AL}|}{|\mathcal{L}|}$, and i=1–|*p*–*a*|, where $|\mathcal{L}|$ is the number of lines in $\mathcal{MPA}$, whereas, $\mathcal{PL}$ and $\mathcal{AL}$ were defined above.

It is easy to prove from Proposition 3 that $|\mathcal{PL}|+|\mathcal{AL}|=|\mathcal{L}|$, $p,a,i\in[0,1]$ and therefore $p+a=1$.

Example 3: The MPA in Figure 2 satisfies $|\mathcal{L}|=7$, $|\mathcal{PL}|=4$, $|\mathcal{AL}|=3$, then its DTG is the triad $\left(\frac{4}{7},1-\left|\frac{4}{7}-\frac{3}{7}\right|,\frac{3}{7}\right)=\left(\frac{4}{7},\frac{6}{7},\frac{3}{7}\right)$. The pure projective and affine spaces also have their DTG, e.g., in the left of Figure 2 the projective plane has DTG= (1,0,0), and for the affine plane in the left of Figure 3, we have DTG= (0,0,1).

Proposition 4: The algebraic axioms of projective and affine spaces are NeutroAlgebras for the MPA.

Proof. As we saw, there is an equivalent algebraic representation of both, the projective and the affine spaces, so from this point of view the algebraic representation of MPAs are also NeutroAlgebras.□

Remark 1: For example, the projective spaces satisfy that the intersection of two different 2-dimensional vector spaces (two lines) in the same 3-dimensional vector space (the plane) is always a 1-dimensional vector space (one point). These properties are deducted from the axioms of projective spaces. But in the MPA, some intersections of two 2-dimensional spaces are a 1-dimensional space (if they are in $\mathcal{PL}$) and others are empty (if they are parallels in $\mathcal{AL}$). Therefore, for some pairs of 2-dimensional spaces this algebraic property is satisfied and for others not and this is a NeutroProperty, then in the axiomatic of projective spaces there is at least a NeutroAxiom for MPA.

Also, the property such that given a 2-dimensional space l and a 1-dimensional space P which is not a subspace of l, there exists a 2-dimensional space s such that P is a subspace of s and $l\cap s=\varnothing$, this is a NeutroProperty for MPA, so, there is at least a NeutroAxiom of affine spaces for MPA.

See that although we are dealing with NeutroGeometries, the vector spaces associated with them are classic finite vector spaces.

The second question is that given a finite NeutroVector Space if we can associate it with a Finite NeutroGeometry. The answer is not simple, that is why it will be the subject of study for future works. Here we will only give a brief explanation on this topic. The properties of vector space that are often used in the theory of finite geometries are fundamentally based on the vector dimension of spaces and subspaces. That is why within the concepts of vector space it is necessary to highlight the concept of a basis of the space and as a consequence the concepts of generator vectors of the space and linearly independent vectors.

It would be necessary to develop a theory of NeutroVector Spaces, and the NeutroAxioms that would influence the concept of the basis, taking into account that each element of the NeutroSpace would have to be represented as an independent linear combination of the elements of a basis.

Another additional matter is if the NeutroVector Space is so due to the field on which it is defined, i.e., it is defined over a NeutroField. In some finite geometry texts, the authors demand weaker conditions for the scalars, for example, that they be only skew-fields. If within the NeutroVector Space the axioms related to commutativity of the sum between vectors or the product by a scalar are converted into NeutroAxioms, it is very likely that the definition of the basis will not be affected, because it would be enough to perform these operations only by the right or by the left and the definition of dimension would not be affected.

In case that associativity was not fulfilled, it would be enough to take the elements in a certain order and the dimension would depend on this order, thus, this would not be a great difficulty if the order of its elements were specified in the basis.

A more important problem would be if the axioms of the existence of a neutral vector or the opposite for each vector of the space become NeutroAxioms. The first of them would be a more complex problem since in some cases there would be a neutral and in others not, and this vector is fundamental to define linear independence. That is why, for these NeutroAxioms, the existence of bases for a subset of the space would be fulfilled and it would not be fulfilled for another, depending on where the neutral element exists and where it does not.

GENERALIZATION OF THE INCIDENCE MATRIX IN FINITE GEOMETRIES

Another way of generalizing finite geometries in the plane is the processing of their incidence matrices. Each plane is represented by an incidence matrix and each incidence matrix represents a plane, although this does not mean that the plane and the matrix are either projective or affine geometries. From the existence of a projective plane of a certain order, the existence of its equivalent affine plane can be deduced and vice versa, that is why from the non-existence of projective planes of order n = 6 or n = 10 the non-existence of affine planes of these orders can be deduced.

A solution of this problem is to obtain planes that are neither affine nor projective, but which can be mixed, not necessarily MPA. For this purpose, *Mixed Approximate Planes* are defined for the first time.

Definition 12: A *Mixed Approximate Plane* is a plane whose incidence matrix $A=\left(a_{ij}\right)_{0\le i,j\le n}$ satisfies one of the solutions of the equation $\min_{a_{ij}\in\{0,1\}} AA^{T}-nI-J$.

Where A^T is the transpose of A, I is the identity matrix, J is the matrix of all 1's, n is the order of the plane, and ||•|| is a norm.

This is the consequence of the well-known equations such that the incidence matrix represents a projective plane if it satisfies the following conditions:

(i) A has constant row sum n+1,
(ii) A has constant column sum n+1,
(iii) The inner product of two distinct rows is 1,
(iv) The inner product of two distinct columns is 1.

The conditions corresponding to rows can be reduced to one single matrix equation:

$$AA^{T}=nI+J$$

The condition shown in Definition 12 is a binary programming problem and the equation $AA^T - nI - J = 0$ represents a projective plane, where 0 means the null matrix.

For example, let us study the incidence matrix of the two planes in Figure 5, considering that they started from a projective plane so that A is a square matrix and the previous calculation makes sense.

For the figure on the left we have:

$$A_L = \begin{bmatrix} 1 & 0 & 0 & 0 & 0 & 0 & 1 & 0 & 0 & 0 & 1 & 1 & 0 \\ 0 & 0 & 0 & 0 & 0 & 0 & 0 & 0 & 0 & 1 & 1 & 0 & 1 \\ 0 & 1 & 0 & 0 & 0 & 0 & 0 & 1 & 0 & 0 & 0 & 1 & 1 \\ 0 & 0 & 0 & 1 & 1 & 0 & 1 & 0 & 0 & 0 & 0 & 0 & 1 \\ 1 & 0 & 0 & 0 & 1 & 0 & 0 & 1 & 0 & 0 & 0 & 0 & 0 \\ 0 & 1 & 0 & 0 & 0 & 0 & 1 & 0 & 1 & 0 & 0 & 0 & 0 \\ 1 & 1 & 0 & 1 & 0 & 0 & 0 & 0 & 0 & 1 & 0 & 0 & 0 \\ 0 & 0 & 1 & 1 & 0 & 0 & 0 & 0 & 0 & 0 & 0 & 1 & 0 \\ 0 & 0 & 0 & 1 & 0 & 0 & 0 & 1 & 1 & 0 & 1 & 0 & 0 \\ 0 & 0 & 0 & 0 & 1 & 0 & 0 & 0 & 1 & 1 & 0 & 1 & 0 \\ 0 & 0 & 1 & 0 & 0 & 0 & 1 & 1 & 0 & 1 & 0 & 0 & 0 \\ 0 & 1 & 1 & 0 & 1 & 0 & 0 & 0 & 0 & 0 & 1 & 0 & 0 \\ 1 & 0 & 1 & 0 & 0 & 0 & 0 & 0 & 1 & 0 & 0 & 0 & 1 \end{bmatrix}$$

Whereas the figure on the right has the following incidence matrix:

$$A_R = \begin{bmatrix} 1 & 0 & 0 & 0 & 0 & 0 & 1 & 0 & 0 & 0 & 1 & 1 & 0 \\ 0 & 0 & 0 & 0 & 0 & 0 & 0 & 0 & 0 & 1 & 1 & 0 & 1 \\ 0 & 1 & 0 & 0 & 0 & 0 & 0 & 1 & 0 & 0 & 0 & 1 & 1 \\ 0 & 0 & 0 & 1 & 1 & 0 & 1 & 0 & 0 & 0 & 0 & 0 & 1 \\ 1 & 0 & 0 & 0 & 1 & 0 & 0 & 1 & 0 & 0 & 0 & 0 & 0 \\ 0 & 1 & 0 & 0 & 0 & 0 & 1 & 0 & 1 & 0 & 0 & 0 & 0 \\ 1 & 1 & 0 & 1 & 0 & 0 & 0 & 0 & 0 & 1 & 0 & 0 & 0 \\ 0 & 0 & 0 & 1 & 0 & 0 & 0 & 0 & 0 & 0 & 0 & 1 & 0 \\ 0 & 0 & 0 & 1 & 0 & 0 & 0 & 1 & 1 & 0 & 1 & 0 & 0 \\ 0 & 0 & 0 & 0 & 1 & 0 & 0 & 0 & 1 & 1 & 0 & 1 & 0 \\ 0 & 0 & 0 & 0 & 0 & 0 & 1 & 1 & 0 & 1 & 0 & 0 & 0 \\ 0 & 1 & 0 & 0 & 1 & 0 & 0 & 0 & 0 & 0 & 1 & 0 & 0 \\ 1 & 0 & 0 & 0 & 0 & 0 & 0 & 0 & 1 & 0 & 0 & 0 & 1 \end{bmatrix}$$

So, we have:

$$A_L A_L^T = \begin{bmatrix} 4 & 1 & 1 & 1 & 1 & 1 & 1 & 1 & 1 & 1 & 1 & 1 & 1 \\ 1 & 3 & 1 & 1 & 0 & 0 & 1 & 0 & 1 & 1 & 1 & 1 & 1 \\ 1 & 1 & 4 & 1 & 1 & 1 & 1 & 1 & 1 & 1 & 1 & 1 & 1 \\ 1 & 1 & 1 & 4 & 1 & 1 & 1 & 1 & 1 & 1 & 1 & 1 & 1 \\ 1 & 0 & 1 & 1 & 3 & 0 & 1 & 0 & 1 & 1 & 1 & 1 & 1 \\ 1 & 0 & 1 & 1 & 0 & 3 & 1 & 0 & 1 & 1 & 1 & 1 & 1 \\ 1 & 1 & 1 & 1 & 1 & 1 & 4 & 1 & 1 & 1 & 1 & 1 & 1 \\ 1 & 0 & 1 & 1 & 0 & 0 & 1 & 3 & 1 & 1 & 1 & 1 & 1 \\ 1 & 1 & 1 & 1 & 1 & 1 & 1 & 1 & 4 & 1 & 1 & 1 & 1 \\ 1 & 1 & 1 & 1 & 1 & 1 & 1 & 1 & 1 & 4 & 1 & 1 & 1 \\ 1 & 1 & 1 & 1 & 1 & 1 & 1 & 1 & 1 & 1 & 4 & 1 & 1 \\ 1 & 1 & 1 & 1 & 1 & 1 & 1 & 1 & 1 & 1 & 1 & 4 & 1 \\ 1 & 1 & 1 & 1 & 1 & 1 & 1 & 1 & 1 & 1 & 1 & 1 & 4 \end{bmatrix}$$

and

$$A_R A_R^T = \begin{bmatrix} 4 & 1 & 1 & 1 & 1 & 1 & 1 & 1 & 1 & 1 & 1 & 1 & 1 \\ 1 & 3 & 1 & 1 & 0 & 0 & 1 & 0 & 1 & 1 & 1 & 1 & 1 \\ 1 & 1 & 4 & 1 & 1 & 1 & 1 & 1 & 1 & 1 & 1 & 1 & 1 \\ 1 & 1 & 1 & 4 & 1 & 1 & 1 & 1 & 1 & 1 & 1 & 1 & 1 \\ 1 & 0 & 1 & 1 & 3 & 0 & 1 & 0 & 1 & 1 & 1 & 1 & 1 \\ 1 & 0 & 1 & 1 & 0 & 3 & 1 & 0 & 1 & 1 & 1 & 1 & 1 \\ 1 & 1 & 1 & 1 & 1 & 1 & 4 & 1 & 1 & 1 & 1 & 1 & 1 \\ 1 & 0 & 1 & 1 & 0 & 0 & 1 & 2 & 1 & 1 & 0 & 0 & 0 \\ 1 & 1 & 1 & 1 & 1 & 1 & 1 & 1 & 4 & 1 & 1 & 1 & 1 \\ 1 & 1 & 1 & 1 & 1 & 1 & 1 & 1 & 1 & 4 & 1 & 1 & 1 \\ 1 & 1 & 1 & 1 & 1 & 1 & 1 & 0 & 1 & 1 & 3 & 0 & 0 \\ 1 & 1 & 1 & 1 & 1 & 1 & 1 & 0 & 1 & 1 & 0 & 3 & 0 \\ 1 & 1 & 1 & 1 & 1 & 1 & 1 & 0 & 1 & 1 & 0 & 0 & 3 \end{bmatrix}$$

Then, $\|A_L A_L^T - nI - J\|_{Frob} = 4$ and $\|A_R A_R^T - nI - J\|_{Frob} = 5.8310$, where $\|\bullet\|_{Frob}$ is the Frobenius norm. See that $\|A_L A_L^T - nI - J\|_{Frob} < \|A_R A_R^T - nI - J\|_{Frob}$ because A_L is nearer than A_R to be a projective plane.

An interesting problem is to find the matrix satisfying $\min_{a_{ij}\in\{0,1\}}\|AA^T - 10I - J\|$, where A is a square matrix of order 111 of an MPA plane. Then, we can interpret this A like the most approximate MPA plane to the projective plane of order 10.

Incidence matrices can be used to calculate the DTG, see the following proposition:

Proposition 5: Let A be an incidence matrix of an MPA plane, if m is the number of rows in AA^T having at least one 0, i.e., m=|R|, where

$$R = \left\{ \prod_j d_{ij} = 0 : D = (d_{ij}), D = AA^T \right\}.$$

Then the triad $a = \frac{m}{n^2+n+1}$, p=1–a, and i=1– |a–p| are the components of the DTG of the plane.

Proof. If two lines in the plane are parallels, the inner product between their rows in the incidence matrix is 0. Therefore, the row where at least one element of the matrix AA^T has a 0 is due to there is a parallelism between two lines. On the other hand, n^2+n+1 is the number of lines in the projective plane, so, $a = \frac{m}{n^2+n+1}$. Evidently, p=1–a and i=1– |a–p|.□

Example 4: For the plane in Figure 5 on the left we have m = 4 and $n^2+n+1=13$, so, from the matrix $A_L A_L^T$, $DTG = \left(\frac{9}{13}, \frac{8}{13}, \frac{4}{13}\right)$.

Additionally, the plane of Figure 5 on the right satisfies m = 7 and $DTG = \left(\frac{6}{13}, \frac{12}{13}, \frac{7}{13}\right)$ using the matrix $A_R A_R^T$.

The incidence matrices also allow us to deal with the indeterminacy of the plane, when the domain of the incidence matrix is extended from {0,1} to {0,1,i}, and i is the symbol to represent indeterminacy. For example, the following matrix:

$$A = \begin{bmatrix} 1 & 0 & 0 & 0 & 0 & 0 & 1 & 0 & 0 & 0 & i & 1 & 0 \\ 0 & 0 & 0 & 0 & 0 & i & 0 & 0 & 0 & i & i & 0 & i \\ 0 & 1 & 0 & 0 & 0 & 0 & 0 & 1 & 0 & 0 & 0 & 1 & i \\ 0 & 0 & 0 & 1 & 1 & 0 & 1 & 0 & 0 & 0 & 0 & 0 & i \\ 1 & 0 & 0 & 0 & 1 & i & 0 & 1 & 0 & 0 & 0 & 0 & 0 \\ 0 & 1 & 0 & 0 & 0 & i & 1 & 0 & 1 & 0 & 0 & 0 & 0 \\ 1 & 1 & 0 & 1 & 0 & 0 & 0 & 0 & 0 & i & 0 & 0 & 0 \\ 0 & 0 & 1 & 1 & 0 & i & 0 & 0 & 0 & 0 & 0 & 1 & 0 \\ 0 & 0 & 0 & 1 & 0 & 0 & 0 & 1 & 1 & 0 & i & 0 & 0 \\ 0 & 0 & 0 & 0 & 1 & 0 & 0 & 0 & 1 & i & 0 & 1 & 0 \\ 0 & 0 & 1 & 0 & 0 & 0 & 1 & 1 & 0 & i & 0 & 0 & 0 \\ 0 & 1 & 1 & 0 & 1 & 0 & 0 & 0 & 0 & 0 & i & 0 & 0 \\ 1 & 0 & 1 & 0 & 0 & 0 & 0 & 0 & 1 & 0 & 0 & 0 & i \end{bmatrix}$$

Represents projective, affine, and MPA planes at the same time, when i means a value in $\{0, 1\}$ we do not know exactly which is it. When all of them are ones, it is a projective plane, if all of them are zeros, it is an affine plane, when some of them are zeros and others are ones, it is another kind of plane including MPA.

We call this type of matrices, *neutrosophic incidence matrices*, which represent bipartite graphs with neutrosophic values. Moreover, these geometries could model networks whose connections are not always 100% true or 100% false, so, we can extend the elements of the incidence matrix to the domain $[0,1] \cup i$ were a value $v \in (0,1)$ means there is not a completely true or a completely false link between the two elements connected by v.

APPLICATIONS

This section contains an approach to the possible applications of MPA in cryptography in two subjects, secret sharing and authentication systems.

Secret Sharing

Suppose there is a group of n participants who have the key to reveal a secret. To reveal it, the following two conditions must be satisfied (Klein & Storme, 2011):

1. Collectively, any group of t participants can reconstruct the secret.
2. Less than t participants cannot reconstruct the secret.

A solution to this problem is the model of the t-threshold scheme, where a set $P_0,P_1,\ldots,P_n$ of n + 1 three to three non-collinear points is selected. A hyperplane π is fixed from the projective space of dimension t and order q. A line *l* is taken from the space such that $\pi \cap l = P0$.

Each ith point $P_i (1 \leq i \leq n)$ is assigned to the ith participant. The information that P_0 is contained in *l* is public. However, we do not know which is the hyperplane π.

If a set U of participants associated with some points tries to access the system then they span a subspace C and it is checked if $C \cap l = P_0$. As can be deduced, if $U \neq \pi$, either because U is a subset of $\{P_1,\ldots,P_n\}$ with cardinality t-1 or less, or because U contains an element that does not belong to $\{P_1,\ldots,P_n\}$, then the condition is not satisfied.

An application of MPAs in secret sharing can be inferred from the plane of Figure 5 on the left. Suppose that the secret is reconstructed from more than one point, as defined in (Beutelspacher, 1990). In this case, the line *BCEK* is taken as the one that contains the set of three points *B*, *C*, and *E* that constitute the secret points. Let us suppose we have three independent sets of evaluators, each one of them is constituted by two people represented by two points of the three parallel lines, *AEH*, *BGI*, and *CDL*. From the first line the participants are represented by the points *A* and *H*; from the second line by the points *G* and *I*; from the third line by the points *D* and *L*. If the two participants for each line reach an agreement, then they generate the line and therefore the third point of each of them is included, and when making the intersection with the line *BCEK* we obtain the three points that are used for the secret.

Another variant is to select one or more parallel lines and additionally another one corresponding to the projective part of the MPA, this guarantees, in addition to the independence of the participants for each line, also a certain veto power, since the participants represented by parallel lines need two participants to generate the line, while the others have more possible combinations of the agreement by having lines with one more element. This gives more importance to the participants represented by the parallel lines.

This idea can be generalized to higher dimensional spaces, not only the plane, especially when we want more participants. For example, if we want a room for one more participant, we should increase the dimension of the space and also the order. As well, the possible combinations of cases are richer than with pure projective spaces, this is easy to understand by analyzing the plane that appears in Figure 5 on the right.

Authentication Systems

In an *authentication system,* it is assumed that one person issues a message M for another person. Faced with the danger that an attacker would exchange the message for another one with a malicious objective, the sender and the receiver secretly agree on both a key *K* and an authentication function f that only they know. Then, the sender sends the message encoded with the agreed key and the output is f(M,*K*).

When the sender sends the message M along with f(M,*K*), the receiving person receives a message M' and an authenticator A'. Then, he (she) calculates $f(M', K)$ and accepts the received message as correct if $f(M', K)$ matches A', otherwise, he (she) realizes that the message was adulterated. Formally this is expressed as follows:

Definition 13: A *message authentication code* is a 4-tuple $(\mathcal{S}, \mathcal{A}, \mathcal{K}, \mathcal{E})$ constituted by the following elements (Klein & Storme, 2011):

- $\mathcal{S}$ is a finite set of elements called *messages*.
- $\mathcal{A}$ is a finite set of *authentication tags*.
- $\mathcal{K}$ is a finite set of *keys*.
- $\forall K \in \mathcal{K}\ \exists e_K \in \mathcal{E},\ e_K : \mathcal{S} \to \mathcal{A}$ called *authentication rule*.

A geometric solution to this problem uses a projective plane of order n, let us call it π, and let *l* be a fixed line contained in π. In this model, the points of *l* are taken as messages or elements of $\mathcal{S}$ and the points of the affine plane $\pi\backslash l$ are taken as keys. The authentication tag is obtained from the *authentication rule* $e_K(s)$ which is defined as the line in π that joins the message (point) $s \in \mathcal{S}$ with the key $K \in \pi\backslash l$.

Suppose that an attacker wants to replace the original message *s* by s' without knowing the secret key $K \in \pi\backslash l$, only knowing that the messages are in l. For this, he (she) must guess the line $e_K(s')$ that joins s' with a *K* contained in a line that is not *l*. We know that in the plane there are other n lines containing M and therefore the probability of guessing the correct one is $\frac{1}{n}$. Furthermore, it follows from the properties of the affine planes that $|\mathcal{K}| = n^2$, that is, the possible number of possible keys is the square of the order of π, and the probability of finding the correct key is $\frac{1}{\sqrt{|\mathcal{K}|}}$, this is what is called a *perfect system*.

However, this model has two limitations (Beutelspacher, 1990):

- There is a much smaller number of possible messages compared to the number of possible keys.
- If the attacker gets to know that two messages have been authenticated with the same key *K*, then he (she) can know *K*.

Regarding the first limitation, one of its possible causes is the dichotomy between projective and affine planes that exists in classical finite geometries. In the approach proposed in this chapter, it is possible to define the messages as points of more than one line; they could be one line and part of another, several lines and part of another, or several complete lines.

For the solution of the second limitation, the messages are defined in hyperplanes instead of lines. In this case, they can also be extended to the theory of MPA spaces.

DISCUSSION

Finite Geometry is a point of confluence of various subjects of mathematics, such as the orthogonal Graeco-Latin squares, geometry, combinatory, finite vector spaces, graphs, among others. They are also related to other subjects such as the computational proof of combinatorial theorems, code theory, and cryptography.

One aspect of finite geometries that has motivated the research on this topic is the non-existence of finite projective planes of certain orders, which logically extends to projective spaces. There are examples of this for small orders such as n = 6 and n = 10. This last case has attracted more attention from researchers since it satisfies the necessary conditions of the Bruck-Ryser Theorem ($10\equiv2(\text{mod}4)$ and $n=1^2+3^2$), not so n = 6. In the case n = 10, the demonstrations have been carried out with the help of computational tools, which can raise doubts about whether the final results of non-existence are definitive or not. Moreover, the solution of mathematical theorems in a computational way is a controversial issue in mathematical philosophy, since it is not difficult for the coded algorithm to contain errors that can lead to a false result.

The theory introduced in this chapter is far from being considered as exhaustive in what could be developed on finite geometries. There are multiple research works on the subject of finite geometries with projective planes and affine planes, as well as their extensions to spaces. We hope that the ideas introduced in this work will overlap with classical theories, in two senses. First, Finite NeutroGeometries naturally extend all the results obtained in classical finite geometries, including possible applications in theories of codes or cryptography. Secondly, the structures and methods obtained from Finite NeutroGeometries serve as support for the development of Finite Geometries.

For example, within the theory of finite geometries, although there are no projective planes of certain orders, it is possible to work with mixed geometries as close as possible to them. Mixed finite geometries by themselves could generate interest in the modeling of networks that cannot be represented by pure projective or pure affine planes, although the results of classical theory can logically be extended to study new types of geometries that contain a mixture of various geometries. Already F. Smarandache and other authors had indicated the convenience of using mixed geometries since the universe does not correspond to a pure specific type of geometric model. That is why we could also think of a possible application of this new theory in problems of physics that are modeled utilizing affine planes.

See that in the previous paragraph a reference is made to mixed finite geometries instead of MPA, this is because even MPA could be further generalized to obtain other geometries with other axioms, for example, that some elements of space contain characteristics of hyperbolic geometry.

With the emergence of NeutroAlgebras, the existence of a theory of neutrosophic models is demanded, as there is a theory of fuzzy models that generalize the classical model theory (Hájek, 1998). This would be a great step forward since Model Theory seen in its most intuitive sense can be considered a provider of semantics to the total, partial algebras, and NeutroAlgebras.

Another interesting topic that remains an open problem is seeking for conditions that must meet finite NeutroAlgebras to define a finite NeutroGeometry. This topic was discussed in a previous section due to NeutroGeometries are algebraically modeled by using finite vector spaces.

CONCLUSION

NeutroGeometry is a theory that emerged from Neutrosophy and NeutroAlgebras. NeutroGeometries are geometric structures that contain at least one axiom or its definition of structure, which is both partially true and partially false, and neither of them is totally false. Some examples had arisen in previous decades when F. Smarandache defined mixed geometries where part of the space is Euclidean, another part is hyperbolic and a third one is elliptical. This chapter offers for the first time a brief introduction to finite geometries in the framework of NeutroGeometries. A finite geometry contains a finite set of points and lines, and they can have dimensions greater than 2. In this chapter, a Finite NeutroGeometry is defined as a finite geometric structure where at least one axiom is fulfilled by a part of the elements and other elements do not satisfy it or them. The Mixed Projective-Affine (MPA) geometries were defined as an example of Finite NeutroGeometry, where some elements of the space satisfy elliptical properties typical of projective spaces and others satisfy

Euclidean properties typical of affine spaces. Some properties of these new mixed geometries are demonstrated, such as the existence of projective, affine, or mixed subspaces. An algorithm is also introduced to obtain mixed spaces from a projective or an affine one. It provides a way to quantify MPA spaces in the form of triads. In the case of planes, some properties are given from the incidence matrix; specifically this matrix is linked with bipartite neutrosophic graphs. Some ideas are given about possible applications of this theory in cryptography. We hope that this chapter opens up many perspectives in the development of the vast theory of finite geometries and that it finds a large number of applications for solving problems in physics, code theory, cryptography, Latin squares, and even pure mathematics.

REFERENCES

Beutelspacher, A. (1990). Applications of finite geometry to cryptography. In G. Longo, M. Marchi, & A. Sgarro (Eds.), *Geometries, Codes and Cryptography* (pp. 161–186). Springer Verlag. doi:10.1007/978-3-7091-2838-1_6

Cameron, P. J. (2000). *Projective and Polar Spaces.* Queen Mary and Westfield College.

Dembowski, P. (1997). Finite Geometries. Berlin: Springer.

Dissett, L. A. (2000). *Combinatorial and Computational Aspects of Finite Geometries* [Unpublished Ph.D. Thesis]. University of Toronto.

Hájek, P. (1998). Metamathematics of Fuzzy Logic. Dordrecht: Springer Science+Business Media. doi:10.1007/978-94-011-5300-3

Hirschfeld, J. W. P. (1979). *Projective Geometries over Finite Fields.* Clarendon Press.

Iseri, H. (2002). Partially Paradoxist Smarandache Geometries. *Smarandache Notions Journal, 13*(1-2-3), 5-12.

Kärteszi, F. (1976). *Introduction to finite geometries.* American Elsevier Publishing Company, Inc.

Kiss, G., & Szonyi, T. (2020). *Finite Geometries.* CRC Press Taylor & Francis Group.

Klein, A., & Storme, L. (2011). Applications of finite geometry in coding theory and cryptography. In *Information Security, Coding Theory and Related Combinatorics* (pp. 38–58). IOS Press.

Kou, Y., Lin, S., & Fossorier, M. P. C. (2001). Low-Density Parity-Check Codes Based on Finite Geometries: A Rediscovery and New Results. *IEEE Transactions on Information Theory*, *47*(7), 2711–2736. doi:10.1109/18.959255

Lam, C. W. H. (1991). The Search for a Finite Projective Plane of Order 10. *The American Mathematical Monthly*, *98*(4), 305–318. doi:10.1080/00029890.1991.12000759

Lam, C. W. H., Thiel, L., & Swiercz, S. (1989). The Non-Existence of Finite Projective Planes Of Order 10. *Canadian Journal of Mathematics*, *41*(6), 1117–1123. doi:10.4153/CJM-1989-049-4

O'Keefe, C. M. (1993). Applications of finite geometries to information security. *Australasian Journal of Combinatorics*, *7*, 195–212.

Rabounski, D. (2010). Smarandache Spaces as a New Extension of the Basic Space-Time of General Relativity. *Progress in Physics*, *2*, L1–L2.

Smarandache, F. (2020a). Introduction to NeutroAlgebraic Structures and AntiAlgebraic Structures (revisited). *Neutrosophic Sets and Systems*, *31*, 1–16.

Smarandache, F. (2020b). NeutroAlgebra is a Generalization of Partial Algebra. *International Journal of Neutrosophic Science*, *2*(1), 8–17. doi:10.54216/IJNS.020103

Smarandache, F. (2021). NeutroGeometry & AntiGeometry are alternatives and generalizations of the Non-Euclidean Geometries. *Neutrosophic Sets and Systems, 46*(2021), 456-476.

Zangalho-Raposo, A. P. (2014). *Geometrias Finitas* [Unpublished Master Thesis]. University of Évora, Évora.

Zumbragel, J. (2016). *Designs and Codes in Affine Geometry.* arXiv e-prints arXiv -1605.03789.

ADDITIONAL READING

Akça, Z., & Altıntas, A. (2021). A Note on Fano Configurations in the Projective Space PG(5,2). *Konuralp Journal of Mathematics*, *9*(1), 190–192.

Al-Seraji, N. A. M., & Jafar, Z. S. (2020). Applications of the Projective Plane in Coding Theory. *Journal of Southwest Jiaotong University*, *55*(1), 1–8. doi:10.35741/issn.0258-2724.55.1.2

Ali, A. T. (2010). Special Smarandache Curves in the Euclidean Space. *International Journal of Mathematical Combinatorics*, *2*, 30–36.

Braun, D., Magaud, N., & Schreck, P. (September 2018). *Formalizing Some "Small" Finite Models of Projective Geometry in Coq*. Paper presented at the 13th International Conference on Artificial Intelligence and Symbolic Computation (AISC'2018), Suzhou, China. 10.1007/978-3-319-99957-9_4

Glynn, D. G. (2010). Theorems of Points and Planes in Three-Dimensional Projective Space. *Journal of the Australian Mathematical Society*, *88*(1), 75–92. doi:10.1017/S1446788708080981

Hirschfeld, J. W. P. (1985). *Finite Projective Spaces of Three Dimensions*. Clarendon Press.

Rottey, S. (2015). *Characterising substructures of finite projective spaces* [Unpublished PhD Thesis]. Vrije Universiteit Brussel, Brussel.

Singh, P. K. (2021). Anti-Geometry and NeutroGeometry Characterization of Non-Euclidean Data. *Journal of Neutrosophic and Fuzzy Systems*, *1*(1), 24–33. doi:10.54216/JNFS.010102

Smarandache, F. (2020). *Neutroalgebra is a generalization of partial algebra*. Infinite Study.

KEY TERMS AND DEFINITIONS

Affine Geometry: Geometric structure such that given a line and an external point, there is a unique line parallel to the aforementioned one, which contains the point.

Finite Field: Algebraic structure consisting of a finite set of elements and two internal operations + and ×, such that + and × are commutative groups except for 0 in ×; 0 is the neutral element of +; 1 is the neutral element of ×; + and × are distributive each other; 0 has no inverse for ×.

Finite Geometry: Geometric system with only a finite number of points.

Finite Vector Space: Vector space over a finite field.

Incidence Matrix: A matrix whose every row represents every line of the finite plane and every column represents every point, or vice versa, such that the element of the matrix is 1 if there is an incidence relation between the corresponding pair of line and point, and it is 0 otherwise.

Incidence Structure: Symmetric relation among points and lines, where if a line l contains a point P it is said that l is incident with P and P is incident with l.

Mixed Geometry: Also called *Smarandache geometry*, it is the geometry where some elements satisfy the axiomatic of some type of geometry, and other elements satisfy other types of geometries, all in the same space. E.g., there are geometric models which are partially Euclidean, partially hyperbolic, and partially elliptic.

NeutroGeometry: A geometric structure whose definition or at least one axiom is simultaneously satisfied by some elements and not satisfied by others.

Principle of Duality: Principle satisfied by the projective spaces where the theorems which link some geometric elements are also valid if they are interchanged. E.g., in the finite projective plane every theorem which links the terms points, lines, and the incidence is valid where the two firsts of them are interchanged.

Projective Geometry: Geometric structure where there is non-parallelism of the lines in the same plane. Moreover, every pair of planes meets in a line, and so on.

Chapter 4
An Introduction to NeutroHyperstructures on Some Chemical Reactions

Fakhry Asad Agusfrianto
Universitas Negeri Jakarta, Indonesia

Madeleine Al Tahan
https://orcid.org/0000-0002-4243-7935
Abu Dhabi University, UAE

Yudi Mahatma
https://orcid.org/0000-0002-3917-3529
Universitas Negeri Jakarta, Indonesia

ABSTRACT

Algebraic hyperstructure is a generalized form of the ordinary algebraic structures. Research on algebraic hyperstructures can focus on the theory and its applications. In this chapter, the discussion will focus on the application of algebraic hyperstructures in chemistry. Algebraic hyperstructures play a role in investigating the types of structures found in a chemical reaction. Furthermore, the concept of algebraic hyperstructures is associated with the concept of NeutroAlgebra so that a new type of structure called NeutroHyperstructures is obtained. In this chapter the types of NeutroHyperstructures contained in some chemical reactions will be investigated along with the application of these NeutroHyperstructures.

DOI: 10.4018/978-1-6684-4740-6.ch004

INTRODUCTION

Algebraic hyperstructures are the generalized form of algebraic structures. Algebraic hyperstructures were introduced by Marty (1934). Since then, much research on hyperstructures has been carried out both in theory and in its application. Further, Vougiouklis (1996) generalized hyperstructures into structures called H_v-structures. Research has been conducted related to the application of hyperstructures in the fields of natural sciences, such as physics (Nezhad, Nejad, Nadjafikhah, & Davvaz, 2012), biology (Al-Tahan & Davvaz, 2017) and chemistry. This chapter focuses on the application of hyperstructures in chemistry. Hyperstructures play a role in analyzing the types of structures contained in chemical reactions. research that has been carried out so far analyzes the types of hyperstructures contained in dismutation reactions (Davvaz, Nezhad, & Benvidi, 2012) and redox reactions (Davvaz, Nezhad, & Mazloum-Ardakani, 2014). Besides that, investigations were also carried out on the types of hyperstructures found in electrochemical cells (Al-Tahan., & Davvaz, 2018), ozone depletion reactions (Chung, 2019), and salt formation (Heidari, Mazaheri, & Davvaz, 2019).

On the other hand, Smarandache in 2020 introduced the concept of NeutroAlgebra as a generalization of partial algebra (Smarandache, 2020). Motivated by this, neutro's concept of hyperstructures was defined as NeutroHyperstructures by Al-Tahan et al (2021) and Ibrahim and Agboola (2021). Motivated by previous studies related to analyzing the types of hyperstructures contained in a chemical reaction, this paper will investigate the types of NeutroHyperstructures contained in several chemical reactions for which the type of hyperstructures has been analyzed, such as redox reactions, electrochemical cells, and ozone depletion reaction.

BACKGROUND

Algebraic Hyperstructures are generalizations of algebraic structures. This concept was introduced by Marty in 1934 (Marty, 1934). This concept has various applications in the fields of physics, biology, chemistry, and cryptography.

Let M be a non-empty set. A mapping $\boxplus : M \times M \to P^*(M)$ where $P^*(M)$ is the entire nonempty subset of M, called *hyperoperation* in M. The pair $(M, \boxplus)$ is called a *hypergroupoid.* A *hypergroupoid* $(M, \boxplus)$ is called a *semihypergroup* if for every $a,b,c \in M$, $(a \boxplus b) \boxplus c = a \boxplus (b \boxplus c)$. A *hypergroupoid* $(M, \boxplus)$ is called a *quasi-hypergroup* if for every $m \in M$, $m \boxplus M = M \boxplus m = M$. This condition is called the *reproduction axiom*. A hypergroupoid $(M,\boxplus)$ is called a *hypergroup* if it is a *semihypergroup* and *quasi-hypergroup* (Davvaz & Leoreanu-Fotea, 2022).

Example 1. Let $M=\{a,b,c,d\}$, define a *hyperoperation* "$\boxplus$" as follows:

Table 1. $(M,\boxplus)$

$\boxplus$	*a*	*b*	*c*	*d*
a	*a*	$\{a,b\}$	$\{a,b,c\}$	*M*
b	$\{a,b\}$	*B*	$\{b,c\}$	*M*
c	$\{a,b,c\}$	$\{b,c\}$	*c*	$\{c,d\}$
d	*M*	*M*	$\{c,d\}$	*d*

It can be checked that $(M,\boxplus)$ satisfies all conditions for the *semihypergroup* and *quasi-hypergroup*. Since Table 1 is symmetric, then $(M,\boxplus)$. is a *commutative hypergroup*.

Now, Let *M* be the non-empty set, and $\boxplus: M\times M \rightarrow, P^{*}(M)$ is *hyperoperation*. The operation "$\boxplus$" is said to be weak associative if for every *a,b,c*∈*M*, $(a\boxplus b)\boxplus c\cap a\boxplus(b\boxplus c)\neq\varnothing$. A *hypergroupoid* $(M,\boxplus)$. is said to be H_v-*semigroups* if the operation "$\boxplus$" satisfies the weak associative properties. If $(M,\boxplus)$ satisfies the reproduction axiom, then $(M,\boxplus)$ is called a H_v-*groups* (Davvaz & Vougioklis, 2019)

Example 2. Let *N*= $\{p,q,r,t\}$. Define hyperoperation "$\boxdot$" as follows:

Table 2. $(N,\boxdot)$

$\boxdot$	*p*	*q*	*r*	*s*	*t*
p	$\{q,r\}$	$\{p,q\}$	$\{p,r\}$	$\{p,s\}$	$\{p,t\}$
q	$\{p,q\}$	*q*	$\{q,r\}$	$\{q,s\}$	$\{q,t\}$
r	$\{p,r\}$	$\{q,r\}$	*r*	$\{r,s\}$	$\{r,e\}$
s	$\{p,s\}$	$\{q,s\}$	$\{r,s\}$	*s*	$\{s,t\}$
t	$\{p,t\}$	$\{q,t\}$	$\{r,t\}$	$\{s,t\}$	*t*

It can be checked that $(N, \boxdot)$ satisfies all conditions for the *semihypergroup* and *quasi-hypergroup*. Since Table 2 is symmetric, then $(N, \boxdot)$ is a *commutative hypergroup*.

Furthermore, Al-Tahan et al. (2021) and Ibrahim and Agboola (Ibrahim & Agboola, 2021) Generalized the concept of hyperstructures to the concept of NeutroHyperstructures.

Definition 3. (Al-Tahan et al., 2021). Let G be a nonempty set and "$\boxminus$" be a *hyperoperation* in G. Then the operation "$\boxminus$" is called a *NeutroHyperoperation* in G if some (or all) of the following conditions are satisfied with

$$(T, I, F) \notin \{(1,0,0),(0,0,1)\}.$$

1. There exists $p,q \in G$ with $p \boxminus q \subseteq G$ (degree of truth "*T*")
2. There exists $p,q \in G$ with $p \boxminus q \subsetneq G$ (degree of falsity "*F*")
3. There exists $p,q \in G$ with $p \boxminus q$ is indeterminate in G (degree of indeterminacy "*I*")

Definition 4. (Al-Tahan et al., 2021). Let G be a nonempty set and "$\boxminus$" is a *hyperoperation* in G. Then "$\boxminus$" is called *AntiHyperoperation* in G if for every $p,q \in G$, $p \boxminus q \subsetneq G$.

Definition 5. (Al-Tahan et al., 2021). Let G is a nonempty set and "$\boxminus$" is a *hyperoperation* in G. Then "$\boxminus$" is called *NeutroAssociative* in G if there exists $p,q,r,x,y,z,a,b,c \in G$ that satisfies some (or all) of the following conditions with $(T,I,F) \notin \{(1,0,0),(0,0,1)\}$.

1. $p \boxminus (q \boxminus r) = (p \boxminus q) \boxminus r$ (degree of truth "*T*")
2. $x \boxminus (y \boxminus z) \neq (x \boxminus y) \boxminus z$ (degree of falsity "*F*")
3. $a \boxminus (b \boxminus c)$ is indeterminate or $(a \boxminus b) \boxminus c$ is indeterminate (degree of indeterminacy "*I*")

Definition 6. (Al-Tahan et al., 2021). Let G be a nonempty set and "$\boxminus$" be a hyperoperation in G. Then "$\boxminus$" is called *NeutroWeakAssociative* in G if there exists $a,b,c,x,y,z,p,q,r \in G$ that satisfies some (or all) of the following conditions with $(T,I,F) \notin \{(1,0,0),(0,0,1)\}$.

1. $[a \boxminus (b \boxminus c)] \cap [(a \boxminus b) \boxminus c] \neq \varnothing$ (degree of truth "*T*")
2. $[x \boxminus (y \boxminus z)] \cap [(x \boxminus y) \boxminus z] = \varnothing$ (degree of falsity "*F*")

3. $p \boxminus (q \boxminus r)$ is indeterminate or $(p \boxminus q) \boxminus r$ is indeterminate (degree of indeterminacy "*I*")

Definition 7. (Al-Tahan et al., 2021). Let *G* is a nonempty set and $"\boxminus"$ is a *hyperoperation* in *G*. Then $(G, \boxminus)$ is called *NeutroHypergroupoid* if $"\boxminus"$ is *NeutroHyperoperation*, *NeutroSemihypergroup* if $"\boxminus"$ is *NeutroAssociative* but not *AntiHyperoperation*, and *Neutro H_v-semigroup* if $"\boxminus"$ is *NeutroWeakAssociative* but not *AntiHyperoperation*.

Definition 8. (Ibrahim, M.A., & Agboola, A.A.A., 2021). A *NeutroHypergroupoid* $(G, \boxminus)$ is called a *NeutroHypergroup* if *G* is a *NeutroSemihypergroup* and there exists $e,f,g \in G$ that satisfies some (or all) of the following conditions with $(T,I,F) \notin \{(1,0,0),(0,0,1)\}$.

1. $e \boxminus G = G \boxminus e = G$ (degree of truth "*T*")
2. $f \boxminus G \neq G \boxminus f \neq G$ (degree of falsity "*F*")
3. $g \boxminus G$ and $G \boxminus g$ indeterminate (degree of indeteminacy "*I*")

Example 9. Let *H*= $\{a,d,u\}$, define a hyperoperation $"\boxdot"$ as follows.

Table 3. ($H, \boxdot$)

$\boxdot$	*a*	*d*	*u*
a	$\{a,d\}$	*u*	*d*
d	$\{d,u\}$	*d*	*d*
u	$\{a,u\}$	*u*	*u*

It can be checked that $(H, \boxdot)$ satisfies all conditions for *NeutroSemihypergroup* and *reproduction axiom* for *NeutroHyperstructures*. Then, $(H, \boxdot)$ is a *NeutroHypergroup*.

CHEMICAL NEUTROHYPERSTRUCTURES ON SOME REDOX REACTIONS

According to redox reaction of Astatine in Al-Tahan and Davvaz (2019), let *P*= $\{At, At_2, At^+, At^{5+}\}$ and write $p=At$, $q=At_2$, $r=At^+$, and $s=At^{5+}$. Next, define the operation

"$\oplus$" as the most positive spontaneous reaction that occurs between two elements in *P*. Table 3 is obtained.

Table 4. $(P,\oplus)$

$\oplus$	p	q	r	s
p	p	$\{p,r\}$	$\{p,r\}$	r
q	$\{p,r\}$	$\{p,r\}$	$\{p,r\}$	r
r	$\{p,r\}$	$\{p,r\}$	r	$\{p,r\}$
s	r	r	$\{r,s\}$	s

Theorem 10. $(P,\oplus)$ is a *NeutroSemihypergroup.*

Proof. It is clear that a hyperoperation "$\oplus$" is not an *AntiHyperoperation* since there is no exist *x,y*∈*P* such that $x\oplus y \subsetneq P$. Next, for *p*∈*P*,

$(p\oplus p)\oplus p = p\oplus(p\oplus p) = p$ and for *q,s*∈*P*,

$(q\oplus s)\oplus s = r\oplus s = \{p,r\}$ and $q\oplus(s\oplus s) = q\oplus s = r$

so $(q\oplus s)\oplus s \neq q\oplus(s\oplus s)$.

Thus, $(P,\oplus)$ is a *NeutroSemihypergroup.*

Remark 11. $(P,\oplus)$ is not a *Neutro H_v-semigroup* since according to Al-Tahan and Davvaz (2019), $(P,\oplus)$ is a *H_v-semigroup*. In other words, the weak associative property is always satisfied.

Theorem 12. Let $Q=\{p,q,s\}\subseteq P$. Then, $(Q,\oplus)$ is a *NeutroSubsemihypergroup.*

Proof. Since

$(p\oplus p)\oplus p = p\oplus(p\oplus p) = p$ and $(q\oplus s)\oplus s \neq q\oplus(s\oplus s)$,

then $(Q,\oplus)$ is a *NeutroSubsemihypegroup.*

Similarly, $(Q'=\{p,r,s\},\oplus)$ and $(Q''=\{q,r,s\},\oplus)$ are *NeutroSubSemihypergroup.*

Now, According redox reaction of Tellurium in Al-Tahan and Davvaz (2019), let $A=\{Te^{6+}, Te^{4+}, Te, Te^{2-}\}$ and write $w=Te^{6+}$, $x=Te^{4+}$, $y=Te$, and $x=Te^{2-}$. Next, define a hyperoperation "$\otimes$". as the most positive spontaneous reaction that occurs between two elements in *P*. Table 5 is obtained.

Table 5. $(A,\otimes)$

$\otimes$	w	x	y	z
w	w	$\{w,x\}$	x	$\{y,z\}$
x	$\{w,x\}$	x	$\{x,y\}$	y
y	x	$\{x,y\}$	y	$\{y,z\}$
z	$\{y,z\}$	y	$\{y,z\}$	z

Theorem 13. $(A,\otimes)$ is a *NeutroSemihypergroup*

Proof. It is clear that a hyperoperation "$\otimes$" is not an *AntiHyperoperation* since there is no exist $x,y \in A$ such that $x\otimes y \subsetneq A$. . Next, for $w \in A$,

$(w\otimes w)\otimes w = w\otimes(w\otimes w) = w$ and for $x,y,z \in P$,

$$(x\otimes y)\otimes z = \{x,y\}\otimes z = \{y,z\} \text{ and } x\otimes(y\otimes z) = x\otimes\{y,z\} = \{x,y\}$$

so $(x\oplus y)\oplus z \neq x\oplus(y\oplus z)$.

Thus, $(A,\otimes)$ is a *NeutroSemihyperoup.*

Remark 14. $(A,\otimes)$ is not a *Neutro H_v-semigroup* since according to Al-Tahan and Davvaz (2019), $(A,\otimes)$ is a *H_v-semigroup*. In other words, the weak associative property is always satisfied.

Theorem 15. Let $B=\{x,y,z\}\subseteq A$. Then, $(B,\otimes)$ is a *NeutroHypergroup.*

Proof. For $y \in B$,

$$y\otimes(y\otimes y) = (y\otimes y)\otimes y \text{ and } x\otimes(x\otimes z) = \{x,y\} \neq (x\otimes x)\otimes z = y.$$

So, $(B,\otimes)$ is Neutro*Semihypergroup*. Furthermore, for $y \in B$,

$y \otimes B = B \otimes y = B$ and for $z \in B$, $z \otimes B = B \otimes z = \{y, z\} \neq B$.

Thus, $(B, \otimes)$ is a *NeutroHypergroup*.

Similarly, $(B' = \{w, y, z\}, \otimes)$ is also a *NeutroHypergroup*. Now, according to Al-Tahan and Davvaz (2019, 2022) for redox reaction of Bismuth, Silver, Antimony, Iron, Platinum, and Americium since hyperstructures of Them is isomorphic to hyperstructures of Tellurium, then, *Theorem 13* and *Theorem 15* is satisfied for the redox reaction of Them.

Furthermore, According to the redox reaction on Vanadium (Chung, 2014), let

$$D = \{V, V^{2+}, V^{3+}, VO^{2+}, VO_2^+\}$$

and write

$$t = V, u = V^{2+}, v = V^{3+}, w = VO^{2+}, x = VO_2^+.$$

Next, define a hyperoperation "$\odot$" as the most positive spontaneous reaction that occurs between two elements in *D*. Table 6 is obtained.

Table 6. $(D, \odot)$

$\odot$	t	u	v	W	x
t	t	$\{t,u\}$	u	$\{u,v\}$	$\{u,w\}$
u	$\{t,u\}$	u	$\{u,v\}$	v	$\{v,w\}$
v	u	$\{u,v\}$	v	$\{v,w\}$	w
w	$\{u,v\}$	v	$\{v,w\}$	w	$\{w,x\}$
x	$\{u,w\}$	$\{v,w\}$	w	$\{w,x\}$	x

Theorem 17. $(D, \odot)$ is a *NeutroSemihypegroup*

Proof. It is obvious that a *hyperoperation* "$\odot$" is not an *AntiHyperoperation* since there is no exist *a,b*∈*D* such that $a \odot b \subsetneq D$. Next, for *t*∈*D*,

$$t \odot (t \odot t) = (t \odot t) \odot t$$

and for $u,v,w \in D$, $u \odot (v \odot w) = \{u,v\} \neq (u \odot v) \odot w = \{v,w\}$.

Thus, $(D,\odot)$ is a *NeutroSemihypegroup*.

Remark 18. $(D,\otimes)$ is not a *Neutro* H_v*-semigroup* since according to Chung (2014), $(D,\otimes)$ is a H_v*-semigroup*. In other words, the weak associative property is always satisfied.

Theorem 19. Let $F = \{t,u,v\} \subseteq D$. Then, $(F,\odot)$ is a *NeutroHypergroup*.

Proof. For $u \in F$,

$$u \odot (u \odot u) = (u \odot u) \odot u = u$$

and for $t,v \in F$, $t \odot (v \odot v) = u \neq (t \odot v) \odot v = \{u,v\}$.

Then, $(F,\odot)$ is a *NeutroSemihypergroup*. Next, for $u \in F$,

$u \odot F = F \odot u = F$ and for $t \in F$, $t \odot F = F \odot t \neq F$.

Thus, $(F,\odot)$ is a *NeutroHypergroup*.

Similarly, $(\boldsymbol{F'} = \{\boldsymbol{u},\boldsymbol{v},\boldsymbol{w}\},\otimes)$ and $(\boldsymbol{F''} = \{\boldsymbol{t},\boldsymbol{w},\boldsymbol{x}\},\otimes)$ are also *NeutroHypergroups*.

CHEMICAL NEUTROHYPERSTRUCTURES FOR ELECTROCHEMICAL CELLS

According to Al-Tahan and Davvaz (2018), there are three cases of Electrochemical Cells. The three cases are Galvanic Cells, Electrolytic Cells, and Galvanic/Electrolytic Cells.

Case 1. *Galvanic Cells*. Based on Al-Tahan and Davvaz (2018), let X is metal with larger electronegativity on the first half-cell and Y is metal with smaller electronegativity. The following redox reaction is obtained.

$$X^{n+} + ne^- \rightarrow X$$

$$Y \rightarrow Y^{m+} + me^-$$

If the above reactions are combined, obtained

$$nY + mX^{n+} \rightarrow mX + nY^{m+}$$

Now, let $F=\{X,Y,X^{n+},Y^{m+}\}$ and define a hyperoperation "$\boxdot$" as a possible reaction between two elements in F. Write $X=p$, $Y=q$, $X^{n+}=r$, $Y^{m+}=s$. Table 7 is obtained.

Table 7. $(F,\boxdot)$

$\boxdot$	**p**	**q**	**r**	**s**
p	p	$\{p,q\}$	$\{p,r\}$	$\{p.s\}$
q	$\{p,q\}$	q	$\{p,s\}$	$\{q,s\}$
r	$\{p,r\}$	$\{p,s\}$	r	$\{r,s\}$
s	$\{p,s\}$	$\{q,s\}$	$\{r,s\}$	s

Theorem 20. $(F,\boxdot)$ is a *NeutroHypergroup*

Proof. It is obvious that $(F,\boxdot)$ is not an AntiHyperoperation. Next, for $p\in F$,

$$p\boxdot(p\boxdot p)=(p\boxdot p)\boxdot p=p$$

and for $q,r\in F$, $q\boxdot(r\boxdot r)=\{p,s\}\neq(q\boxdot r)\boxdot r=\{p,r,s\}$.

So, $(F,\boxdot)$ is a *NeutroSemihypergroup*. Now, for $p\in F$,

$$p\boxdot F=F\boxdot p=F \text{ and for } q\in F,\ q\boxdot F=F\boxdot q\neq F.$$

Thus, $(F,\boxdot)$ is a *NeutroHypergroup*.

Theorem 21. Let $F_1=\{q,r,s\}\subseteq F$. Then, $(F_1,\boxdot)$ is a *NeutroSubhypergroup*

Proof. Similar to proof in *Theorem 20*.

Case 2. *Electrolityc Cells.* Based on Al-Tahan and Davvaz (2018), let P is metal with larger electronegativity on first half-cell and Q is metal with smaller electronegativity. The following redox reaction is obtained

$$P\rightarrow P^{n+}+ne^{-}$$

$$Q^{m+}+me^{-}\rightarrow Q$$

If the above reactions are combined, the following redox reaction is obtained

$$mP+nQ^{m+}\rightarrow nQ+mP^{n+}$$

Now, let G= {P, Q, P^{n+}, Q^{m+}} and define a hyperoperation "$\boxminus$" as a possible reaction between two elements in G. Write $P=k$, $Q=l$, $P^{n+}=m$, $Q^{m+}=n$. Table 8 is obtained.

Table 8. $(G,\boxminus)$

$\boxminus$	k	l	m	n
k	k	$\{k,l\}$	$\{k,m\}$	$\{l,m\}$
l	$\{k,l\}$	l	$\{l,m\}$	$\{l,n\}$
m	$\{k,m\}$	$\{l,m\}$	m	$\{m,n\}$
n	$\{l,m\}$	$\{l,n\}$	$\{m,n\}$	n

Theorem 22. $(G,\boxminus)$ is a *NeutroHypergroup*.

Proof. It is clear since $(G,\boxminus)\cong(F,\boxdot)$.

Theorem 23. Let G_1= $\{k,m,n\}$. Then, $(G_1,\boxdot)$ is a *NeutroHypergroup*.

Proof. Similar to proof in *Theorem 20*.

Case 3. *Galvanic/Electrolytic Cells*. Based on Al-Tahan and Davvaz (2018), let K= $\{R,S,R^{n+},S^{m+}\}$ and define a hyperoperation $\ominus$ as possible reactions result between x and y either in a galvanic cell or in an electrolytic cell with $x,y\in K$. write $R=a$, $S=b$, $R^{n+}=c$, and $S^{m+}=d$. Table 9 is obtained.

Table 9. $(K,\ominus)$

$\ominus$	a	b	c	d
a	a	$\{a,b\}$	$\{a,c\}$	$\{b,c\}$
b	$\{a,b\}$	b	$\{a,d\}$	$\{b,d\}$
c	$\{a,c\}$	$\{a,d\}$	c	$\{c,d\}$
d	$\{b,c\}$	$\{b,d\}$	$\{c,d\}$	d

Remark 24. $(K,\ominus)$ is not a *NeutroHypergroup*. Since there is no element in $\{a,b,c,d\}$ such that $(K,\ominus)$ satisfies falsify or indeterminate condition on reproduction axiom.

Theorem 25. $(K,\ominus)$ is a *NeutroSemihypergroup.*

Proof. It is obvious that $(K,\ominus)$ is not an *AntiHyperoperation.* Next, for some $a \in K$,

$$a\ominus(a\ominus a)=(a\ominus a)\ominus a=a$$

and for $b,c \in K$, $b\ominus(c\ominus c)=\{a,d\}\neq(b\ominus c)\ominus c=\{a,c,d\}$.

Thus, $(K,\ominus)$ is a *NeutroSemihypergroup.*

Theorem 26. Let $K_1=\{a,b,c\}$. Then, $(K_1,\ominus)$ is a *NeutroHypergroup*

Proof. Similar to proof in *Theorem 20.*

Similarly, $(K_2=\{a,b,d\},\ominus)$ and $(K_3=\{a,c,d\},\ominus)$ are also *NeutroHypergroups.*

CHEMICAL NEUTROHYPERSTRUCTURES FOR OZONE DEPLETION REACTIONS

According to Chung (2019), ozone depletion by freon gas $CFCl_3$ has chemical reaction in three steps:

$$CFCl_3 \rightarrow CFCl_2 + Cl$$

$$Cl + O_3 \rightarrow ClO + O_2$$

$$ClO + O_3 \rightarrow Cl + 2O_2$$

Now, let $A=\{CFCl_3, CFCl_2, Cl, ClO, O_2, O_3\}$ and write $u=CFCl_3$, $v=CFCl_2$, $w=Cl$, $x=ClO$, $y=O_2$, $z=O_3$ define a *hyperoperation* "$\boxdot$" as a possible reaction between two elements in A. If for every $x,y \in A$, x and y is not reacted, according to Chung (2019), then $x\boxdot y=\{x,y\}$ for every $x,y \in A$. Based on Chung 2019), Table 10 is obtained.

Table 10. $(A, \boxdot)$

$\boxdot$	u	v	w	x	y	z
u	$\{v,w\}$	$\{u,v\}$	$\{u,w\}$	$\{u,x\}$	$\{u,y\}$	$\{u,z\}$
v	$\{u,v\}$	v	$\{v,w\}$	$\{v,x\}$	$\{v,y\}$	$\{v,z\}$
w	$\{u,w\}$	$\{v,w\}$	w	$\{w,x\}$	$\{w,y\}$	$\{x,y\}$
x	$\{u,x\}$	$\{v,x\}$	$\{w,x\}$	x	$\{x,y\}$	$\{w,z\}$
y	$\{u,y\}$	$\{v,y\}$	$\{w,y\}$	$\{x,y\}$	y	$\{y,z\}$
z	$\{u,z\}$	$\{v,z\}$	$\{w,z\}$	$\{w,z\}$	$\{y,z\}$	z

Theorem 27. $(A, \boxdot)$ is a *NeutroHypergroup*

Proof. It is clear that $(A, \boxdot)$ is not an *AntiHyperoperation*. Now, For $v \in A$,

$$v \boxdot (v \boxdot v) = v \boxdot (v \boxdot v)$$

and for $u,w \in A$, $w \boxdot (u \boxdot u) = \{v, w\} \neq (w \boxdot u) \boxdot u = \{u, v, w\}$.

Then, $(A, \boxdot)$ is a *NeutroSemihypergroup*. Next, for $u \in A$,

$$u \boxdot A = A \boxdot u = A \text{ and for } w \in A, w \boxdot A = A \boxdot w = \{u, v, w, x, y\} \neq A.$$

Thus, $(A, \boxdot)$ is a *NeutroHypergroup*.

Theorem 28. Let $B = \{u, v, x\} \subseteq A$. Then, $(B, \boxdot)$ is a *NeutroSubhypergroup*.

Proof. Similar to *Theorem 27*

It is clear that

$$(\{u,v,x\},\boxdot),(\{u,v,x\},\boxdot),(\{u,v,z\},\boxdot),(\{u,w,x\},\boxdot),(\{u,w,y\},\boxdot),(\{u,x,y\},\boxdot),$$
$$(\{u,y,z\},\boxdot),(\{v,w,x\},\boxdot),(\{v,w,z\},\boxdot),(\{v,x,z\},\boxdot),(\{w,x,z\},\boxdot),(\{x,y,z\},\boxdot)$$

is a *NeutroSubhypergroup*.

Theorem 29. Let $C = \{u, v, w, z\} \subseteq A$. Then, $(C, \boxdot)$ is a *NeutroSubhypergroup*.

Proof. Similar to *Theorem 27*

It is clear that

$$(\{u,w,x,y\},\boxdot),(\{u,w,x,y\},\boxdot),(\{u,w,x,z\},\boxdot),$$
$$(\{u,w,y,z\},\boxdot),(\{u,x,y,z\},\boxdot),(\{v,w,x,z\},\boxdot),(\{w,x,y,z\},\boxdot)$$

is a *NeutroSubhypergroup*.

Theorem 30. Let $\boldsymbol{D}=\{u,v,w,x,z\}\subseteq \boldsymbol{A}$. Then, $(\boldsymbol{D},\boxdot)$ is a *NeutroSubhypergroup*.

Proof. Similar to *Theorem 27*

It is clear that $(\{v,w,x,y,z\},\boxdot)$ is a *NeutroSubhypergroup*

FUTURE RESEARCH DIRECTIONS

The results in this chapter may be considered as a base for any possible study in the field of Chemical NeutroHyperstructures. The authors suggest some research ideas in this direction.

1. Investigate types of NeutroHyperstructures on another chemical hyperstructures.
2. Generalized these results for redox reaction for elements with four oxidation states.
3. Find Another Application of NeutroHyperstructures in Biology and Physics.

CONCLUSION

Based on the explanation above, it has been shown the type of NeutroHyperstructures for both redox reactions, especially for Astatine, Tellurium, Bismuth, and Vanadium, electrochemical cells, and ozone depletion.

Funding

This research received no specific grant from any funding agency in the public, commercial, or not-for-profit sectors

ACKNOWLEDGMENT

The authors are grateful to the referees for their valuable comments and suggestions.

REFERENCES

Al-Tahan, M., & Davvaz, B. (2017). Algebraic Hyperstructures Associated to Biological Inheritance. *Mathematical Biosciences*, *285*, 112–118. doi:10.1016/j.mbs.2017.01.002 PMID:28108295

Al-Tahan, M., & Davvaz, B. (2018). Weak Chemical Hyperstructures Associated to Electrochemical Cells. *Iranian Journal of Mathematical Chemistry*, *9*(1), 65–75.

Al-Tahan, M., & Davvaz, B. (2019). Chemical hyperstructures for Astatine, Tellurium and for Bismuth. *Bull. Comput. Appl. Math*, *7*(1), 9–25.

Al-Tahan, M., & Davvaz, B. (2022). Chemical Hyperstructures for Element with Four Oxidation States. *Iranian Journal of Mathematical Chemistry*, *13*(2), 85–97.

Al-Tahan, M., Davvaz, B., Smarandache, F., & Anis, O. (2021). On some NeutroHyperstructures. *Symmetry*, *13*(4), 535. doi:10.3390ym13040535

Chung, S. C. (2014). Chemical Hyperstructures for Vanadium. *Journal of Chungcheong Mathematical Society*, *27*(2), 309–317. doi:10.14403/jcms.2014.27.2.309

Chung, S. C. (2019). Chemical Hyperstructures For Ozone Depletion. *Journal of Chungcheong Mathematical Society*, *32*(4), 491–508.

Davvaz, B. & Leoreanu-Fotea. (2022). *Hypergroup Theory.* World Scientific.

Davvaz, B., Nezhad, A. D., & Benvidi, A. (2012). Chemical Hyperalgebra: Dismutation Reactions. *MATCH Commun.Math.Comput.Chem*, *67*(1), 55–63.

Davvaz, B., Nezhad, A. D., & Mazloum-Ardakani, M. (2014). Chemical Hyperalgebra: Redox Reactions. *MATCH Commun.Math.Comput.Chem*, *71*(2), 323–331.

Heidari, D., Mazaheri, D., & Davvaz, B. (2019). Chemical Salt Reactions as Algebraic Hyperstructures. *Iranian Journal of Mathematics Chemistry*, *10*(2), 93–102.

Ibrahim, M. A., & Agboola, A. A. A. (2020). Introduction to NeutroHyperGroups. *Neutrosophic Sets and System*, *38*, 15–32.

Marty, F. (1934). Sur une generalization de la notion de groupe. *Proceedings of the 8th Congress on Mathmatics, Scandinaves*, 45–49.

Nezhad, A. D., Nejad, S. M. M., Nadjafikhah, M., & Davvaz, B. (2012). A Physical Example of Algebraic Hyperstructures: Leptons. *Indian Journal of Physics and Proceedings of the Indian Association for the Cultivation of Science*, *86*(11), 1027–1032. doi:10.100712648-012-0151-x

Smarandache, F. (2020). Introduction to NeutroAlgebraic structures and AntiAlgebraic structures (revisited). *Neutrosophic Sets and System*, *31*, 1–16.

Smarandache, F. (2020). NeutroAlgebra is a generalization of partial algebra. *International Journal of Neutrosophic Science*, *2*, 8–17. doi:10.54216/IJNS.020103

Vougiouklis, T. (1996). -groups defined on same set. *Discrete Mathematics*, *155*(1-3), 259–265. doi:10.1016/0012-365X(94)00390-5

ADDITIONAL READING

Al-Tahan, M., & Davvaz, B. (2017). N-ary Hyperstructures Associated to The Genotypes of F_2. *International Journal of Biomathematics*, *10*(8), 1750118. doi:10.1142/S1793524517501182

Al-Tahan, M., & Davvaz, B. (2019). A New Relationship Between Intuitionistic Fuzzy Sets and Genetics. *Journal of Classification*, *36*(3), 494–512. doi:10.100700357-018-9276-8

Douglas, B., McDaniel, D. H., & Alexander, J. J. (1983). *Concepts and Models of Inorganic Chemistry* (2nd ed.). John Wiley & Sons, Inc.

Chapter 5
Use of Soft and Neutrosophic Sets for a Mathematical Representation of the Ethical Rules

Michael Gr. Voskoglou
https://orcid.org/0000-0002-4727-0089
Graduate Technological Educational Institute of Western Greece, Greece

Joachim Feuerstein
University of Applied Sciences, Berlin, Germany

Evangelos Athanassopoulos
Technical University of Athens, Greece

ABSTRACT

Soft and neutrosophic sets are used in this chapter as tools for introducing a multi-valued logic for ethics. The introduction of a multi-valued logic in ethics is not a new idea, but there is not any integrated proposal about this reported in the literature until now. The target is not to add another theory about ethics, but instead to create a new basis enabling a modern approach to the subject. The chapter starts by examining the role of human logic and statistical thinking for the creation and evolution of ethics. A brief historical account of the development of ethics follows with emphasis to the moral dilemmas, the existence of which motivates the application of a multi-valued logic. The basic information about fuzzy sets, fuzzy logic, soft sets, and neutrosophic sets, needed for the understanding of the rest of the chapter, are also presented before using soft and neutrosophic sets for a mathematical representation of the ethical rules and in extension of the moral theories.

DOI: 10.4018/978-1-6684-4740-6.ch005

INTRODUCTION

The terms *ethics* and *morality* are frequently used interchangeably, loosely meaning a framework of rules and behaviors about "good" and "bad" or "right" and "wrong" conduct. The etymology of the word ethics is connected to the Greek word "ήθος" which means character (with the wide sense of the word), whereas the root of the word morality comes from the Latin word "mores" which means customs. Many people, however, think of morality as something which is personal and ethics as the standards of "good" and "bad" or "right" and "wrong" distinguished by a certain community or social setting.

For example, assume that your local community considers that sex before marriage is immoral, but you have no strong feelings about it. In this case your personal morality contradicts the ethics of your community. Ethics also has the meaning of a deontological code for those working in medicine, business or law, whereas morality is connected to Christianity for many people, since moral theology is prominent in the church. In general, the distinction between the two terms is not so clear.

The arising question, in any case, is what is "good" and "bad" or "wrong" and "right". The several answers given through the centuries to this question, which played a very important role in the formation of human behavior and history, are in most cases inconsistent or even contradictory to each other and differ from time to time and from society to society. In fact, the humans are characterized by deep differences among each other, the result of which is the lack of unity and the absence of a universal will (Koestler, 1979). The development, therefore, of a universal system of principles, rules and laws for the right human behavior, or in other words of a universally acceptable theory of ethics, seems to be impossible or, at least, very difficult.

The previous problem becomes even harder due to the fact that the various theories about an ideal system of ethics, developed from the ancient years until nowadays, are based on the principles of the classical *bivalent logic (BL)*, of Aristotle (384-322 BC). As a result, every human action or behavior is characterized only as "good" or "bad", without taking into account how good or bad it is. This creates frequent moral dilemmas and impasses, which will be discussed later in this paper. A possible solution to this problem is the introduction of a multi-valued logic in ethics, where degrees of truth are used. This is not actually a new idea, but as far as we know, apart from some general suggestions (e.g. Kosko, 1993) or specific examples (e.g. Papaioannou, 2013), nobody has attempted until now to propose a concrete and general way to materialize it in practice.

In this chapter we propose the use of *soft sets (SSs)* and *neutrosophic sets (NSs)*, introduced by Molodstov (1999) and Smarandache (1995) respectively, as tools for the introduction of a multi-valued logic in ethics. It must be emphasized, however, that our target is not to add, in the already existing long catalogue, a new theory about ethics, but simply to propose a new basis that enables a modern approach to the subject.

The rest of the chapter is formulated as follows: In the next section the influence of human logic for the creation and evolution of ethics is examined. The third section is devoted to a brief historical account of the development of ethics with emphasis on the *moral dilemmas* that motivate the application of a multi-valued logic to ethics. In the fourth section basic information about *fuzzy sets (FSs)*, *fuzzy logic (FL)*, NSs and SSs, which is needed for the understanding of the rest of the paper, is presented. The use of SSs and NSs for a mathematical representation of the ethical rules, and in extension of the ethical theories, is attempted in the fifth section and the chapter closes with some hints for further research and the final conclusions, presented in its last two sections.

CREATION AND EVOLUTION OF ETHICS

Life on our planet is in continuous evolution. According to the anthropologists' estimations the first anthropoids appeared on Earth 3.5 million years ago, whereas the *Homo-Sapiens* appeared about 500 thousand years ago, after the discovery of the fire. During the last 200 thousand years a shocking fact took place, the rapid – in comparison to the whole time of existence of humans on Earth - development of the *neocortex* of the human brain, which has no precedent in the history of biological evolution. The development of neocortex made humans the only creatures on the planet being able to think, to design and organize their actions and generally to affect the evolution of the whole life on the Earth. Humans acquired "world-view" about their planet and the whole universe, about their origin and their future. This world-view unfolded from the worship of animals and many Gods' to the variety of the religious and philosophical views of the contemporary societies (Koestler, 1979, Timmer, 1994, Razis, 1989).

Humans lived 98% of the total time of their existence on Earth in small groups as hunters and food collectors moving from place to place. During the last 10-12 thousand years, however, they discovered agriculture and since then they stopped moving around and organized in cities, societies and nations. The golden period of ancient Greece, which took place some thousand years later (started around 600 BC), is connected to the first accumulated increase of knowledge in human history. Centuries of "spiritual numbness" followed until the "explosion in intelligence and informatics" (Nicolis, 1991), started with Newton's scientific revolution during the 17th century and approaching its peak in our days. During the last 300 years the curve of increase of human knowledge transformed from exponential to a rocket-like form.

The recent findings of the magnetic tomography verified that the human brain consists of three parts (Figure 1).

Figure 1. A simplified representation of the anatomy of the human brain

- The first part (black colored in Figure 1) is the center of the *instincts* and the *automated functions* of the human body and corresponds to the brain of the first reptiles that moved from the sea to the earth.
- The second part (grey colored in Figure 1), referred to as the *cerebral cortex,* is the center of the *emotions* and corresponds to a second cortex created in the brain of the upper class of mammals.
- The third part (white colored in Figure 1) is the *neo-cortex*, which is the center of the *human logic*.

As we have already seen, the neocortex was created too rapidly. As a result, the other two parts of the human brain, having not enough time to be transformed, remained almost unchanged. This means that the human instincts and emotions are almost the same as the instincts and emotions of the animals. The difference is that human logic can manage instincts and emotions, a fact which does not happen to animals.

The previous one, however, is only a simplified presentation of the anatomy of the human brain. In reality, its three parts are not strictly separated from each other. This explains why one can find some traces of simple logic to certain kinds of animals, like monkeys, dogs, etc.

It must be noted that the five senses (touch, sight, hearing, smell and taste), which preexisted in humans before the development of the neocortex, remain connected to the two older parts of the brain. As a result, the reactions from the instincts and emotions appear first, whereas the elaboration of them by logic is always delayed. This is a crucial point for the understanding of the principles of emotional intelligence.

The *amygdala* is a group of almond-shaped cells near the base of the brain, which helps the definition and regulation of human emotions. The amygdala also activates the *fight-or-flight response*, which can help people in immediate physical danger to react quickly, without any initiative by them, for their safety and security. When the amygdala senses danger, it signals the human organization to pump stress hormones preparing the body either to fight for survival or to flee for safety. This response helped early humans to avoid being injured or killed in the frequently occurring dangerous situations at that time, whereas today it usually appears in cases of fear, anxiety, aggression and anger. The fight-or-flight response was termed by Goleman (2007) as the *amygdala hijack* in resemblance to the speedy action of pirates when attacking ships.

Figure 2 illustrates in a simplified way how logic influences human emotions to generate moral and immoral actions or behaviors.

Figure 2. Influence of logic to emotions for creating moral and immoral human actions

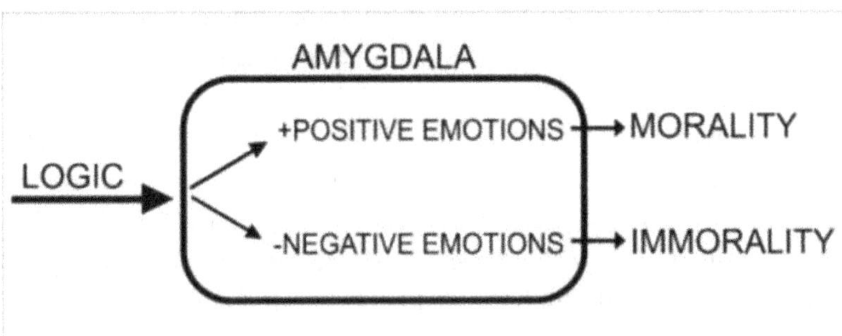

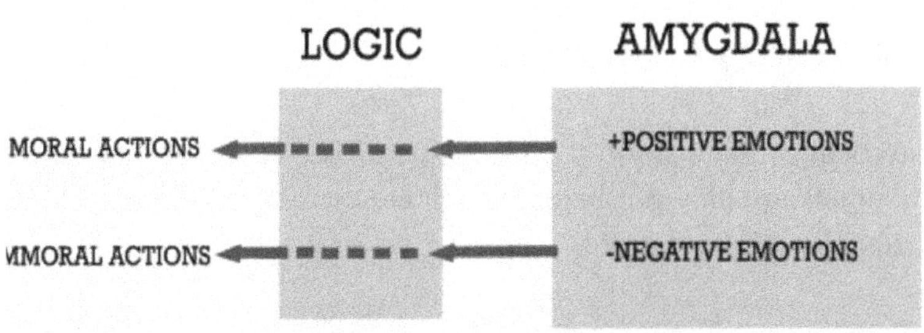

In the upper part of Figure 2 the process starts from logic, which combined, through the amygdala, with positive emotions results in moral actions, whereas combined with negative emotions results in immoral actions. As it is shown in the lower part of Figure 2, however, this process is frequently inverted starting from the individual's personal emotions and resulting, through logic, to moral or immoral actions respectively. Logic plays, therefore, a central role in the creation of both ethical and unethical human behavior.

The lower part of Figure 2 indicates further that the moral or immoral actions of an individual depend upon his/her personality. Consequently, the ethical rules in a human society must be imposed in a strict way by the corresponding authorities, a fact which obviously has to do with the process illustrated in the upper part of Figure 2. Ethics, in other words, is inevitably connected to the systems of justice of human societies.

Logic combines *statistical* and *critical thinking* for the creation of the ethical rules of human society. Let us consider, for example, the following possible scenario about the origin of the known from the Bible and the Moses' Ten Commandments ethical rule "Thou shalt not kill". Some thousand years ago, when humans lived in small groups as hunters, some of them, affected by their negative emotions (lower part of Figure 2), started to kill other hunters for stealing their games of hunting. In this way, however, the number of hunters could be dangerously decreased, so that the other people in the groups could remain without food (statistical thinking). This made the leaders of the groups, with the help of critical thinking, decide and agree to issue a common order that killing of other people is forbidden and will be strictly punished (upper part of Figure 2).

Statistics, of course, which laid its foundations as a science only during the Renaissance thanks to the work of John Graunt (1620-1674) who provided a framework for the modern demography (Voskoglou & Athanassopoulos, 2020) was a completely unknown topic at that time. Statistical thinking, however, preexisted in human nature from ages through the instinct of "following the herd".

In conclusion, human ethics, but human immorality too, were created and developed through the influence of logic (critical and statistical thinking) on human emotions.

A BRIEF HISTORICAL ACCOUNT

A detailed account of the history of ethics is out of the scope of this paper. Here some central points are only presented, which are considered to be useful for the better understanding of the rest of the paper.

From the beginning of human history morality was connected to religion and philosophy (Razis, 1990). The ancient Greek philosopher Socrates (470-399 BC) was the first who attempted to give a definition of ethics and his axiom "Virtue is knowledge" became famous. Aristotle, who followed, formulated the eternal problems of ethics which remain unchanged until nowadays. The Stoic philosophers divided philosophy to three branches: Logic, physics and ethics. This division was acceptable until the era of Renaissance (14th to 17th century) and was adopted by the famous German philosopher Immanuel Kant (1724-1804), who influenced so much the formulation of the modern ethics of the West Societies. Kant supported the existence of two levels of reality: The level of *phenomena* connected to science and the level of *intellectuals'* connected to ethics. The former is created by the human mind, whereas the latter surpasses the human intelligence corresponding to a higher spiritual reality that supports the moral and religious human life (Encyclopedia Britannica, 1971).

Ethics is one of the forces that formulated human societies. Thrasybulus in Plato's "Republic" argued that the ethical rules are created for serving the leader or the group which governs a country. In this way Plato preceded the Marxist philosophers, who believed that the ethical rules are a product of human nature and serve the needs of the human societies at a certain time and place. The opposite view is that the ethical rules come from an absolute authority, beyond the human's existence, like Plato's "absolute good", Stoics' "physical law", Thomists' "Devine law", Hegel's "absolute idea" and Kant's "a priori ethical law" (Razis, 1990). The existence, however, of certain basic ethical principles which are generally acceptable, at least by the Western societies, is a view supported until nowadays (Encyclopedia Britannica, 1971). .

Although there is much disagreement about what exactly constitutes a *moral theory*, roughly speaking one may say that the object of such a theory is morality as a normative system (Driver, 2022). The philosophical theories about the nature and origins of morality are broadly divided in two classes: *Moral realism*, according to which there are true moral statements reporting objective facts and *moral anti-realism* supporting the view that moral statements do not attempt to describe the world's reality, but rather to express an emotion or a command deriving from people's personal motives and practice (Capouthier, 2004).

The *moral dilemmas* reported during the 20th century disputed the old moral theories, like *Aristotle's ethics* (Aristotle, 350 BC), *Kantian ethics* (Sullivan, 1994), Bentham's *utilitarianism* (Bentham, 1789), etc. A moral dilemma is a situation in which a person has moral reasons to do two (or more) actions, he/she can do each of them, but not both (or all) of them. No matter what he/she does, he/she will do something wrong, thus seems condemned to moral failure. The *problem of the trolley* (Thomson, 1976) is probably the most well-known example of a moral dilemma. One of the many variations of this dilemma is illustrated in Figure 3.

Figure 3. A variation of the trolley's problem

The mother in Figure 3 is looking to the train, which cannot be stopped, approaching with high speed to her son, who is trapped on the rail. The only thing that she can do is to use the bar in order to direct the train to the other side of the rail, where unfortunately two other people are also trapped. What must she do?

At that time the poor mother has no time to think (amygdala hijack). Most probably, therefore, her motherly instinct will push her to turn the train to the other side for saving the life of her son, but killing the other two people. According to the ethical theory of utilitarianism, however, whose basic principle is to prescribe actions maximizing the well-being and happiness of all the individuals affected by them, this is an immoral action.

The previous dilemma, in which the moral principle of not killing other people conflicts with itself, indicates that BL is not suitable for ethics. The next dilemma of the *searching assassin* (Sandel, 2009, pp. 187-195), however, in which the moral principle of not killing conflicts to the obligation, according to Kant (Sullivan, 1994) of telling always the truth, marks out more emphatically the need of introducing degrees of truth to the moral principles. This happens, because assassination is much more serious than telling a lie, since its consequences, i.e. the loss of a human life, cannot be corrected.

According to the dilemma of the searching assassin, stated by the contemporary to Kant French philosopher B. Constant, an individual who searches for his enemy on the purpose of killing him, comes to the house where his enemy is hidden, and asks the owner and friend of his enemy where he can find him. Which should be the right answer for the owner of the house? Kant commented on this dilemma with a written report of about 10 pages, insisting that the right answer is to tell the truth, because this is a sacred ethical obligation. Such kind of dilemmas were the reason for starting to make evident to a number of ethical philosophers the need of introducing a multi-valued logic in ethics, with Sir W.D. Ross being a characteristic example (Ross, 1930).

MATHEMATICAL BACKGROUND

Fuzzy Sets and Logic

The Aristotle's BL, which is based on the "*Principle of the Excluded Middle*" (a proposition is either "true" or "not true"), was for many centuries in the center of human reasoning, affecting enormously the development of science and civilization and, in extension, the formulation of the human ethics.

Opposite views, however, appeared also early in the history of humanity, supporting the existence of a third area between "true" and "not true"; e.g. by Buddha Siddhartha Gautama, India, around 500 BC, by Plato (427-377 BC), more recently by the Marxist philosophers, etc. Integrated ideas about multi-valued logics appeared, at any case, only during the early 1900s by Lukasiewicz, Tarski and others (Athanassopoulos & Voskoglou, 2020, Section 2). According to Lukasiewicz's "*Principle of Valence*", in reality there exist intermediate truth-values between the two extremities "true" and "not true".

The electrical engineer L. Zadeh, Professor of Computer Science at the University of California, Berkeley, in order to tackle the partial truths, extended the classical concept of crisp set to the concept of *FS* by generalizing the characteristic function of crisp sets to the *membership function* of FSs, as follows (Zadeh, 1965):

Definition 1: A FS F on the universe U is of the form

$$F = \{(x, m(x)): x \in U\} \tag{1}$$

In formula (1) m: U→ [0,1] is the membership function of F and m(x) is called the *membership degree* of x in F. The greater m(x), the more x satisfies the property of F.

Based on the notion of FS, Zadeh introduced the FL (Zadeh, 1973), in which the propositions possess infinitely many truth-values in the interval [0, 1]. FSs and FL are also used for handling the existing in real world uncertainty. FL generalizes and completes the traditional BL, reducing to it when no uncertainty exists in the corresponding situation. For general facts on FSs and FL we refer to (Klir & Folger, 1988).

Neutrosophic Sets

Following Zadeh FSs, various generalizations of the FS theory have been proposed during the last years for a more effective management of all types of the existing in real world uncertainty. A brief description of the most important among these theories can be found in (Voskoglou, 2019).

On the purpose of simulating the existing imprecision in human reasoning, K.T. Atanassov, Professor at the Bulgarian Academy of Sciences in Sofia, enriched Zadeh's membership degree by adding the degree of *non-membership* n(x) and defined the notion of *intuitionistic fuzzy set (IFS)*, which is a generalization of FS, as follows (Atanassov, 1986):

Definition 2: An IFS A in the universe U is of the form

$$A = \{(x, m(x), n(x)): x \in U, 0 \leq m(x) + n(x) \leq 1\} \quad (2)$$

In formula (2) m: U→ [0,1] is the membership function and n: U→ [0,1] is the non-membership function of A.

The Romanian-American mathematician and writer F. Smarandache, motivated by the various neutral situations appearing in real life - like <friend, neutral, enemy>, <positive, zero, negative>, <small, medium, high>, <win, draw, defeat>, etc. – introduced the degree of *indeterminacy* (or *neutrality*) of the elements of the universal set U in a subset of U and extended the concept of IFS to the concept of *neutrosophic set* (NS). The term neutrosophic is the result of the synthesis of the words "neutral" and "sophia", which means in Greek language "wisdom". According to Smarandache, neutrosophy means the knowledge of the neutral thought.

In this work we need only the simplest version of the concept of NS, which is defined as follows (Smarandache, 1995):

Definition 3: A *single valued NS (SVNS)* A in U is of the form

$$A = \{(x,T(x),I(x),F(x)): x \in U, T(x),I(x),F(x) \in [0,1], 0 \leq T(x)+I(x)+F(x) \leq 3\} \quad (3)$$

In formula (3) T(x), I(x), F(x) are the degrees of *truth* (or membership), indeterminacy (or neutrality) and *falsity* (or non-membership) of x in A respectively, called the *neutrosophic components* of x. For simplicity, we write A<T, I, F>. Indeterminacy is understood to be everything which is between the opposites of truth and falsity (Smarandache, 2021).

For example, let U be the set of the players of a football team and let A be the SVNS of the good players of U. Then each player x of U is characterized by a *neutrosophic triplet* (t, i, f) with respect to A, with t, i, f in [0, 1]. For instance, x(0.7, 0.1, 0.4) ∈ A means that there is a 70% belief that x is a good player, a 10% doubt about it and a 40% belief that x is not a good player. In particular, x(0,1,0) ∈ A means that we do not know absolutely nothing about x's affiliation with A.

When the sum T(x) + I(x) + F(x) of the neutrosophic components of x ∈ U in a SVNS A in U is <1, then x leaves room for *incomplete* information, when is equal to 1 for *complete* information and when is greater than 1 for *paraconsistent* (i.e. contradiction tolerant) information. A SVNS may contain simultaneously elements leaving room to all the previous types of information. For general facts on SVNSs we refer to (Wang et al., 2010).

Soft Sets

A serious disadvantage of the FS theory, and of all its generalizations involving the use of membership functions (e.g. IFSs, NSs, etc.), is that there is no exact rule for defining the membership function properly. The methods used for this are usually statistical or intuitive/empirical. Moreover, the definition of the membership function is not unique depending on the "signals" that each observer receives from the environment. For example, defining the FS of "tall men" one could consider all men having heights greater than 1.9 meters as tall and another one all those having heights greater than 2 meters. Consequently, the first observer would assign membership degree 1 to all men with heights between 1.9 and 2 meters, whereas the second one would assign to them membership degrees less than 1. Analogous differences are logical to appear also to the membership degrees of all the other men. Thus, the only restriction for the definition of the membership function is that it must be compatible with the common logic; otherwise the resulting FS does not give a reliable description of the corresponding real situation. This could happen, for instance, if in the previous example men with heights less than 1.6 meters possessed membership degrees ≥0.5.

Several theories related to FSs have been proposed for overcoming this difficulty, like the *interval-valued FSs* (Dubois & Prade, 2005) in which the membership function is replaced by a mapping from the universe U to the set of closed subintervals of [0, 1], the *grey systems/numbers* (Deng, 1982) in which the definition of a membership function is not necessary, the *rough sets* (Pawlak, 1991) using a pair of sets which give the lower and the upper approximation of the original crisp set, etc.

The Russian mathematician D. Molodstov introduced the notion of *soft set (SS)* for tackling the uncertainty in a parametric manner, as follows (Molodstov, 1999):

Definition 4: Let E be a set of parameters, let A be a subset of E, and let f be a map from A into the power set P(U) of the universe U. Then the SS (f, A) in U is defined by

$$(f, A) = \{(e, f(e)): e \in A\} \tag{4}$$

For example, let U= $\{H_1, H_2, H_3\}$ be a set of houses and let E = $\{e_1, e_2, e_3\}$ be the set of the parameters e_1=cheap, e_2=wooden and e_3= beautiful. Let us further assume that H_1, H_2 are the cheap houses, H_3 is the beautiful house and H_2, H_3 are the wooden houses. Then, a map f: E P(U) is defined by $f(e_1)=\{H_1, H_2\}$, $f(e_2)=\{H_2, H_3\}$ and $f(e_3)=\{H_3\}$. Therefore, the SS (f, E) in U is the set of the ordered pairs (f, E) = $\{(e_1, \{H_1, H_2\}), (e_2, \{H_2, H_3\}), (e_3, \{H_3\})\}$.

In other words, a SS is a parameterized family of subsets of U. The term “soft” is related to the fact that the form of (f, A) depends on the parameters of A.

A FS in U with membership function y = m(x) is a SS in U of the form (f, [0, 1]), with f(α)={x U: m(x) α}, for each α in [0, 1]. The concept of SS is, therefore, a generalization of the concept of FS. For general facts on SSs we refer to (Maji, Biswas & Roy, 2003).

REPRESENTATION OF THE ETHICAL RULES BY SOFT AND NEUTROSOPHIC SETS

As we already have mentioned in our Introduction, the implementation of the principles of FL in ethics is not a new idea. To the best of our knowledge, however, no attempt has been reported in the literature until now to present an integrated model for the materialization of this idea. Here, we show that, under the light of FL, or another multi-valued logic, the ethical rules of a human community can be represented by SSs or NSs. As a result, a moral theory, being a collection of ethical rules, could be considered as being a collection of SSs, or NSs. This consideration, which reduces the ethical dilemmas drastically, could be proved in future as being a new basis for a modern approach to ethics.

Let us go back, for example, to Moses’ rule “Thou shalt not kill”, which is more or less connected nowadays to the ethical beliefs of most people, regardless their religion. According to BL, therefore, assassination is an immoral action. The question, however, is: “Have all assassinations the same degree of immorality?” Is it the same, for instance, an assassination designed with every detail and performed under conditions of complete soberness (case A_1), with an assassination performed under a “boiling soul” condition (case A_2), or with the assassin being in defense trying to protect his/her life (case A_3), or even with the instinctive reaction of the mother (case A_4) in the moral dilemma with the train (Figure 3)? Consider also a neutral observer in place of the mother, who prefers to save the lives of the two people leaving the child to be killed (case A_5). The answer to this question is obviously negative, which shows that BL is not suitable to tackle successfully such kind of situations.

In order to overcome this difficulty, let us consider the set U of all kinds of assassinations as the universal set of the discourse. We introduce the following parameters: e_1 = immoral, e_2 = immoral, but with lightning, e_3 = neutral and e_4 = moral. Set E = {e_1, e_2, e_3, e_4} and let f: E → P(U) be the function assigning to each parameter of E the subset of U consisting of all kinds of assassinations characterized by this parameter. Then the case A_1 belongs to the set f(e_1), A_2 to f(e_2), A_3 and A_4 could belong to f(e_3), A_5 could belong to f(e_4), etc. Consequently, the SS (f, E) = {(e, f(e)): e ∈ E} represents mathematically the ethical rule "Thou shalt not kill".

Of course, this assignment still leaves some doubts and space to the personal beliefs of the observer. One may consider, for example, that A_3 and/or A_4 belong to f(e_2), or even that A_5 belongs to f(e_3). The choice of the parameters is not unique too; one may add more parameters or change some of the previous ones (e.g. the parameter e_3). In any case, the use of SSs gives much better conclusions than the application of the BL. This approach can be improved further, if the choice of the parameters and the assassinations characterized by each one of them is decided not by a single person, but by a group (the wider possible) of specialists. In this way the previous choices will be statistically more objective. This skepticism could be also useful to the judges and the jurymen for reaching the right decisions in the courts.

In an analogous way one can represent any other ethical rule, with the universe U and the set of parameters E varying from case to case. Consequently, since a moral theory is a collection of ethical rules, such a theory could be mathematically represented by a collection of SSs.

An alternative option is to work with NSs instead of SSs. This is useful when one has doubts about the credibility of the parameters assigned to the elements of U, like it happens in the previous cases A_3 and A_4 of assassinations. In our example, for instance, one could consider the SVNS of the "compatible" to the common ethics assassinations and characterize with neutrosophic triplets its elements as follows: A_1 (0, 0, 1), A_2 (0.2, 0.5, 0.7), A_3 (0.6, 0.8, 0.2), A_4 (0.7, 0.9, 0.2), A_5 (0.9, 0.3, 0), etc. The credibility of this option, of course, still depends on the objectivity of the choice of the corresponding neutrosophic triplets. On the basis of this option, a moral theory could be represented by a collection of NSs.

As a second example, we go back to the ethical dilemma of the "searching assassin" mentioned in the third section of this chapter. How moral could one characterize the possible answers of the owner of the house to the candidate assassin? Let us imagine some of these answers: A_1: I don't know him, A_2: I have not seen him during the last few days, A_3: He passed from here for a few minutes and then he left, A_4: He used to be here some time ago, but now he has left, A_5: He is hiding in the basement, etc. Obviously only the last of the previous answers is true. Consequently, with respect to the Kantian ethics and the principles of BL, this is the unique moral answer. The common logic, however, says that it is better to tell a lie for saving a life. Considering, therefore, the SS (f, E) = $\{(e, f(e)): e \in E\}$in the universal set of all possible answers and with respect to the same with the previous example set E of parameters, one could put the previous answers A_1-A_4 to the set $f(e_2)$ and the answer A_5 to the set $f(e_1)$! Alternatively one could introduce the following neutrosophic triplets in the SVNS of the moral answers: $A_1(0.3, 0.5, 0.6)$, $A_2(0.4, 0.6, 0.4)$, $A_3(0.5, 0.7, 0.3)$, $A_4(0.8, 0.7, 0.2)$, $A_5(0.2, 0, 0.8)$.

In conclusion, a moral theory could be represented either as a collection of SSs or as a collection of NSs. The latter is useful when one has doubts about the credibility of the parameters assigned to the elements of the universal set. Both methods have their weaknesses, because they depend on the subjective feelings of the user, but they are better than the application of the principles of BL.

FUTURE RESEARCH DIRECTIONS

The previous two methods of using SSs and NSs to introduce multi-valued logics in ethics could be proved as a new basis for a modern approach of the subject in future. Further theoretical and empirical research is needed, however, for improving the two methods and applying them on the existing moral theories, in order to obtain safer results and conclusions.

CONCLUSION

The discussion performed in this paper leads to the following conclusions:

- There is not a clear distinction in the literature between the terms of morality and ethics. Morality is usually referred to as something personal, whereas ethics are the standards of "good" and "bad" distinguished by a certain community. The ethical rules of a community must be imposed strictly by the corresponding authorities, since humans are characterized by deep differences among each other, the result of which is the lack of unity and the absence of a universal will. Ethics is inevitably connected, therefore, to the social systems of justice of human society.
- The creation and evolution of ethics came as a result of the influence of human logic (critical and statistical thinking) on human emotions. The introduction of a multi-valued logic in ethics, seems to be necessary, since the conclusions of the BL create many moral dilemmas and impasses.

- In this work the use of SSs and NSs was proposed for a mathematical representation of the ethical rules. Thus, a moral theory, being a collection of ethical rules, could be represented as a collection of SSs or NSs. To the best of our knowledge, this is the first integrated proposal for implementing FL (or another multi-valued logic) in ethics. This proposal, which reduces drastically the moral dilemmas, although it still leaves some doubts and space to the personal beliefs of the user, is much better than the outcomes of BL, especially when the choice of the corresponding parameters/neutrosophic triplets is statistically correct. Further research is needed, however, for obtaining safer results and conclusions.

Funding

This research received no specific grant from any funding agency in the public, commercial, or not-for-profit sectors.

REFERENCES

Aristotle. (1908). Nicomachean Ethics (W. D. Ross, Trans.). Clarendon Press.

Atanassov, K. T. (1986). Intuitionistic Fuzzy Sets. *Fuzzy Sets and Systems*, *20*(1), 87–96. doi:10.1016/S0165-0114(86)80034-3

Athanassopoulos, E., & Voskoglou, M. (2020). A Philosophical Treatise on the Connection of Scientific Reasoning with Fuzzy Logic. *Mathematics*, *8*(6), 875. doi:10.3390/math8060875

Bentham, J. (1789). An Introduction to the Principles of Morals and Legislation. Dover Publications Inc.

Chapouthier, G. (2004). To what extent is moral judgment natural? *European Review (Chichester, England)*, *12*(2), 179–183. doi:10.1017/S1062798704000171

Deng, J. (1982). Control Problems of Grey Systems. *Systems & Control Letters*, 288–294.

Driver, J. (2022). Moral Theory. The Stanford Encyclopedia of Philosophy.

Dubois, D., & Prade, H. (2005), Interval-Valued Fuzzy Sets, Possibility Theory and Imprecise Probability. *Proceedings EUSFLAT-LFA,* 314-319.

Encyclopedia Britannica. (1971). *History of Ethics - Ethics Comparative*. William Benton.

Goleman, D. (2007). Emotional Intelligence: Why It Can Matter More Than IQ (10th ed.). Random House USA Inc.

Klir, G. J., & Folger, T. A. (1988). *Fuzzy sets, Uncertainty and Information*. Prentice-Hall.

Koestler, A. (1979). *The Ghost in the Machine*. Hutchinson.

Kosko, B. (1993). *Fuzzy thinking: The new science of fuzzy logic*. Hyperion.

Maji, P. K., Biswas, R., & Roy, A. R. (2003). Soft Set Theory. *Computers & Mathematics with Applications (Oxford, England)*, *45*(4-5), 555–562. doi:10.1016/S0898-1221(03)00016-6

Molodtsov, D. (1999). Soft set theory – first results. *Computers & Mathematics with Applications (Oxford, England)*, *37*(4-5), 19–31. doi:10.1016/S0898-1221(99)00056-5

Nicolis, J. S. (1991). *Chaos and Information Processing*. World Scientific Publishers. doi:10.1142/0984

Nodelman, (Eds.). (2022). https://plato.stanford.edu/archives/fall2022/entries/moral-theory

Papaioannou, Y. (2013). *Fuzzy Logic in Science, Law and Ethics, 2013*. Available online: https://blogs.elpais.com/atomium-culture/2013/01/fuzzy-logic-in-science-law-and-ethics.html

Pawlak, Z. (1991). *Rough Sets: Aspects of Reasoning about Data*. Kluer Academic Publishers. doi:10.1007/978-94-011-3534-4

Razis, D. V. (1989). Modern Cassandras and our survival as a species - a new role for medicine? *Journal of the Royal Society of Medicine*, *82*(10), 575–576. doi:10.1177/014107688908201001 PMID:2810296

Razis, D. V. (1990). Ethics and the Universality of Morality. *Quality Assurance in Health Care*, 2(3-4), 369–373. doi:10.1093/intqhc/2.3-4.369 PMID:1983258

Ross, W. D. (1930). The Right and the Good. Oxford University Press.

Sandel, M. J. (2011). *Justice. What is the Right Thing to do?* (A. Kioupkiolis, Trans.). Polis Editions.

Smarandache, F. (1998). *Neutrosophy / Neutrosophic Probability*. Set, and Logic.

Smarandache, F. (2021). Indeterminacy in Neutrosophic Theories and their Applications. *International Journal of Neutrosophic Science, 15*(2), 89–95. doi:10.54216/IJNS.150203

Sullivan, R. J. (1994). *An Introduction to Kant's Ethics*. Cambridge University Press. doi:10.1017/CBO9780511606151

Thomson, J. J. (1976). Killing, Letting Die, and the Trolley Problem. *The Monist, 59*(2), 204–217. doi:10.5840/monist197659224 PMID:11662247

Timmer, E. (1994). Medical ethics and evolution. *Journal of the Royal Society of Medicine, 87*(5), 250–252. doi:10.1177/014107689408700502 PMID:8207716

Voskoglou, M. G., & Athanassopoulos, E. (2020). Statistical Thinking in Problem Solving. *American Journal of Educational Research, 8*(10), 754–761.

Voskoglou, M. G. (2019). Generalizations of Fuzzy Sets and Related Theories. In M. Voskoglou (Ed.), *An Essential Guide to Fuzzy Systems* (pp. 345–352). Nova Science Publishers.

Wang, H., Smarandanche, F., Zhang, Y., & Sunderraman, R. (2010). Single Valued Neutrosophic Sets. *Review of the Air Force Academy (Brasov), 1*(16), 10–14.

Zadeh, L. A. (1965). Fuzzy Sets. *Information and Control, 8*(3), 338–353. doi:10.1016/S0019-9958(65)90241-X

Zadeh, L. A. (1973). Outline of a new approach to the analysis of complex systems and decision processes. *IEEE Transactions on Systems, Man, and Cybernetics, 3*(1), 28–44. doi:10.1109/TSMC.1973.5408575

ADDITIONAL READING

Atanassov, K. T. (1999). *Intuitionistic Fuzzy Sets*. Physica-Verlag. doi:10.1007/978-3-7908-1870-3

Cuong, B. C. (2014). Picture Fuzzy Sets. *Journal of Computer Science and Cybernetics, 30*(4), 409–420.

Dubois, D., & Prade, H. (1980). *Fuzzy Sets and Systems: Theory and Applications*. Academic Press.

Farid, F., Saeed, M., & Ali, M. (2019). Representation of Soft Set and its Operations by Bipartite Graph. *Sci. Inquiry Rev., 3*(3), 30–42. doi:10.32350ir.33.03

Gentili, P. L. (2021). Establishing a New Link between Fuzzy Logic, Neuroscience and Quantum Mechanics through Bayesian Probability: Perspectives in Artificial Intelligence and Unconventional Computing. *Molecules (Basel, Switzerland)*, *26*(19), 5987. doi:10.3390/molecules26195987 PMID:34641530

Kharal, A., & Ahmad, B. (2011). Mappings on soft classes. *New Mathematics and Natural Computation*, *7*(3), 471–481. doi:10.1142/S1793005711002025

Maji, P. K., Roy, A. R., & Biswas, R. (2002). An Application of Soft Sets in a Decision Making Problem. *Computers & Mathematics with Applications (Oxford, England)*, *44*(8-9), 1077–1083. doi:10.1016/S0898-1221(02)00216-X

Sakawa, M. (1993). *Fuzzy Sets and Interactive Multiobjective Optimization*. Plenum Press. doi:10.1007/978-1-4899-1633-4

Smarandache, F. (2016). Subtraction and Division of Neutrosophic Numbers. *Critical Review*, *13*, 103–110.

Tripathy, B. K., & Arun, K. R. (2016). Soft Sets and Its Applications. In J. S. Jacob (Ed.), *Handbook of Research on Generalized and Hybrid Set Structures and Applications for Soft Computing* (pp. 65–85). IGI Global. doi:10.4018/978-1-4666-9798-0.ch005

Voskoglou, M., Broumi, S., & Smarandache, F. (2022). A Combined Use of Soft and Neutrosophic Sets for Student Assessment with Qualitative Grades. *Journal of Neutrosophic and Fuzzy Systems*, *4*(1), 15–20. doi:10.54216/JNFS.040102

Voskoglou, M. G. (2011). *Stochastic and Fuzzy Models in Mathematics Education, Artificial Intelligence and Management*. Lambert Academic Publishing.

Voskoglou, M. G. (2017). *Finite Markov Chain and Fuzzy Logic Assessment Models: Emerging Research and Opportunities, Create Space Independent Publishing Platform*. Amazon.

Voskoglou, M. G. (2019). Methods for Assessing Human-Machine Performance under Fuzzy Conditions. *Mathematics*, *7*(3), 230. doi:10.3390/math7030230

Voskoglou, M. G. (2022). Fuzziness, Indeterminacy and Soft Sets: Frontiers and Perspectives. *Mathematics*, *10*(20), 3909. doi:10.3390/math10203909

Zadeh, L. A. (1996). Fuzzy logic = computing with words. *IEEE Transactions on Fuzzy Systems*, *4*(2), 103–111. doi:10.1109/91.493904

KEY TERMS AND DEFINITIONS

Amygdala Hijack: The amygdala is a group of almond-shaped cells near the base of the brain, which helps the definition and regulation of human emotions. The amygdala activates the fight-or-flight response, which can help people in immediate physical danger to react quickly, without any initiative by them, for their safety and security. This was termed by Goleman in 2007 as the amygdala hijack in resemblance to the speedy action of pirates when attacking ships.

Ethics: The standards of "good" and "bad" or "right" and "wrong" distinguished by a certain community or social setting.

Fuzzy Logic (FL): A logic based on the concept of FS that, in contrast to the classical bi-valued logic of Aristotle (right-wrong), characterizes cases with multiple values.

Fuzzy Set (FS): A FS, say A, is characterized by its membership function $y = m_A(x)$ defined on the universal set of the discourse U and taking values in the interval [0, 1], thus assigning to each element x of U a membership degree $m_A(x)$ with respect to A. The closer is $m_A(x)$ to 1, the better x satisfies the characteristic property of A. A crisp subset of U is a FS in U with membership function its characteristic function.

Indeterminacy: Indeterminacy is understood to be everything which is between the opposites of truth and falsity.

Moral Dilemma: A moral dilemma is a situation in which a person has moral reasons to do two (or more) actions, he/she can do each of them, but not both (or all) of them. No matter what he/she does, he/she will do something wrong, thus seems condemned to moral failure.

Morality: A personal framework of rules and behaviors about "good" and "bad" or "right" and "wrong" conduct.

Single-Valued Neutrosophic Set (SVNS): The concept of SVNS is an extension of the concept of FS. A SVNS, say A, in the universe U is defined as $A = \{(x,T(x),I(x),F(x)): x \in U,\ T(x),I(x),F(x) \in [0,1],\ 0 \leq T(x)+I(x)+F(x) \leq 3\}$, where T(x), I(x), F(x) are the degrees of truth (or membership), indeterminacy (or neutrality) and falsity (or non-membership) of x in A respectively, called the neutrosophic components of x. When the sum $T(x) + I(x) + F(x) < 1$, then x leaves room for incomplete information, when is equal to 1 for complete information and when is greater than 1 for paraconsistent (i.e. contradiction tolerant) information. A SVNS may contain simultaneously elements leaving room to all the previous types of information.

Soft Set (SS): Let E be a set of parameters, let A be a subset of E, and let f be a map from A into the power set P(U) of the universe U. Then the SS (f, A) in U is defined by $(f, A) = \{(e, f(e)): e \in A\}$.

Chapter 6
Neutro Geometric Topology and Its Examples

Somen Debnath
https://orcid.org/0000-0003-3585-0868
Umakanta Academy, India

Prem Kumar Singh
https://orcid.org/0000-0003-1465-6572
Gandhi Institute of Technology and Management, India

ABSTRACT

Neutro geometric topology is obtained by combining the neutro topology and the neutro geometry. Similarly, the hybridization of anti-topology and the anti-geometry produced anti-geometric topology. Considering this knowledge, in this chapter, neutro geometric topology is studied. Undoubtedly the classical geometric topology provides the foundation for introducing the neutro geometric topology. Also, some basic properties and results associated with neutro geometric topology are obtained and proved. Moreover, to justify the proposed study, some examples are given. The future direction of the proposed study is mentioned at the end of the chapter.

INTRODUCTION

Topology made to study the surfaces of different objects whose properties remains invariant under homeomorphism. It is useful to study those objects whose geometrical properties are preserved under continuous deformation like stretching and bending. Due to which, it is known as "rubber-sheet geometry". Topology can be regarded as a new branch of mathematics and it has some other subbranches like general topology

DOI: 10.4018/978-1-6684-4740-6.ch006

(Sierpinski, 2020), geometric topology (Martelli, 2016), combinatorial topology (Weisstein), differential topology (Brocker& Janich, 1982), algebraic topology (Munkers, 2018), etc. On the other hand, simple geometry, or Euclidean geometry is the study of the shapes, sizes, and relative positions of 2D and 3D objects, based on axioms, postulates, and theorems. Topology and geometry seem to be akin to one another in many areas. There are many instances where geometric properties of objects and its study requires other things (attributes) like color, smell, melting point, etc. To understand these instances precisely the study of their topological properties is required. In this way, the topology helps in building the relationship among the physical and logical properties of an object. The above discussion shows that, geometry cannot answer many questions among the pair of objects and its instances which can be answered via topology. It is one of the reasons which gave flexibility to apply topology in numerous fields of science and technology (Adhikari, 2016). Cheng-Ming, (1985) extended the classical topology by introducing the fuzzy topological spaces. Sostak et al. (1996) studied the basic structures of fuzzy topology which is studied via Rough set theory base topological space by Lashin et al. (2005). Sarkar (1981) defined the L-fuzzy topological spaces (Sarkar, 1981) followed by Mondal and Samanta (2011) et al. presented topology via interval-valued intuitionistic fuzzy sets (Mondal & Samanta, 2001). Hur et al. (2014) introduced the intuitionistic fuzzy topological spaces for dealing the hesitant uncertainty. Recently, Pythagorean fuzzy topological spaces is introduced by Olgun et al. (2019) whereas Lee et al. (2022) introduced the hesitant fuzzy topological spaces for dealing with the hesitant or indeterminacy part. It gives a way to introduce a new geometry for dealing with the uncertainty and indeterminacy in the objects and its instances. This paper focuses on introducing NeutroGeometry Topological space.

Fuzzy logic (Zadeh, 1988) is introduced for dealing with the uncertainty in words for knowledge representation tasks. It means fuzzy logic is one of the ways for approximate reasoning that is essential for rational decision-making under the uncertain scenarios. It can be considered as one of the useful tools to control multi-valued linguistic variables by using fuzzy quantifiers. The problem arises when the uncertainty associated with object and instances exists independently when compared to true and false membership values. To deal with these types of uncertainty, Smarandache introduced the neutrosophic logic (Smarandache, 1998) based on an ordered triple $\langle t\%, i\%, f\% \rangle$, where $t\%$ is the percentage of truth degree, $i\%$ is the percentage of indeterminate degree and $f\%$ is the percentage of false degree. Smarandache (2019) promoted neutro algebra and anti-algebra as an extension of classical algebra called Neutroalgebra (Smarandache, 2020a). It introduced that if there is an entity say $\langle P \rangle$ then corresponding to $\langle P \rangle$ there exists two more entities called $\langle anti\ P \rangle$ and $\langle neut\ P \rangle$ whereas $\langle P \rangle$ and $\langle anti\ P \rangle$ are two opposite ones and $\langle neut\ P \rangle$ is something between $\langle P \rangle$ and $\langle anti\ P \rangle$. Thus, the structure (triplet) ($\langle P \rangle$,

$\langle neut\ P\rangle$, $\langle anti\ P\rangle$) is formed due to the tri-section of a non-empty space, an axiom, a proposition, an operation, or any other concept that is close to human cognition. Due to the neutrosophication of statistical data, it has been noticed that some data are informative, some are non-informative, and the remaining may be informative or non-informative i.e., unknown. There is no such concept that exists in a particular domain that is beyond the triplet. In a broader sense, neutrosophication of uncertain information provides a structure that helps us to boost our analytical thinking lucidly so that it ensures to reach the ultimate goal. In another paper, Smarandache presented neutro algebra as an extension of partial algebra (Smarandache(b), 2020). Inspired by the neutro algebraic structure and the anti-algebraic structure, some recent works are highlighted in the following: neutro-BCK-algebra and its applications are investigated (Smarandache&Hamidi, 2020); Rezaei et al. studied the neutro-BE-algebra and anti-BE-algebra and their basic properties (Rezaei & Smarandache, 2020); Sahin et al. extended the classical metric space by introducing the neutro-metric spaces (Sahin, Kargin & Altun, 2021); Sahin et al. investigated the neutro-G modules and anti-G modules (Sahin, Kargin, & Smarandache, 2021); neutro-Lie algebras and anti-Lie algebras (Sahin, Kargin, &Uz, 2021); Smarandache et al. presented the neutro quadruple groups (Smarandache et al., 2021); Al-Tahan et al. introduced the neutrohyperstructures (Al-Tahan, Davvaz, Smarandache&Anis, 2021); Rezaei et al. given applications related to neutrosemihypergroups and anti semihypergroups (Rezaei, Smarandache & Mirvakili, 2021).

Euclid geometry deals with the properties, rules, operators, axioms, postulates, etc which are true for every element in a homogeneous space. By critical investigation, it has been observed that there exist some spaces that are not homogeneous where rules are not equally applicable to all the elements due to the vague information. To tackle such conflicting information, non-Euclidean geometry (Coxeter, 1942) and its data sets (Singh, 2022)are introduced. Thus the non-Euclidean geometry has the potential to address some special situation that is ruled out by Euclidean geometry. For example, the surface of a sphere is non-Euclidean in the sense that the angle sum of a triangle on a sphere is not equal to 180°. Hyperbolic geometry is also known as Lobachevskian geometry is a non-Euclidean geometry that is responsible for rejecting Euclid's fifth postulate (Lobachevsky, 2010). In elliptic or Riemannian geometry also, Euclid's parallel postulate does not hold. The Riemannian geometry fails the second postulate also which gives a new way to deal with non-Euclidean data sets (Singh, 2022). The reason is the failure of fifth postulates, or its removal is considered as Neutral Geometry (Dillon, 2018). However, Smarandache introduced the paradoxist geometry (Smarandache, 1969; Smarandache, 1969) by simultaneous construction of partial Euclidean and partial non-Euclidean space called as NeutroGeometry (Popov, 1996; Chimienti & Bencze, 1998a, 1998b;

Kuciuk & Antholy, 2003; Bhattacharya, 2004). This paper focused on dealing with NeutroGeometry, Anti-Geometry, and its topology.

The introduction of multi-space (Mao, 2007; Smarandache& Cheng-Wei, 2012) or other spaces (Bal, Singh & Ahmad, 2022; Singh, Ahmad, Bal & Aswad, 2021) which exists beyond three dimensions given a way to understand its geometry. It becomes popular among the researchers as they found it as an effective mathematical theory (Stillwell, 1993; Balachandran, 1991), has philosophical (Singh, 2021) and logical influence (Singh, 2021), to address uncertainty (Singh, 2021) in a more accurate manner in numerous subjects (Bal, Singh & Ahmad, 2022; Bal, Singh & Ahmad, 2022; Pandey, Ojha, Singh, Dwivedi&Bergey, 2016). After that, the introduction of the neutro-structures and anti-structures was recognized as a more general structure because of their successful utilization in different areas (Morandi, 1992), whereas the classical mathematical structures are unable to make any decision (Martelli, 2021). We know that topology is an integral part of mathematics, and its ignorance makes us dull in mathematics. Thus, the classical topological concept has been instrumental to investigate several scientific studies, for details see the references given in (Chen & Kikuchi, 2001; Sahin, Kargin&Yucel, 2021; Smarandache, 2021; Singh, 2021; Singh, 2022; Granados, 2022). Sahin et al. (Sahin, Kargin&Yucel, 2021) have extended the classical topological concept by introducing the neutro-topological space and anti-topological space. In their work, they have shown the significance of introducing the neutro-structure and anti-structure via topology to improve the classical-topological structure. In geometry perspective, Smarandache introduced neutro-geometry and anti-geometry to generalize the non-Euclidean geometry (Smarandache, 2021). Moreover, the characterization of non-Euclidian data via neutro-geometry and anti-geometry has been carried out by Singh (Singh, 2021; Singh, 2022) followed in (Granados, 2022). Based on the above literature review, the author is trying to combine the neutro-topology and neutro geometry to produce a neutro-geometric topology and in a similar way, with a combination of anti-topology and anti-geometry responsible to form an anti-geometric topology. Furthermore, various properties and results based on neutro-geometric topology and anti-geometric topology are investigated in a proper way. A comparison of the existing topological structures and the proposed structure is shown below to make sense.

The rest of the chapter is organized as follows: In the next section, some basics about neutrosophic triplet and law is given. Section 3 contains some examples for the neutrosophic Triples Axioms and their notions are given followed by Acknowledgements and conclusions.

BASIC CONCEPTS

Definition: (Smarandache, 2020) (Neutrosophic Triplet, Axioms, and Laws)

According to neutrosophy, if there is a concept, say $\langle\wp\rangle$ then its opposite is denoted by $\langle anti\wp\rangle$ and between $\langle\wp\rangle$ and $\langle anti\wp\rangle$ there is $\langle neut\wp\rangle$. Also, $\langle non\wp\rangle$ implies not $\langle\wp\rangle$ and it is defined by

$$\langle non\wp\rangle = \langle neut\wp\rangle \cup \langle anti\wp\rangle$$

Thus, $(\langle\wp\rangle, \langle neut\wp\rangle, \langle anti\wp\rangle)$ indicates a neutrosophic triplet for $\langle\wp\rangle$.

Neutrosophic Triplet Axioms

Let X be a universe of discourse and $A \subseteq X$, $A \neq \varnothing$. Also, let λ be an axiom defined on A. Then, we consider the following:

1. If the axiom λ is defined for all the elements of A, then λ is known as a classical axiom.
2. If the axiom λ holds for some elements of A and do not hold for others then it is called neutro axiom.
3. If the axiom do not hold for any element of A then it is known as anti axiom.

Therefore, ⟨*axiom, neut axiom, anti axiom*⟩ form a neutro axiom triplet where *neut axiom* ∪ *anti axiom* = *non axiom* and *neut axiom* ∩ *anti axiom* = ∅.

Remark: It is to be noted that axiom means 100% true, neut axiom means it is partially true and partially false, and anti axiom means it is 100% false. So, the neutro axiom triplet enables to study of all possible types of elements that may not be encouraged by the classical axiom.

Neutrosophic Triplet Laws

In real-science, there exist some laws or rules which may be true from one's opinion and the same law may be treated as false according to another opinion. Also, some laws are unpredictable whether they are true/false. Furthermore, there are many conflicting areas where the people's opinion varies, and thus to address such issues, neutrosophic laws or neutro laws serve the purpose.

Neutrosophic Triplet Associativity

Let X be a universal set and $P \subseteq X$, $P \neq \varnothing$ and '•'is a binary operation. Then,

I. $\forall x,y,z \in P$, $x \bullet (y \bullet z) = (x \bullet y) \bullet z$ (Classical)
II. There exists at least one triplet $x_1, y_1, z_1 \in P$ such that $x \bullet (y \bullet z) \neq (x \bullet y) \bullet z$(neutro)
III. For any triplet $x_2, y_2, z_2 \in P$, $x \bullet (y \bullet z) \neq (x \bullet y) \bullet z$(anti)

Similarly, (b) Neutrosophic Triplet Commutativity (c) Neutrosophic Triplet Unity (d) Neutrosophic Triplet Inverse,etc have been defined.

Definition: (Sierpinski, 2020; Martelli, 2016; Weisstein, n.d.; Balachandran, 1991) (Classical Topology)

Let G be a non-empty set and T be a family of all subsets of G. Then, the pair (G,T) is called a topological space where T is a topology on G if the following properties hold:

A1) $\varnothing$, $G \in T$
A2) Arbitrary union of elements of T belongs to T.
A3) Finite intersection of elements of T belongs to T.

Definition: (Smarandache, 1969; Popov, 1996) (Neutro-Topological Space)

Let H be a non-empty set and T_N be a collection of subsets of H. Then, (H,T_N) is said to be a neutro-topological space whereas T_N is a neutro-topology on H if at least one of the following properties hold:

B1) $(\varnothing, H \in T_N)$ or $\left(\varnothing \in T_N, H \notin T_N\right)$ or $\left(\varnothing \notin T_N, H \in T_N\right)$

B2) $\left(\bigcup_{i \in I} s_i \in T_N\right) or \left(\bigcup_{j \in I} r_j \notin T_N\right)$ (arbitrary union)

B3) $\left(\bigcap_{k=1}^{n} p_k \in T_N\right) or \left(\bigcap_{l=1}^{n} q_l \notin T_N\right)$ (finite intersection)

Definition: (Chimienti & Bencze, 1998a, 1998b; Granados, 2022) (Anti-Topological Space)

Let T_A be a collection of subsets of a non-empty set F. Then the pair (F,T_A) is called an anti-topological space where T_A is an anti-topology on F if the following conditions are satisfied:

C1) $\varnothing, F \notin T_A$

C2) $\bigcup_{m \in I} u_m \notin T_A$ (arbitrary union)

C3) $\bigcap_{r=1}^{n} v_r \notin T_A$ (finite intersection)

Definition: (Smarandache, 2021; Singh, 2021, 2022; Granados, 2022) (Geometric Neutrosophic Triplet)

From an algebraic point of view, we consider a neutrosophic triplet as (⟨*Algebra, NeutroAlgebra, AntiAlgebra*⟩) is introduced to generalize the classical algebraic concept. Motivating from this algebraic structure, to generalize the non-Euclidean geometry, a geometric neutrosophic triplet (⟨*Geometry, NeutroGeometry, AntiGeometry*⟩) is formed. Moreover, in a classical geometry (Euclidean) all the concepts (axioms, postulates, operator, property, etc.) are 100% true, while in Neutro Geometry there exists at least one Neutro concept (i.e. no Anti concept) and, in Anti Geometry, on a given space there is at least one Anti concept. Therefore, in a given geometric space (which is not necessarily homogeneous or Euclidean) one can consider a neutrosophic triplet as (⟨(1,0,0), (T,I,F), (0,0,1)⟩) which covers all possible geometric concepts that exist in the real-world.

Definition: (Singh, 2022) (Geometric Neutosophication and Geometric Antisophication)

Let $\langle G \rangle$ be a classical geometric structure on a space. Due to the neutrosophication on $\langle G \rangle$, a neutro geometry is produced whereas antisophication on $\langle G \rangle$ produced anti geometry. Therefore, neutrosophication and antisophication processes are responsible to trisect the classical geometric space into three subspaces.

For example, in an arbitrary geometric space, a concept denoted by $\langle H \rangle$ means this geometric concept is absolutely true and it is denoted by $\langle H \rangle_{(1,0,0)}$. Also, by $\langle neut\ H \rangle$ means the geometric concept is partially true $\langle T\%$ true$\rangle$, partially indeterminate $\langle I\%$ in determinate$\rangle$, and partially false $\langle F\%$ false$\rangle$ and it is denoted by $\langle neut\ H \rangle_{(T,I,F)}$, and by $\langle anti\ H \rangle$ means the geometric concept is absolutely false and it is denoted by $\langle anti\ H \rangle_{(0,0,1)}$. Thus, the structure of the tri-section of the geometric space is denoted by $(\langle H \rangle, \langle neut\ H \rangle, \langle anti\ H \rangle)$ where these three subsections may or may not be disjoint but their union will form the whole geometric space.

Definition: (Sahin et al., 2021; Smarandache, 2021) (Manifold, Neutro Manifold, Anti Manifold)

In a classical sense, a manifold (Todd) is a topological space with the property that, near each point has a neighborhood that is homeomorphic to an open unit ball of n-dimensional Euclidean space (R^n). In addition, 1-dimensional manifolds include lines and circles, whereas 2-dimensional manifolds are called surfaces (e.g., sphere, torus, klein bottle, etc.). Manifold plays an important role in geometry as it enables to descriptionof complicated structures with an aid of well-defined topological properties. Another important aspect of utilizing manifolds to study calculus is called differential manifolds. Also, the Riemannian metric on a manifold is useful to measure distances and angles. Thus, for practical knowledge of a manifold, calculus and topology play a significant role.

A neutro manifold is a topological space with the property that, near each point, on the small scales, resembles the neutro geometric space (A space where there exists at least one neutro axiom but no anti axiom).

An anti manifold is a topological space with the property that, near each point, on the small scales, resembles the anti geometric space (A space where there exists at least one anti axiom).

Definition: (Martelli, 2016) (Geometric Topology)

Geometric topology is the study of topological manifolds via geometric structure and mappings between them. It is used in a restrictive sense based on Euclidean or Non-Euclidean Space [29]. It gives a way to characterize the data into true, false or uncertain regions.

NEUTRO GEOMETRIC TOPOLOGICAL SPACE

Recently Singh (Singh, 2022) paid attention to dealing with the non-Euclidean data sets and their uncertainty (Singh, 2022). The problem arises when the data sets contain both Euclidean and Non-Euclidean Geometry for the representation (Smarandache, 2021; Singh, 2021; Singh, 2022; Granados, 2022). To deal with them the NeutroGeometrical data and its characterization a method is presented as follows (Singh, 2022):

1. Let us suppose, two elements $x_1, x_2 \in X$ can be represented via non-Euclidean geometry. The operator $x_1 \circ x_2 \subseteq X$ provides a new element that exists in Non-Euclidean space. It can be represented as true regions as (1, 0, 0).

2. Let us suppose, two elements $x_1, x_2 \in X$ in non-Euclidean geometry. The operator provides $x_1 \circ x_2 \subsetneq X$. It means the new element does not exist in the Non-Euclidean geometry. It can be represented as a false region (0, 0, 1).
3. Let us suppose, two elements $x_1, x_2 \in X$ in non-Euclidean geometry. The operator $x_1 \circ x_2$ generates a new element whose position is uncertain that Euclidean or Non-Euclidean, AntiGeometry. This type of saddle space can be defined as the indeterminacy (*t, i, f*) region of NeutroGeometry.

The precise representation and characterization of these types of data and their pattern analysis are one of the crucial tasks. There are several examples of NeutroGeometry data and its characterization:

Example 1. (Neutro Friend): The friendship can be characterized via NeutroGeometry as follows:

1. **Real Friend (1, 0, 0):** The person is connected with us from school or college can be characterized as real friend. They support each other with loyalty.
2. **Anti-Friend (0, 0, 1):** The person is connected only for profit rather than any reality.
3. **Neutro Friend (*t, i, f*):** These type of people are uncertain. They connect with us on virtual or social media to make friendship. In this case their act is indeterminant.

Example 2. (Gender Biased Law): The gender biased law for sexual harassment or other cases is one of the data set which can be characterized via NeutroGeometry as follows:

1. **Women Law (1, 0, 0):** The women never used to be considered as victim in any type of sexual harassment, crime or either in marriages. This type of act done by police can be characterized as Women law. It means the women are always right or the law is supportive for women in several countries. However it cannot be considered same in case for men.
2. **Men-Law (0, 0, 1):** The same type of sexual harassment, crime or any other case filed by men against women then the police does not support. It means men always considered as wrong.
3. **Neutro Law (*t, i, f*):** There is uncertainty in case of transgender claim same type of act or crime.

Example 3. (Air Pollution): The characterization of Air pollution is one of the major issues for data science researchers. It can be analyzed using NeutroGeometry as follows:

1. True Air (1, 0, 0): The Air quality of given area is suitable for the people.
2. Air Pollution (0, 0, 1): The Air quality of given area is not suitable for the people (0, 0, 1).
3. Neutro Air Pollution (*t, i, f*): The Air pollution is uncertain in the given area which changes based on several parameters. It can be characterized using NeutroGeometry.

Example 4. (Author Characterization): The author characterization is one of the important goals of today's scenario which can be characterized using NeutroGeometry as follows:

1. **Consistent Author (1, 0, 0):** The author hasa number of papers more than the number of co-authors.
2. **Inconsistent-Author (0, 0, 1):** The author hasa number of papers less than the number of co-authors (0, 0, 1).
3. **Neutro Author (*t, i, f*):** The author hasa number of papers almost equal to the number of co-authors. This type of author and its performance can be analyzed using a number of first- author papers, technical papers, or corresponding author papers.

Example 5. (Democracy Analysis): It is observed that recently most of the country claim democracy but do not follow democratic law. These types of countries can be analyzed using NeutroGeometry as follows:

1. **Democracy (1, 0, 0):** The country is fully democratic and everyone participates in voting as well as a decision for the country.
2. **Un-Democratic (0, 0, 1):** The country has no voting right for anyone. It is king rule country or autocratic.
3. **Neutro Democratic (*t, i, f*):** The country situation is uncertain where the country is ruled by both elected leaders, business classes, and other powerful persons.

The problem arises data sets are unknown, and undefined which required human Turiyam consciousness (Singh, 2021; Singh, 2021; Singh, 2021) and its metric (Bal, Singh & Ahmad, 2022; Bal, Singh & Ahmad, 2022; Pandey et al., 2016) to explore the data sets. In the near future, the author's work will focus on dealing the undefined, incomplete, or unknown data sets and their geometry.

CONCLUSION

This chapter focuses on Neutro geometric topology and its illustrative example. Some basic properties associated with neutro geometric topology are discussed with an example. In the near future, authors will focus on dealing with the NeutroGeometrical data and its examples. Also, there is a scope for the researchers to work on Anti geometric topology and its impact on real consequences.

ACKNOWLEDGMENT

Authors thank the editorial team for their valuable time.

REFERENCES

Adhikari, M. R. (2016). *Basic algebraic topology and its applications*. Springer. doi:10.1007/978-81-322-2843-1

Al-Tahan, M., Davvaz, B., Smarandache, F., & Anis, O. (2021). On some neutroHyperstructures. *Symmetry*, *13*(4), 535. doi:10.3390ym13040535

Bal, M., Singh, P. K., & Ahmad, K. D. (2022). A Short Introduction To The Symbolic Turiyam Vector Spaces and Complex Numbers. *Journal of Neutrosophic and Fuzzy Systems*, *2*, 76–87. doi:10.54216/JNFS.020107

Bal, M., Singh, P. K., & Ahmad, K. D. (2022). A short introduction to the concept of symbolic Turiyam matrix. *Journal of Neutrosophic and Fuzzy Systems*, *2*, 88–99. doi:10.54216/JNFS.020108

Bal, M., Singh, P. K., & Ahmad, K. D. (2022). An introduction to the symbolic Turiyam R-modules and Turiyam modulo integers. *Journal of Neutrosophic and Fuzzy Systems*, *2*, 8–19. doi:10.54216/JNFS.020201

Balachandran, A. P. (1991). Classical topology and quantum statistics.*International Journal of Modern Physics B*, *5*(16-17), 2585-2623.

Bhattacharya, S. (2004). A model to the Smarandache geometries. *Journal of Recreational Mathematics*, *33*, 166.

Bröcker, T., & Jänich, K. (1982). *Introduction to differential topology*. Cambridge University Press.

Chen, B. C., & Kikuchi, N. (2001). Topology optimization with design-dependent loads. *Finite Elements in Analysis and Design*, *37*(1), 57–70. doi:10.1016/S0168-874X(00)00021-4

Cheng-Ming, H. (1985). Fuzzy topological spaces. *Journal of Mathematical Analysis and Applications*, *110*(1), 141–178. doi:10.1016/0022-247X(85)90340-3

Chimienti, S. P., & Bencze, M. (1998a). Smarandache paradoxist geometry. *Smarandache Notions Journal*, *9*(1-3), 42.

Chimienti, S. P., & Bencze, M. (1998b). Smarandache anti-geometry. *Smarandache Notions Journal*, *9*(1-3), 48.

Coxeter, H.S.M. (1942). *Non-Euclidean geometry.* University of Toronto Press.

Dillon, M. I. (2018). Neutral geometry. In *Geometry Through History*. Springer. doi:10.1007/978-3-319-74135-2_2

Granados, C. (2022). A note on AntiGeometry and NeutroGeometry and their application to real life. *NeutrosophicSets and Systems*, *49*, 579–593. doi:10.5281/zenodo.6466520

Hur, K., Kim, J. H., & Ryou, J. H. (2004). Intuitionistic fuzzy topological spaces. *The Pure and Applied Mathematics*, *11*, 243–265.

Kuciuk, L., &Antholy, M. (2003). An introduction to the Smarandache geometries. *Infinite Study*.

Lashin, E. F., Kozae, A. M., Khadra, A. A., & Medhat, T. (2005). Rough set theory for topological spaces. *International Journal of Approximate Reasoning*, *40*(1-2), 35–43. doi:10.1016/j.ijar.2004.11.007

Lee, J. G., & Hur, K. (2020). Hesitant fuzzy topological spaces. *Mathematics*, *8*(2), 188. doi:10.3390/math8020188

Lobachevsky, N. (2010). Pangeometry, translator and editor: A. Papadopoulos. Heritage of European Mathematics Series. European Mathematical Society, 4.

Mao, L. F. (2007). An introduction to Smarandache multi-spaces and mathematical combinatorics. *Scientia Magna*, *3*, 54–80.

Martelli, B. (2016). *An introduction to geometric topology.* arXiv preprint arXiv:1610.02592.

Martelli, F. (2021). Topology and complexity of the hydrogen bond network in classical models of water. *Journal of Molecular Liquids, 329*, 115530. doi:10.1016/j.molliq.2021.115530

Mondal, T. K., & Samanta, S. K. (2001). Topology of interval-valued intuitionistic fuzzy sets. *Fuzzy Sets and Systems, 119*(3), 483–494. doi:10.1016/S0165-0114(98)00436-9

Morandi, G. (1992). *The role of topology in classical and quantum physics* (Vol. 7). Springer Science & Business Media. doi:10.1007/978-3-540-46688-8

Munkres, J. R. (2018). *Elements of algebraic topology*. CRC press. doi:10.1201/9780429493911

Olgun, M., Ünver, M., & Yardımcı, Ş. (2019). Pythagorean fuzzy topological spaces. *Complex & Intelligent Systems, 5*(2), 177–183. doi:10.100740747-019-0095-2

Pandey, L. K., Ojha, K. K., Singh, P. K., Singh, C. S., Dwivedi, S., & Bergey, E. A. (2016). Diatoms image database of India (DIDI): A research tool. *Environmental Technology & Innovation, 5*, 148–160. doi:10.1016/j.eti.2016.02.001

Popov, M. R. (1996). The Smarandache non-geometry. *Abstracts of Papers Presented to the American Mathematical Society Meetings, 17*, 595.

Rezaei, A., &Smarandache, F. (2020). On neutro-BE-algebras and anti-BE-algebras. *International Journal of Neutrosophic Science, 4*, 8-15. doi:10.54216/IJNS.040101

Rezaei, A., Smarandache, F., & Mirvakili, S. (2021). Applications of (neutro/anti) sophications to semihypergroups. *Journal of Mathematics, 2021*, 2021. doi:10.1155/2021/6649349

Şahin, M., Kargın, A., & Altun, A. (2021). Neutro-metric spaces. *NeutroAlgebra Theory, I*, 71.

Şahin, M., Kargın, A., & Smarandache, F. (2021). Neutro–G modules and anti–G modules. *NeutroAlgebra Theory, I*, 50.

Şahin, M., Kargın, A., & Uz, M. S. (2021). Neutro-Lie Algebras and Anti-Lie Algebras. *NeutroAlgebra Theory, I*, 101.

Şahin, M., Kargın, A., & Yücel, M. (2021). Neutro-topological space and anti-topological space. *NeutroAlgebra Theory, I*, 16–30.

Sarkar, M. (1981). On L-fuzzy topological spaces. *Journal of Mathematical Analysis and Applications, 84*(2), 431–442. doi:10.1016/0022-247X(81)90179-7

Sierpinski, W. (2020). *General topology*. University of Toronto press.

Singh, P. K. (2021). Turiyam set a fourth dimension data representation. *Zeitschrift für Angewandte Mathematik und Physik*, *9*(7), 1821–1828. doi:10.4236/jamp.2021.97116

Singh, P. K. (2021). Data with Turiyam set for fourth dimension quantum information Processing. *Journal of Neutrosophic and Fuzzy Systems*, *1*, 9–23. doi:10.54216/JNFS.010101

Singh, P. K. (2021). Fourth dimension data representation and its analysis using Turiyam Context. *Journal of Computer and Communications*, *9*(6), 222–229. doi:10.4236/jcc.2021.96014

Singh, P. K. (2021). Anti-geometry and neutrogeometry characterization of non-Euclidean data. *Journal of Neutrosophic and Fuzzy Systems*, *1*, 24–33. doi:10.54216/JNFS.010102

Singh, P. K. (2022). Data with non-Euclidean geometry and its characterization. *Journal of Artificial Intelligence and Technology*, *2*, 3–8. doi:10.37965/jait.2021.12001

Singh, P. K. (2022). Non-Euclidean, anti-geometry and neutrogeometry characterization. *International Journal of Neutrosophic Sciences*, *18*(3), 8–19. doi:10.54216/IJNS.180301

Singh, P. K. (2022). NeutroAlgebra and neutrogeometry for dealing heteroclinic patterns. In *Theory and Applications of NeutroAlgebras as Generalizations of Classical Algebras* (pp. 90–102). IGI Global Publishers. doi:10.4018/978-1-6684-3495-6.ch006

Singh, P. K., Ahmad, K. D., Bal, M., & Aswad, M. (2021). On the symbolic Turiyam rings. *Journal of Neutrosophic and Fuzzy Systems*, *1*, 80–88. doi:10.54216/JNFS.010204

Smarandache, F. (1969). *Paradoxist geometry*. Academic Press.

Smarandache, F. (1969). Mixed noneuclidean geometries. *Infinite Study*.

Smarandache, F. (1998). *Neutrosophy: Neutrosophic probability, set, and logic: analytic synthesis & synthetic analysis*. American Research Press.

Smarandache, F. (2019). Introduction to Neutroalgebraic structures and antialgebraic structures, in advances of standard and nonstandard neutrosophic theories. *Pons Publishing House Brussels, Belgium*, *6*, 240–265.

Smarandache, F. (2020). Introduction to neutro algebraic structures and anti algebraic structures (revisited). *Neutrosophic Sets and Systems*, *31*, 1–16. doi:10.5281/zenodo.3638232

Smarandache, F. (2020). NeutroAlgebra is a Generalization of Partial Algebra. *International Journal of Neutrosophic Science, 2*, 8-17. doi:10.54216/IJNS.020103

Smarandache, F. (2021). Neutro geometry and anti geometry are alternatives and generalizations of the non-euclidean geometries (revisited). *Neutrosophic Sets and Systems*, *46*, 456–477.

Smarandache, F., & Chang-Wei, H. (2012). *Vacuum*. Space-Time, Matter and the Models of Smarandache Geometry.

Smarandache, F., & Hamidi, M. (2020). Neutro-BCK-algebra. *International Journal of Neutrosophic Science*, *8*, 110–117.

Smarandache, F., Rezaei, A., Agboola, A. A. A., Jun, Y. B., Borzooei, R. A., Davvaz, B., ... Mirvakili, S. (2021). On Neutro Quadruple Groups. *Infinite Study*.

Šostak, A. P. (1996). Basic structures of fuzzy topology. *Journal of Mathematical Sciences*, *78*(6), 662–701. doi:10.1007/BF02363065

Stillwell, J. (1993). *Classical topology and combinatorial group theory* (Vol. 72). Springer Science & Business Media. doi:10.1007/978-1-4612-4372-4

Todd, R. (n.d.). Manifold. *MathWorld*. https://mathworld.wolfram.com/Manifold.html

Weisstein, E. W. (n.d.). Combinatorial Topology. *MathWorld*. https://mathworld.wolfram.com/CombinatorialTopology.html

Zadeh, L. A. (1988). Fuzzy logic. *Computer*, *21*(4), 83–93. doi:10.1109/2.53

Chapter 7
"NeutroGeometry Laboratory": The New Software Dedicated to Finite NeutroGeometries

Erick González Caballero
Technological University of Havana, Cuba

ABSTRACT

Mixed projective-affine-hyperbolic (MPAH) planes belong to both the branches of NeutroGeometries and mixed or Smarandache geometries. This kind of plane is a geometric structure containing a finite number of points. Some lines of the MPAH satisfy the axioms of projective planes, other lines satisfy the axioms of affine planes, and others satisfy the axioms of hyperbolic planes. Therefore, each of the axioms of parallelism is partially satisfied. This chapter describes version 1.0 of a new software called "NeutroGeometry Laboratory" coded in Python by the author that is used for the calculation and visualization related to MPAH planes. This software is easy to use by users once they know the theory of finite mixed projective-affine-hyperbolic planes, and particularly, it supports the study of this topic and finite planes in general.

INTRODUCTION

NeutroGeometries are geometric structures where some axioms or definitions are satisfied by a set of elements and simultaneously not satisfied by other elements of the space (Smarandache, 2019, 2020, 2021). The Mixed or Smarandache Geometries are geometric structures where the elements of the geometric space are divided into subspaces, each of which is identified with a specific type of geometry (Iseri, 2002;

DOI: 10.4018/978-1-6684-4740-6.ch007

Mao, 2006; Smarandache, 1969, 2009). These new types of geometries correspond to the need to match geometric theory with the reality of physical space, where Euclidean or non-Euclidean spaces coexist fulfilling more than one geometry.

On the other hand, Finite Geometries are geometric structures based on axioms where there is a finite number of points (Dembowski, 1997; Dissett, 2000; Kiss & Szonyi, 2020). Within this theory there are three types of planes, the projective planes are the most important ones, where there is no parallelism. In the affine planes, for each point outside a line passes one and only one parallel. Additionally, in a Bolyai-Lobachevsky or hyperbolic plane, for each point outside a line passes at least two lines parallel to the aforementioned one (Korchmáros & Sonnino, 2012, 2014).

The Mixed Projective-Affine-Hyperbolic planes (MPAH planes) contain axioms of these three types of finite planes, especially those related to parallelism. That is to say, this type of plane can be divided into three types of lines, namely, some that satisfy the axiom of parallelism of projective planes, others that satisfy the axiom of parallelism of affine planes, and a third set that satisfies the axiom of parallelism of hyperbolic planes. These planes constitute a previous contribution of the author of this chapter to the theory of NeutroGeometries and Mixed Geometries.

It is well known that there are no projective planes of certain orders, where the order of the plane is determined by the number of points contained in each line. On the other hand, it is also known that an affine plane is obtained by eliminating a line and all its points within a projective plane. Much less studied is the theory of Bolyai-Lobachevsky planes and there are algorithms to obtain them from projective planes.

The drawbacks to obtaining projective planes of some orders may be due to their restrictive axiomatic. MPAH planes can be a solution to this problem because it relaxes the traditional axiomatic of finite planes. E.g., a consequence of the well-known theorem of Bruck-Ryser is that projective planes of order 6 do not exist, where the order plus 1 is the number of points contained in every line and also it is the number of lines passing for every point; however we could define an MPAH of order 6.

Another point in favor of finite geometries is their application in topics such as Code Theory, and Cryptography and they are associated with orthogonal Greco-Latin Squares (Bishnoi, Gijsiwijt, D'haesleer, & Potukuchi, 2022; A.L. Horlemann, 2022; A. Ilić, 2022; Neri, Santonastaso, & Zullo, 2023; Szabó, 2022). This ensures that the new theory of MPAH planes can be applied to these types of problems from a new perspective. Thus, now we have an open question that is how can this new theory contribute to the theoretical development of finite geometry and its applications?

For the study of these geometric structures, the author found some difficulty in studying the examples, due to the need for software that was sufficiently simple and that would allow sufficient calculations to be made for studying the finite planes and in particular the MPAH planes. That is the motivation to code the software named

NeutroGeometry Laboratory version 1.0. This software was coded in python 3.6 for Windows, preferably Windows 10 or higher, which includes the visual programming of the Tkinter package and the drawing of graphics with the Turtle package (Lutz, 2011 ; Shipman, 2013). Therefore, this software allows us to carry out the calculations that are proposed in other works related to 2-dimensional Finite NeutroGeometries and also the graphical visualization of these planes. An advantage offered by the software is that it is based on visual programming elements, that is, the user will find buttons, Combo Boxes, editors, etc., and will not have to interact with the python editor console. In addition to the fact that for greater convenience the user is provided with an executable or .exe, so he/she can start the software with just a double click. However, before using the software, the user must study the theory of MPAH planes and the help file; otherwise, the results obtained may be erroneous.

From the software literature for supporting finite geometry, we can cite Zenodo and GAP, which are models of reference for developing our software (Cullinane, 2017; The-GAP-Group, 2022). These software products contain many capabilities that cannot be exhibited in the present software, like versatility. However, so far "NeutroGeometry Laboratory" is the only computational approach to NeutroGeometry or Smarandache Geometry.

Briefly, software supports any kind of calculation in mathematics and other branches like engineering. We hope this software allows us to inspire new results in the novel theory in MPAH planes or others. Also, it can serve as an educational approach to this important subject, which needs visualization and complex calculations by the students on this theme. The software can be used to prove the viability of obtaining MPAH of order 6 or others, and also for the development of practical results applied to Code theory and Cryptography, which are essential subjects in informatics security of today's world.

This chapter is divided into the following structure, a Background section where the main concepts of Mixed Projective-Affine-Hyperbolics planes are explained. Further, a section is developed detailing the use of the new software. The chapter ends with the section dedicated to the Conclusion, the Additional Readings, and Key Words and Definitions. In the Appendix, there are additional examples of finite geometries and the use of this software.

BACKGROUND

Finite Geometry is the branch of geometry where there is a finite number of points and therefore the number of lines, planes, spaces, and hyperspaces is also finite. Evidently, due to their finitude, these are not Euclidean geometries. In this theory, three types of planes are studied namely, the projective planes, the affine planes, and the Bolyai-Lobachevsky or hyperbolic planes.

One of the fundamental concepts is incidence structure:

Definition 1 (Dembowski, 1997): An *incidence structure* is a triple $\boldsymbol{S} = (\mathcal{P}, \mathcal{L}, \mathcal{I})$, where $\mathcal{P}, \mathcal{L}, \mathcal{I}$ are sets satisfying: $\mathcal{P} \cap \mathcal{L} = \varnothing$ and $\mathcal{I} \subseteq \mathcal{P} \times \mathcal{L}$. The elements of $\mathcal{P}$ are called *points*, those of $\mathcal{L}$ *lines* and $\mathcal{I}$ is a symmetric relation called *incidence*.

Let us recall that the Cartesian product $\mathcal{P} \times \mathcal{L}$ is defined as all the ordered pairs (P,l), where P is a point in $\mathcal{P}$, and l is a line in $\mathcal{L}$.

A line is an element of the power set of $\mathcal{P}$ containing at least two points. A point P is incident in a line l and vice versa, a line is incident in the point if P is contained in l and l passes through P. It is preferred the notation of incidence $P\mathcal{I}l$ and $l\mathcal{I}P$, instead of $(P,l) \in \mathcal{I}$ and $(l,P) \in \mathcal{I}$, respectively.

Definition 2 (Dembowski, 1997): The *dual structure* $\bar{\boldsymbol{S}} = (\bar{\mathcal{P}}, \bar{\mathcal{L}}, \bar{\mathcal{I}})$ of the incidence structure $\boldsymbol{S} = (\mathcal{P}, \mathcal{L}, \mathcal{I})$ is defined by $\bar{\mathcal{P}} = \mathcal{L}$, $\bar{\mathcal{L}} = \mathcal{P}$, for $P \in \mathcal{P}$ and $l \in \mathcal{L}$; where $l\bar{\mathcal{I}}P$ if and only if $P\mathcal{I}l$, and $P\bar{\mathcal{I}}l$ if and only if $l\mathcal{I}P$. I.e., the points are interchanged with lines.

Definition 3 (Dembowski, 1997): If $\boldsymbol{S} = (\mathcal{P}, \mathcal{L}, \mathcal{I})$ is an incidence structure, $\mathcal{Q} \subset \mathcal{P}$, and $\mathcal{R} \subset \mathcal{L}$, then a *substructure* of $\boldsymbol{S}$ is the incidence structure $\boldsymbol{T} = (\mathcal{Q}, \mathcal{R}, \mathcal{J})$ with $\mathcal{J} = \mathcal{I} \cap (\mathcal{Q} \times \mathcal{R})$.

Definition 4 (Dembowski, 1997): Let $\boldsymbol{S} = (\mathcal{P}, \mathcal{L}, \mathcal{I})$ be an incidence structure, where points and lines are numbered arbitrarily and fixedly: $P_1, P_2, \ldots, P_m$ and $l_1, l_2, \ldots, l_n$. Then the following two matrices are called *incidence matrices* $C=(c_{ij})$ and $D=(d_{ij})$, where

$$c_{ij} = \begin{cases} 1, & \text{if } P_i\mathcal{I}l_j \\ 0, & \text{otherwise} \end{cases} \text{ and } d_{ij} = \begin{cases} 1, & \text{if } l_i\mathcal{I}P_j \\ 0, & \text{otherwise} \end{cases}.$$

Projective planes satisfy the following axioms:

Definition 5 (Kiss & Szonyi, 2020): A *projective plane* is an incidence structure $\boldsymbol{P} = (\mathcal{P}, \mathcal{L}, \mathcal{I})$ with the following axioms:

1. For each pair of two distinct points P, $Q \in \mathcal{P}$ there is a unique line that is incident with P and Q.
2. Each pair of two distinct lines l, $s \in \mathcal{L}$ has a unique point in common.

3. There are four points, such that no three of them are collinear, i.e., three of them do not belong to the same line. The last axiom is equivalent to the two following axioms:
   a. Every line has at least three points.
   b. There are at least two lines in the plane.

In a projective plane of *order* $n \geq 2$, each line contains $n+1$ points. Also, this type of plane satisfies the *duality principle*, where if a theorem T contains the words "point", "line" and "incidence", and the statement T′ is obtained from T replacing the word "point" by "line" and vice versa, then T′ is also a theorem.

A projective plane of order n satisfies the following properties (Zumbragel, 2016):

1. The number of points in every line is $n+1$.
2. The number of lines passing by only one point is $n+1$.
3. The total number of points in the plane is n^2+n+1.
4. The total number of lines in the plane is n^2+n+1.

It is well known that there are no projective planes of certain orders, as can be deducted from the Bruck-Ryser theorem.

Theorem 1 (Bruck-Ryser) (Bruck & Ryser, 1949): If $n \equiv 1,2 \pmod 4$, then a necessary condition for the existence of a finite projective plane of order n is that integers x, y exist satisfying $x^2+y^2=n$.

The axioms of affine planes are the following:

Definition 6 (Kiss & Szonyi, 2020): An *affine plane* is an incidence structure $A = (\mathcal{P}, \mathcal{L}, \mathcal{I})$ fulfilling the following axioms:

- For each pair of two distinct points P, $Q \in \mathcal{P}$ there is a unique line that is incident with P and Q.
- For each point $P \in \mathcal{P}$ and each line $l \in \mathcal{L}$ with $P \notin l$, there exists a unique line $s \in \mathcal{L}$ such that $P\mathcal{I}s$, and l and s do not meet. This is called *parallelism* in **A** and it is denoted by $l \| s$.
- Three points are not collinear.

As in projective planes, the order of the affine plane is the number of points contained in one line.

An affine plane of order n satisfies the following properties (Zumbragel, 2016):

1. Every line is incident with n points.
2. Every point is incident with $n+1$ lines.
3. The affine plane contains n^2 points.
4. The affine plane contains n^2+1 lines.

Given a projective plane, when a fixed line and all the points contained in this line are removed from the projective plane, then we obtain an affine plane.

Finally, the Bolyai-Lobachevsky or hyperbolic planes are based on the following axioms:

Definition 7 (Raber, 1972): A *Bolyai-Lobachevsky plane* or *Hyperbolic plane* is a geometric structure $\boldsymbol{H} = (\mathcal{P}, \mathcal{L}, \mathcal{I})$ with the following axioms:

- For each pair of two distinct points $P,\ Q \in \mathcal{P}$ there is a unique line that is incident with P and Q.
- For each point $P \in \mathcal{P}$ and each line $l \in \mathcal{L}$ such that $P \notin l$ there exists at least two lines in $\mathcal{L}$ which are incident with P, and these lines and l do not meet.
- If a subset $\mathcal{S}$ of points of $\boldsymbol{H}$ contains three non-collinear points and contains all the points of the lines defined by every pair of points of $\mathcal{S}$, then $\mathcal{S}$ contains all the points of $\boldsymbol{H}$.

A *Smarandache Geometry* (or Hybrid Geometry) is a geometry which has at least one smarandachely denied axiom [or theorem, proposition, etc.], i.e., in the same space the axiom behaves differently (i.e., validated and invalided; or only invalidated but in at least two distinct ways) (Smarandache, 1969).

In *NeutroGeometry*, at least one axiom, definition, theorem, proposition, property, among others, is satisfied by some elements of the space, although not by all. On the other hand, *AntiGeometry* contains an axiom, definition, theorem, proposition, property, among others that are not satisfied by any element of the space (Granados, 2022; Singh, 2022; Smarandache, 2021).

The author of this chapter first defined the finite Mixed Projective-Affine-Hyperbolic (MPAH) planes, which satisfy the following axioms:

Definition 8: A *finite Mixed Projective-Affine-Hyperbolic plane* or a finite MPAH plane is an incidence structure $\mathcal{MPAH} = (\mathcal{P}, \mathcal{L}, \mathcal{I})$, where $\mathcal{L} = \mathcal{L}_{\mathcal{P}} \cup \mathcal{L}_{\mathcal{A}} \cup \mathcal{L}_{\mathcal{H}}$. The elements of $\mathcal{L}_{\mathcal{P}}$ satisfy partially or totally the axiomatic of projective planes, the elements of $\mathcal{L}_{\mathcal{A}}$ satisfy partially or totally the axiomatic of affine planes, and the elements of $\mathcal{L}_{\mathcal{H}}$ satisfy partially or totally the axiomatic of hyperbolic planes. Let us call n the *order of the finite MPAH plane* if $n+1$ is the maximal cardinality of the elements of $\mathcal{L}$.

Let $\rho = (\mathcal{P}, \mathcal{L}, \mathcal{I})$ be a projective plane and $P \in \mathcal{P}$, then $\mathcal{MPAH}_\rho = (\mathcal{P}', \mathcal{L}', \mathcal{I}')$ such that given a fixed-line $l \in \mathcal{L}$ and two fixed points $P_1, P_2 \in l$, $\mathcal{P}' = \mathcal{P} \setminus \{P_1, P_2\}$ and $\mathcal{I}' = \mathcal{I} \cap (\mathcal{P}' \times \mathcal{L}')$, $\mathcal{L}' = \{s \setminus \{P_1, P_2\} : s \in \mathcal{L}\}$ such that it is an incidence structure.

Let $\alpha = (\mathcal{P}, \mathcal{L}, \mathcal{I})$ be an affine plane, then let us select $l_1, l_2 \in \mathcal{L}$ and $l_1 \| l_2$, define $H = \{[l] : l \in \mathcal{L}\} \setminus \{[l_1], [l_2]\}$, where $[l]$ is the parallel class of l. Let us define $\mathcal{MPAH}_\alpha = (\mathcal{P}', \mathcal{L}', \mathcal{I}')$ where for each parallel class in H we add a point in $\mathcal{P}'$ incident with them and we add a line l_3 in $\mathcal{L}'$ containing all those new points.

Remark 1.

1. This procedure can be repeated in other lines, always that at least one line of the plane remains with $n+1$ points, which is the line satisfying the axiomatic of projective planes.
2. The procedure can be extended to more than two points for $\mathcal{MPAH}_\rho$ or more than two lines for $\mathcal{MPAH}_\alpha$ always that at least one line contains $n+1$ points, and every line has at least two points.

Definition 9: Let an $\mathcal{MPAH}$ ($\mathcal{MPAH}_\rho$ or $\mathcal{MPAH}_\alpha$) be a Mixed Projective-Affine-Hyperbolic plane, the *Degree of Type of Geometry* (DTG) is a triad (p,a,h), where

$$p = \frac{|\mathcal{PL}|}{|\mathcal{L}|}, a = \frac{|\mathcal{AL}|}{|\mathcal{L}|}, \text{ and } h = \frac{|\mathcal{HL}|}{|\mathcal{L}|}, |\mathcal{L}|,$$

is the number of lines in $\mathcal{MPAH}$, whereas, $|\mathcal{PL}|, |\mathcal{AL}|$, and $|\mathcal{HL}|$ are the cardinality of the lines with $n+1$ points, with n points, and with $n-1$ points or less, respectively.

Obviously, $p,a,h \in (0,1)$ and $p+a+h=1$.

A finite MPAH plane is an example of NeutroGeometry and also of Mixed Geometry. This is because the axioms of parallelism for the projective, affine and hyperbolic planes are partially satisfied, and in the same geometric plane there are subspaces with characteristics typical of these three types of planes.

THE NEW SOFTWARE CALLED "NEUTROGEOMETRY LABORATORY"

The reader who wishes to experiment with the MPAH planes may encounter a problem, that is the impossibility of easily visualizing and calculating the planes that he/she wishes to study. Obviously, these are new concepts, therefore there is no such software for this end. The main aim of this chapter is to describe the use of the software called "NeutroGeometry Laboratory", which was coded by the author of this chapter to perform the calculations that are usually done on Finite Geometries. Among them, we have the calculation of the incidence matrix, the visualization of the planes obtained, the conversion of a projective plane into an MPAH plane when the user eliminates some points of the original one, and finally when such a plane is obtained, the calculation of its DTG.

Other facilities that the user may need are to save the graphs obtained and the calculus, as well as to load files that he/she may previously have saved to carry out new calculations. Also, it is included a file in pdf format that contains the help that the reader needs to use the software correctly.

Below we detail the characteristics of the software.

This software was coded by Dr. Erick González Caballero, from two book chapters to appear on the subject of Finite NeutroGeometries. Specifically, this software contains the main elements to perform Finite Geometry Planes calculations and plots. It is coded in python 3.6 language for calculus (Lutz, 2011), and its Tkinter (Shipman, 2013) and Turtle packages for visual and graphical programming, respectively. Understanding the theoretical basis of the software is essential for its fruitful use, so we recommend reading both chapters on finite NeutroGeometries. The software is an executable application for Windows 10 or higher. It could work correctly on Windows with versions before 10. The software can be downloaded from the website https://github.com/erickgc5/NeutroGeometry.git.

The main window of the software is shown in Figure 1.

The software has a menu where Workspace options appear referring to the elements at the bottom. Also, we have a Help menu that allows us to access the pdf document and the About.

In Figure 2 we show the elements of the Workspace menu:

Within the menu, the Load workspace submenu allows us to load workspace data previously saved in txt format with a certain syntax that will be explained later. The first line of the txt file is the text to paste into Input Plane, the second line is pasted in Points, and the third one is in Coordinates.

The Save output submenu allows us to save the data in a txt file, that is the data in the output at the bottom of the main window.

Figure 1. The main window of NeutroGeometry Laboratory

Figure 2. Elements of the workspace menu

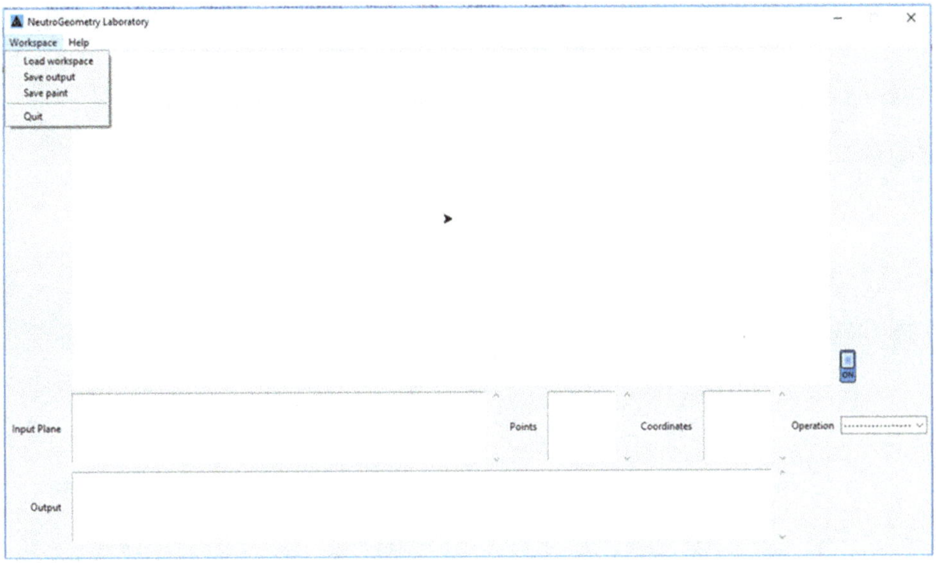

The Save paint submenu is used to save the plotting obtained below in an Encapsulated-Postscript (eps) format.

Finally Quit is another option that the user has for closing the application.

By default, the *on/off* switch button that appears in the middle on the right is in its state *on*. This means that the user will obtain the graphical representation of the calculation that he/she performs. However, when the button is pressed, the status lifts to the state *off*, which will not allow the graphical representation of the planes. The plots will appear on the screen with an arrow in the center; this arrow marks the point of origin. The connection among the points is represented by a smooth curve. Each line is represented by the union of points using a smooth closed curve, that is, two lines are represented by curves with two different colors.

Below there are four places where the text is used:

Input Plane allows us to enter the finite plane from which we will work.

Points allows us to enter the points with which we will work on. They are the additional data depending on the operation, as will be explained further.

Coordinates allows us to enter the coordinates of the points to draw the plane obtained or desired. Keep in mind that the arrow marks the origin and from this, the user determines the coordinates where he/she wants to position the points on the screen, following certain syntax that will be further explained. If the user does not enter this data, the points will be plotted at elliptical positions.

Finally, in the Output the final results of the calculations are achieved according to the operation selected by the user. In Figure 3 there is an example of the syntax used to enter the data:

Figure 3. An example of an uploaded plane, points, and coordinates

In the Input Plane the points of each line are separated by space and each line is separated from the other by ";". In Points the points are separated by spaces and in Coordinates, the coordinates are entered by pairs of integers *(x,y)* and each coordinate is separated from the other utilizing ";". Each point can be represented by letters, numbers, or any set of symbols not separated by spaces. Any other syntax used by the user may be subject to errors in the final results.

Next, the possible operations that can be performed on the input planes appear in the Combo Box, see Figure 4.

Figure 4. Options of the combo box

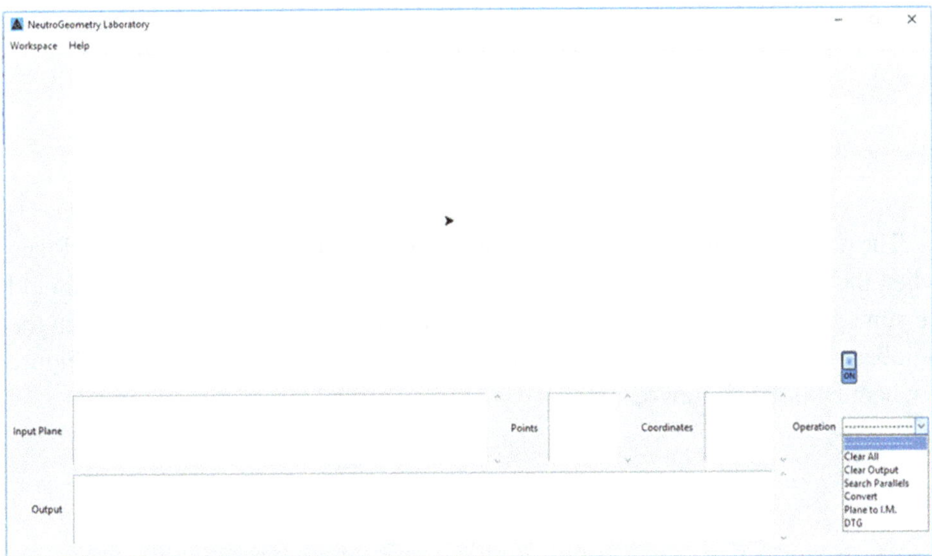

By default, the Combo Box has the option "----------------" which means that there is no option selected yet. Next, Clear All is used to eliminate the input data, the output data, and the previously obtained graphical representation. Clear Output clears both the Output and the drawing on the screen.

The Search Parallels option returns the pairs of parallels in the plane that appears in the Input Plane. For this selection, it is necessary to enter the plane to be analyzed in Input Data and the Points which are all the points that are part of the plane. If there are no parallels, it returns "There is no parallelism". In this option, the plot of the plane returns in red color the lines with parallels, those not having any parallels are colored in blue, see Figure 5.

Figure 5. Example of using the option search parallels

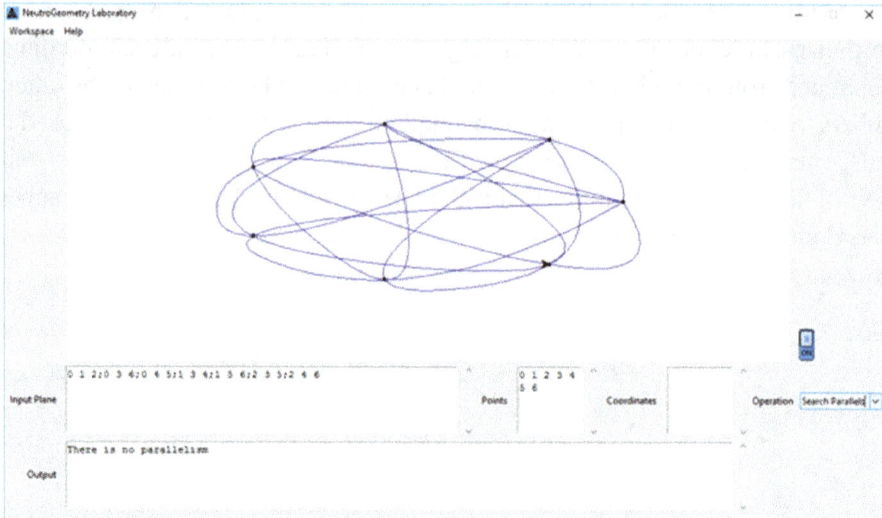

The Convert option results in the conversion of the plane entered in Input Plane when the points that appear in Points are removed. Note that these points should be some but not all of the set of points that are part of the input plane. When the number of removed points is too many, the software can launch an error; see Figure 6 where point 0 is removed.

Figure 6. Example of using the option convert

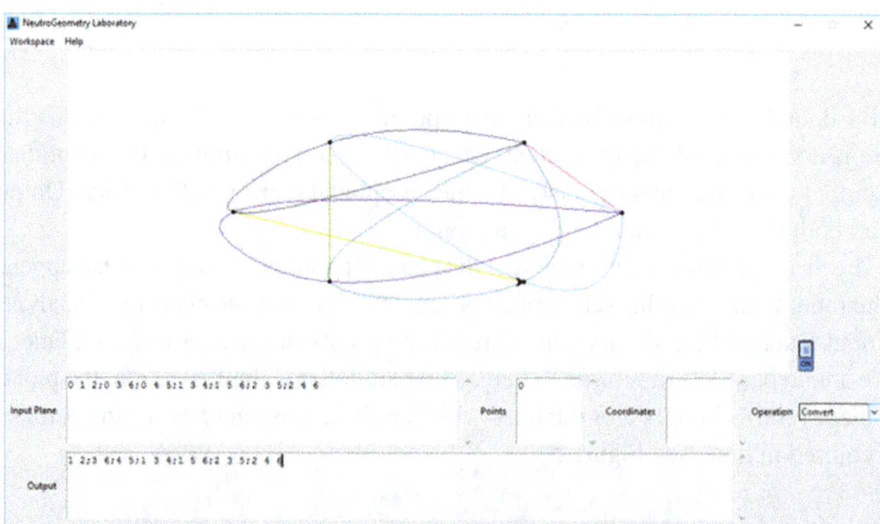

The option Plane to I.M. returns the Incidence Matrix of the plane entered in Input Plane. To do this, in addition to entering the plane, all the points that are part of the plane are entered in **Points**. See Figure 7.

Figure 7. Example of using the option plane to I.M.

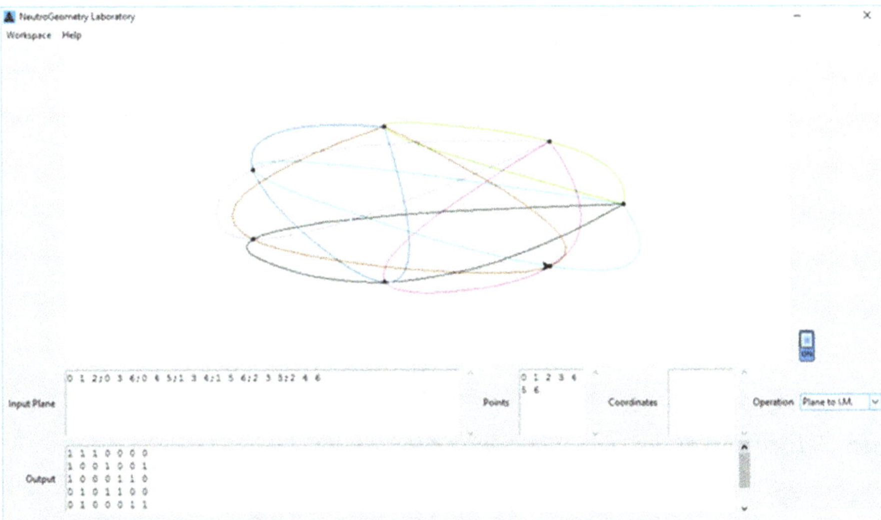

The DTG option for *Degree of Type of Geometry* is used to calculate this triad of indices according to Definition 9. To do this, we need to enter the plane in Input Plane and in Points all the points that are part of the plane. See the example in Figure 8. So far, there are the facilities of the software.

Below, there are some examples of using the software.

Firstly, we exemplify in Figure 9 the calculation of the incidence matrix of the plane that appears in the Input Plane with the switch button in the state *off*:

The other example revisits the same example without including the Coordinates and the *on/off* button is in *on* with the DTG operation. In this case, the software plots a graph with elliptically located points:

In the next example, the projective plane in Figure 10 is converted into another one by eliminating points A and B, see Figure 11. See that we included the Coordinates.

Below in Figure 12, we show one example of a plane with many parallels. The option Search Parallels outputs every pair of parallels in the Input Plane and plots in red color the lines with parallels and in blue color the lines without any parallel.

Figure 8. Example of using the option DTG

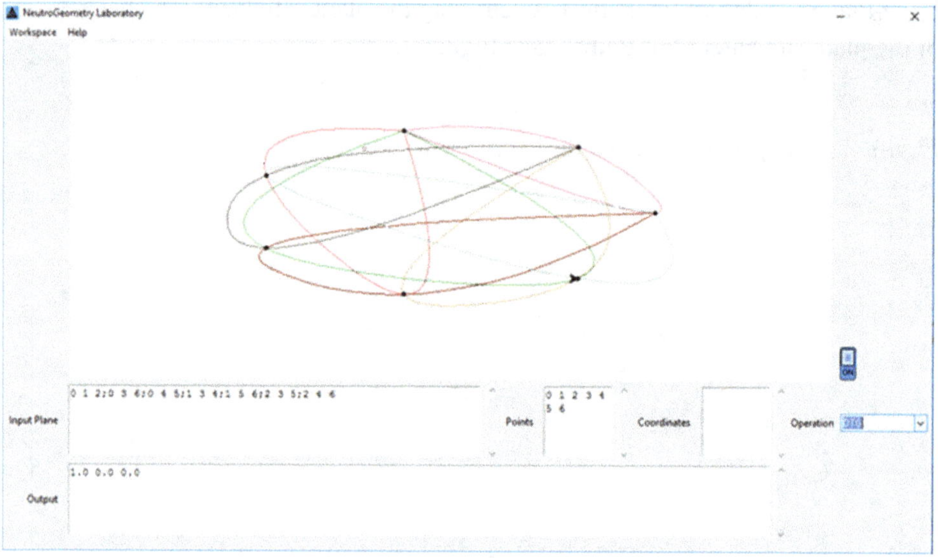

Figure 9. Example of the switch button in the state off

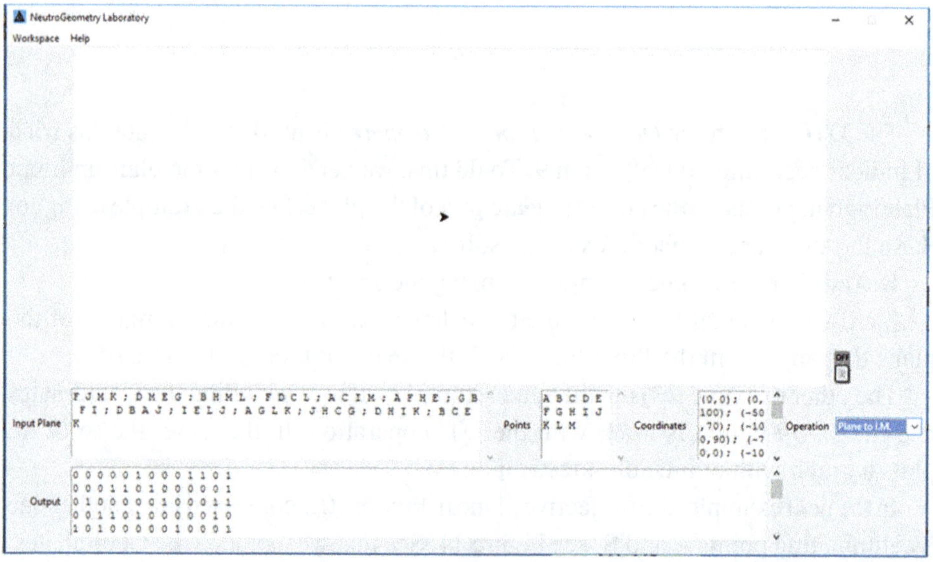

It is also possible to perform calculations with planes of higher orders, as is the case of the projective plane of order 5 which contains 31 points and lines, see Figure 13.

Figure 10. Example of the calculation of the DTG where the switch button is in the state on

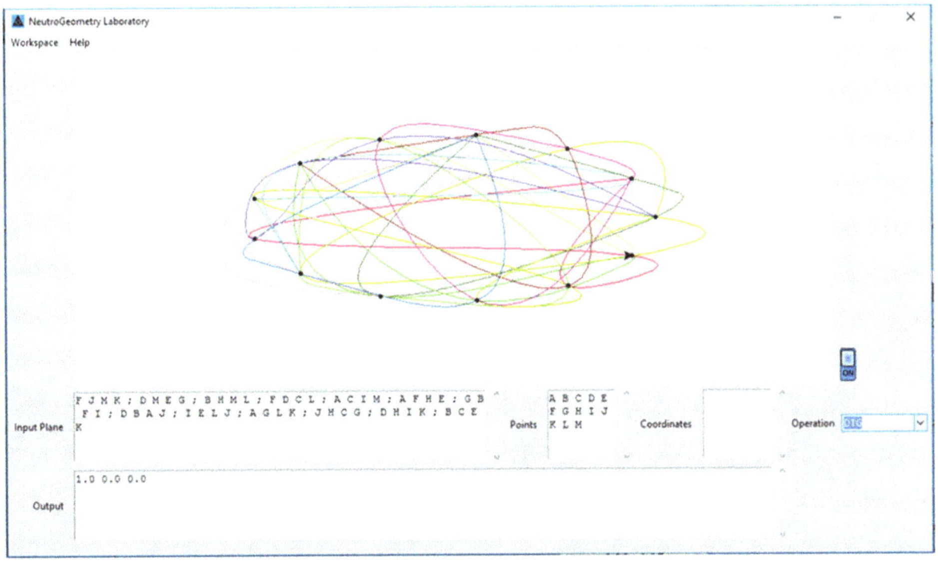

Figure 11. Example of Figure 10 where the user introduces the coordinates of the graphic with the operation convert

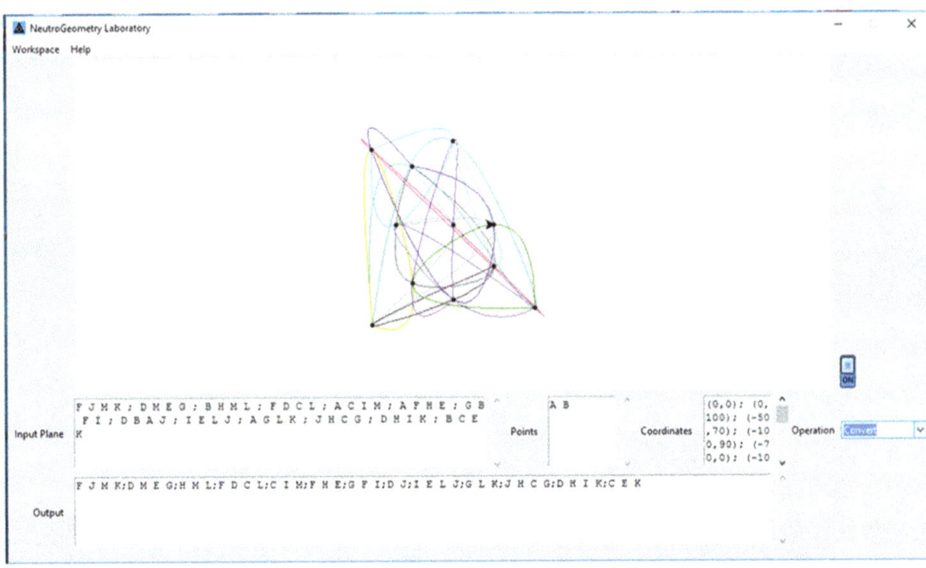

Figure 12. Example for visualizing the parallel lines in red color and the non-parallel lines in blue color

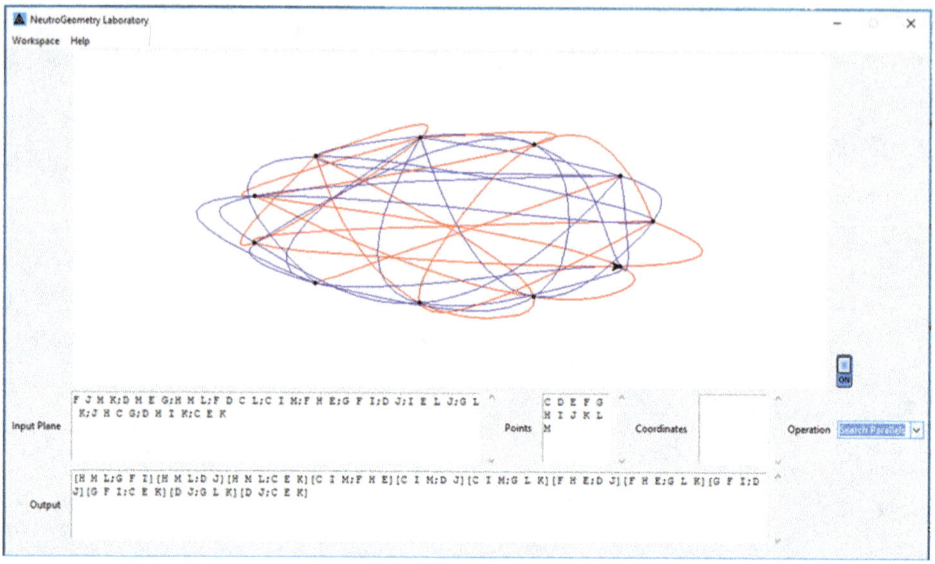

Figure 13. Example of representing the projective plane of order 5, calculating its incidence matrix

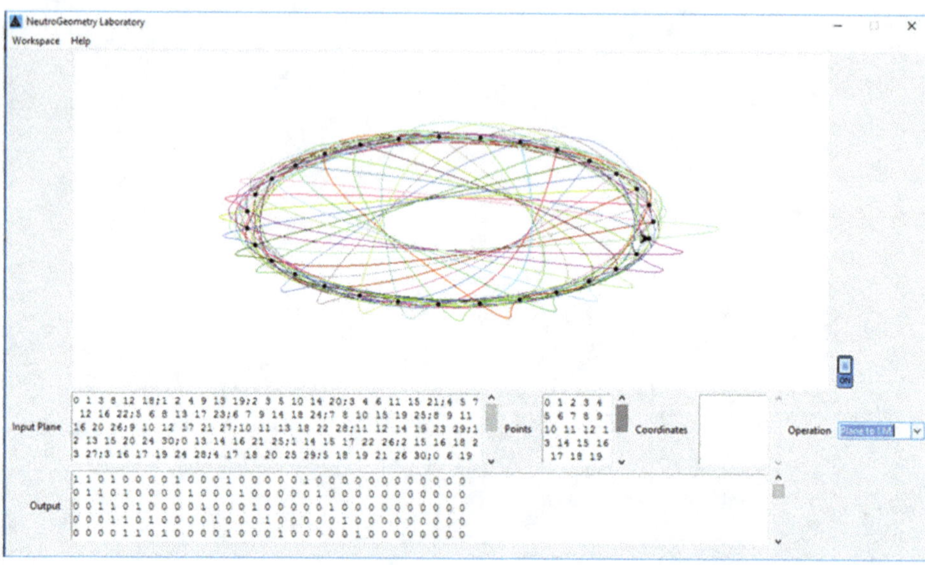

FUTURE RESEARCH DIRECTIONS

The proposed software offers the necessary advantages to perform the main calculations needed to deal with the Mixed Projective-Affine-Hyperbolic planes. However, in future works, the author will intend to improve the software in its appearance, where the user will have more facilities in the visualization of the planes. In addition, taking into account that the software can only be used in Windows, preferably with version 10 or higher, which is an executable with more than 46 megabytes, the author will intend to replace the appearance of the program with the use of other technologies such as Brython. With this technology, the same facilities as the current software would be obtained, but with the characteristics that the user would basically only have an Html file that would require little space in megabytes, it could be run on any operating system and only a Browser would be necessary.

CONCLUSION

This chapter is a summary of how to use the new software called "NeutroGeometry Laboratory 1.0". This software allows the user to comfortably perform the calculations and visualizations that are needed in the case of the Mixed Projective-Affine-Hyperbolic planes, which generalize the classic planes of the theory of Finite Geometries, such as the projective, affine, and hyperbolic planes. These are new types of planes corresponding to the so-called Mixed Geometries or Smarandache Geometries and at the same time to the NeutroGeometries since their parallelism axioms are partially satisfied.

The software allows us to save the results of the obtained calculations in files, to load previously saved planes, to obtain the incidence matrices of a plane, to convert one projective plane into another one, and to calculate the Degree of Type of Geometry. It also allows us to visualize the plane under study, in case the user wants it, and if he/she does not want it, he/she can prevent it by simply pressing an *on/off* button.

REFERENCES

Bishnoi, A., Gijsiwijt, D., D'haesleer, J., & Potukuchi, A. (2022). *Trifferent codes and affine blocking sets*. Paper presented at the Applications of Computer Algebra-ACA 2022.

Bruck, R. H., & Ryser, H. J. (1949). The Nonexistence of Certain Finite Projective Planes. *Canadian Journal of Mathematics*, *1*(1), 88–93. doi:10.4153/CJM-1949-009-2

Brython. (n.d.). http://www.brython.info/

CullinaneS. H. (2017). finite-geometry/galois: Release with Jan. 9, 2016, compressed folder of finitegeometry.org (v1.0). Zenodo. doi:10.5281/zenodo.1038121

Dembowski, P. (1997). Finite Geometries. Berlin: Springer.

Dissett, L. A. (2000). *Combinatorial and Computational Aspects of Finite Geometries* [Unpublished Ph.D. Thesis]. University of Toronto.

Granados, C. (2022). A note on AntiGeometry and NeutroGeometry and their application to real life. *Neutrosophic Sets and Systems*, *49*(1), 579–592.

Horlemann, A. L. (2022). *The Density of Optimal Error-Correcting Codes in Various Metric Spaces*. Paper presented at the Sixth Irsee Conference.

Ilić, J. (2022). *Secret sharing schemes arising from finite geometry* [Unpublished Doctoral thesis]. Univerza na Primorskem, Koper.

Iseri, H. (2002). Partially Paradoxist Smarandache Geometries. *Smarandache Notions Journal, 13*(1-2-3), 5-12.

Kiss, G., & Szonyi, T. (2020). *Finite Geometries*. CRC Press Taylor & Francis Group.

Korchmáros, G., & Sonnino, A. (2012). Finite Bolyai-Lobachevskii Planes. *Acta Mathematica Hungarica*, *134*(4), 405–415. doi:10.100710474-011-0181-4

Korchmáros, G., & Sonnino, A. (2014). Old and Recent Results on Finite Bolyai-Lobachevsky Planes. *Mathematica*, *56*(79), 59–73.

Lutz, M. (2011). Programming Python (4th ed.). O'Reilly.

Mao, L. (2006). *Smarandache Geometries & Map Theory with Applications (I)*. Chinese Branch Xiquan House.

Neri, A., Santonastaso, P., & Zullo, F. (2023). The geometry of one-weight codes in the sum rank metric. *Journal of Combinatorial Theory Series A*, *194*, 105703. doi:10.1016/j.jcta.2022.105703

Raber, N. C. (1972). On Ostrom's Finite Hyperbolic Planes [Unpublished Ph.D.]. The Ohio State University.

Shipman, J. W. (2013). *Tkinter 8.5 reference: a GUI for Python*. New Mexico Tech. Computer Center.

Singh, P. K. (2022). Data with non-Euclidean geometry and its characterization. *Journal of Artificial Intelligence and Technology*, *2*(1), 3–8.

Smarandache, F. (1969). *Mixed noneuclidean geometries*. Infinite Study.

Smarandache, F. (2009). *Degree of Negation of an Axiom*. Infinite Study.

Smarandache, F. (2019). Introduction to NeutroAlgebraic Structures and AntiAlgebraic Structures. In *Advances of Standard and Nonstandard Neutrosophic Theories* (pp. 240–265). Pons Publishing House.

Smarandache, F. (2020). Introduction to NeutroAlgebraic Structures and AntiAlgebraic Structures (revisited). *Neutrosophic Sets and Systems*, *31*, 1–16.

Smarandache, F. (2021). NeutroGeometry & AntiGeometry are alternatives and generalizations of the Non-Euclidean Geometries. *Neutrosophic Sets and Systems, 46*(2021), 456-476.

Szabó, Z. (2022). *Finite Geometric Methods in Quantum Theory* [Unpublished Diploma Thesis]. Budapest University of Technology and Economics.

The-GAP-Group. (2022). *GAP--Groups, Algorithms, and Programming, Version 4.12.2*. https://www.gap-system.org

Zumbragel, J. (2016). *Designs and Codes in Affine Geometry*. arXiv e-prints arXiv -1605.03789.

ADDITIONAL READING

Akça, Z., & Altıntas, A. (2021). A Note on Fano Configurations in the Projective Space PG(5,2). *Konuralp Journal of Mathematics*, *9*, 190–192.

Al-Seraji, N. A. M., & Jafar, Z. S. (2020). Applications of the Projective Plane in Coding Theory. *Journal of Southwest Jiaotong University*, *55*(1), 1–8. doi:10.35741/issn.0258-2724.55.1.2

Ali, A. T. (2010). Special Smarandache Curves in the Euclidean Space. *International Journal of Mathematical Combinatorics*, *2*, 30–36.

Beazley, D., & Jones, B. K. (2013). *Python Cookbook*. O'Reilly.

Korchmáros, G., & Sonnino, A. (2014). Old and Recent Results on Finite Bolyai-Lobachevsky Planes. *Mathematica*, *56*, 59–73.

Singh, P. K. (2021). Anti-Geometry and NeutroGeometry Characterization of Non-Euclidean Data. *Journal of Neutrosophic and Fuzzy Systems*, *1*, 24–33. doi:10.54216/JNFS.010102

KEY TERMS AND DEFINITIONS

Affine Plane: Geometric structure of dimension 2 such that given a line and an external point, there is a unique line parallel to the aforementioned one, which contains the point.

Finite Geometry: Geometric system with only a finite number of points.

Hyperbolic Plane: Also knows as the *Bolyai-Lobachevsky plane*, it is the geometric structure of dimension 2 such that given a line and an external point, there are at least two parallel lines to the aforementioned one, which contain the point.

Incidence Matrix: A matrix whose every row represents every line of the finite plane and every column represents every point, or vice versa, such that the element of the matrix is 1 if there is an incidence relation between the corresponding pair of line and point, and it is 0 otherwise.

Mixed Geometry: Also called *Smarandache Geometry*, it is geometry where some elements satisfy the axiomatic of some type of geometry, and other elements satisfy other types of geometries, all in the same space. For example, there are geometric models which are partially Euclidean, partially hyperbolic, and partially elliptic.

NeutroGeometry: A geometric structure whose definition or at least one axiom is simultaneously satisfied by some elements and not satisfied by others.

Projective Plane: Geometric structure of dimension 2 where there is non-parallelism between the lines.

Python: Is an interpreted programming language whose philosophy emphasizes a syntax that favors readable code. It is a popular multi-paradigm programming language.

Tkinter: Is a graphical library package for the Python programming language. It is considered a standard for the Graphical User Interface (GUI) for Python.

Turtle Graphic: Is a term used in computer graphics as a method to program vector graphics using a relative cursor (the turtle) to Cartesian coordinates.

APPENDIX

This appendix is dedicated to providing the readers with some additional examples for strengthening the reader's knowledge about finite geometries, the MPAH planes, and the use of this software.

First of all, let us study the projective plane of order n=7. Each point is represented by a number between 0 and 56. There is no parallelism in this plane. Every line contains n+1=8 points and every point is contained in n+1=8 lines. Also, the cardinality of the lines and points is the same and it is n^2+ n+ 1= 7^2+ 7+ 1= 57. See Figure 14.

Figure 14. Example for representing the projective plane of order 7

To obtain an affine plane of order 7, we have to eliminate one line and all its points from the projective plane. Let us select the first line in Figure 14 ([0 1 3 13 32 36 43 52]) like the line to remove. The result is seen in Figure 15.

Figure 15. Affine plane of order 7 obtained from the projective plane

See that we used the option Convert and in Points, we input the name of the points to be eliminated in the projective plane.

The result is the following:

[2 4 14 33 37 44 53;2 5 15 34 38 45 54;4 6 16 35 39 46 55;4 5 7 17 40 47 56;5 6 8 18 37 41 48;6 7 9 19 38 42 49;2 7 8 10 20 39 50;8 9 11 21 40 44 51;4 9 10 12 22 41 45;5 10 11 23 42 46 53;6 11 12 14 24 47 54;7 12 15 25 44 48 55;8 14 16 26 45 49 56;9 14 15 17 27 46 50;10 15 16 18 28 47 51;2 11 16 17 19 29 48;12 17 18 20 30 49 53;4 18 19 21 31 50 54;5 14 19 20 22 51 55;6 15 20 21 23 33 56;7 16 21 22 24 34 53;8 17 22 23 25 35 54;2 9 18 23 24 26 55;10 19 24 25 27 37 56;4 11 20 25 26 28 38;5 12 21 26 27 29 39;2 6 22 27 28 30 40;7 14 23 28 29 31 41;4 8 15 24 29 30 42;5 9 16 25 30 31 33;6 10 17 26 31 34 44;7 11 18 27 33 35 45;8 12 19 28 33 34 46;9 20 29 34 35 37 47;10 14 21 30 35 38 48;11 15 22 31 37 39 49;12 16 23 37 38 40 50;17 24 33 38 39 41 51;14 18 25 34 39 40 42;15 19 26 35 40 41 53;16 20 27 41 42 44 54;17 21 28 37 42 45 55;18 22 29 38 44 46 56;19 23 30 39 44 45 47;20 24 31 40 45 46 48;2 21 25 41 46 47 49;22 26 33 42 47 48 50;4 23 27 34 48 49 51;5 24 28 35 44 49 50;6 25 29 45 50 51 53;7 26 30 37 46 51 54;8 27 31 38 47 53 55;9 28 39 48 53 54 56;10 29 33 40 49 54 55;11 30 34 41 50 55 56;2 12 31 35 42 51 56].

Now, let us check that actually this is an affine plane, see Figure 16, where the DTG index is calculated.

Figure 16. DTG index of the affine plane obtained in Figure 15

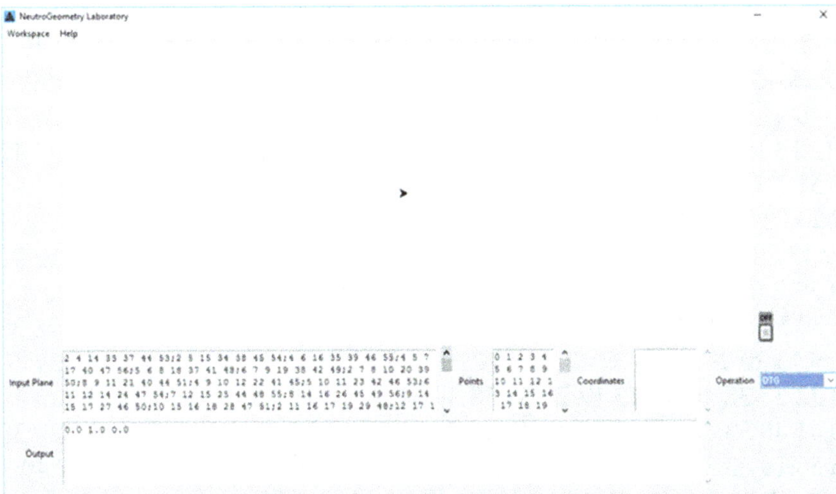

For calculating the DTG we have to input the affine plane in the Input Plane and the list of all its points in Points. The output is the triad (0.1, 1.0, 0.0), which means it is not projective nor hyperbolic but 100% affine.

In the following let us construct an MPAH plane from the projective plane of order 7, see Figure 14. We select two points (e.g., 0 and 1) and we remove them from the projective plane, see Figure 17.

Figure 17. Finite MPAH plane converted from the projective plane

The finite MPAH plane is:

[3 13 32 36 43 52;2 4 14 33 37 44 53;2 3 5 15 34 38 45 54;3 4 6 16 35 39 46 55;4 5 7 17 36 40 47 56;5 6 8 18 37 41 48;6 7 9 19 38 42 49;2 7 8 10 20 39 43 50;3 8 9 11 21 40 44 51;4 9 10 12 22 41 45 52;5 10 11 13 23 42 46 53;6 11 12 14 24 43 47 54;7 12 13 15 25 44 48 55;8 13 14 16 26 45 49 56;9 14 15 17 27 46 50;10 15 16 18 28 47 51;2 11 16 17 19 29 48 52;3 12 17 18 20 30 49 53;4 13 18 19 21 31 50 54;5 14 19 20 22 32 51 55;6 15 20 21 23 33 52 56;7 16 21 22 24 34 53;8 17 22 23 25 35 54;2 9 18 23 24 26 36 55;3 10 19 24 25 27 37 56;4 11 20 25 26 28 38;5 12 21 26 27 29 39;2 6 13 22 27 28 30 40;3 7 14 23 28 29 31 41;4 8 15 24 29 30 32 42;5 9 16 25 30 31 33 43;6 10 17 26 31 32 34 44;7 11 18 27 32 33 35 45;8 12 19 28 33 34 36 46;9 13 20 29 34 35 37 47;10 14 21 30 35 36 38 48;11 15 22 31 36 37 39 49;12 16 23 32 37 38 40 50;13 17 24 33 38 39 41 51;14 18 25 34 39 40 42 52;15 19 26 35 40 41 43 53;16 20 27 36 41 42 44 54;17 21 28 37 42 43 45 55;18 22 29 38 43 44 46 56;19 23 30 39 44 45 47;20 24 31 40 45 46 48;2 21 25 32 41 46 47 49;3 22 26 33 42 47 48 50;4 23 27 34 43 48 49 51;5 24 28 35 44 49 50 52;6 25 29 36 45 50 51 53;7 26 30 37 46 51 52 54;8 27 31 38 47 52 53 55;9 28 32 39 48 53 54 56;10 29 33 40 49 54 55;11 30 34 41 50 55 56;2 12 31 35 42 51 56].

To check that this is an MPAH plane, let us calculate the DTG, see Figure 18.

Figure 18. DTG index of the finite MPAH plane converted from the projective plane of order 7

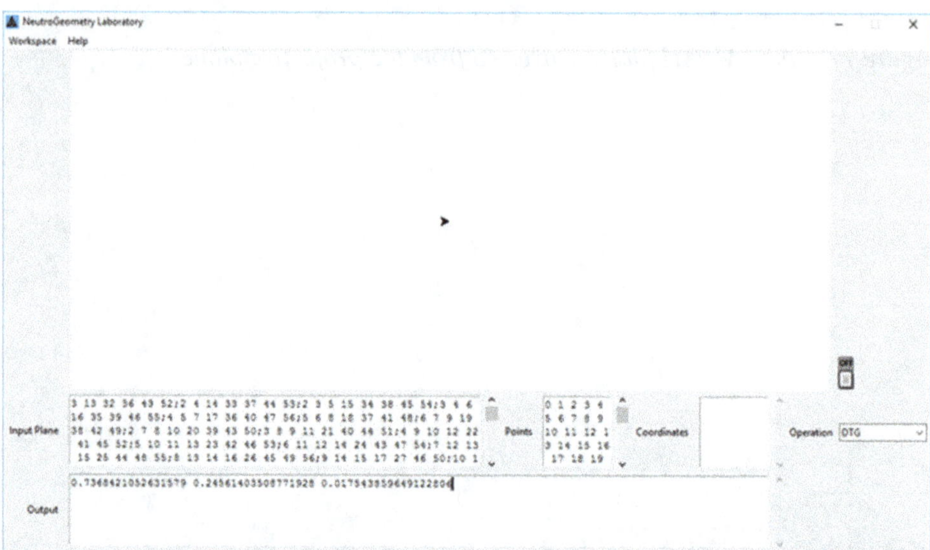

Let us note that now the triad is (0.7368421052631579, 0.24561403508771928, 0.017543859649122806), so this new plane has characteristics of the three types of planes, i.e. it is approximately 73.68% projective, 24.56% affine, and 1.75% hyperbolic. This means that most of the lines in the plane do not have parallels at all, whereas more or less one quarter has only one parallel passing by an external point, and less than 2% has more than 2 parallels by every external point. Let us note that every element is in the interval (0,1) and they sum up to 1.0.

Chapter 8
A Constructive Introduction to Finite Mixed Projective-Affine-Hyperbolic Planes

Erick González Caballero
Technological University of Havana, Cuba

ABSTRACT

NeutroGeometry is the axiomatic approach to geometry from the neutrosophic theory. Neutrosophy is the branch of philosophy that studies neutrality. NeutroGeometry is a geometric structure based on at least one axiom, concept, definition, among others, which is only partially satisfied by the elements of the structure, so they are indeterminate since their definition. In AntiGeometry one of these entities is not satisfied by any element. This chapter introduces the theory and concepts of finite NeutroGeometry, which is geometry based on a finite set of points within the subject of NeutroGeometries in the plane. The author calls them finite mixed projective-affine-hyperbolic planes. Here, finite planes are defined so that lines are divided into three subsets; either they satisfy the axioms of projective planes, the axioms of affine planes, or the axioms of Bolyai-Lobachevsky (also called hyperbolic) planes. They demonstrate the main properties of these planes.

INTRODUCTION

Finite geometry is a branch of geometry closely related to algebra. Several concepts of classical Euclidean geometry are modified since finite geometry deals with geometric structures where there is a finite set of points and therefore it contains a finite set of lines, planes, and so on (Dembowski, 1997; Dissett, 2000; Kiss & Szonyi, 2020).

DOI: 10.4018/978-1-6684-4740-6.ch008

On the other hand, these structures are dependent on the axioms on which they are based and these axioms determine the type of finite geometric structure in question. In dimension 2 there are three main types of structures, they are the projective planes where there is no parallelism between lines (Cameron, 2000), in the affine planes for each point external to the line passes one and only one parallel line (Zumbragel, 2016), while in the Bolyai-Lobachevsky planes or hyperbolic planes through a point outside a line pass at least two lines parallel to the first one (Korchmáros & Sonnino, 2012). The relationship of these geometric structures with algebra is that they are modeled with the support of finite vector spaces, which are vector spaces over finite fields (Hirschfeld, 1979).

Among the three types of finite planes, perhaps the projective planes are the most important of them, of which a large number of books and scientific papers can be found where they are studied exhaustively. These planes are applied in subjects like Code Theory, Cryptography, and are related to orthogonal Latin squares (Beutelspacher, 1990; Klein & Storme, 2011; Zangalho-Raposo, 2014). A useful theorem is the Bruck-Ryser theorem (Bruck & Ryser, 1949), which is a necessary condition for the existence of finite projective planes of certain order n, where n+1 is the number of points that each line contains. This theorem, because it is a necessary condition, allows us to rule out some orders with which there are no finite projective planes, for example, there is no projective plane with an order as small as n = 6. The projective plane of order n = 10 has fascinated some researchers since n = 10 satisfies the necessary conditions of the theorem, but this plane does not exist, which was demonstrated computationally, (Lam, 1991; Lam, Thiel, & Swiercz, 1989).

Less abundant in the literature is the study of affine planes (Zumbragel, 2016). It is well known that to obtain an affine plane from a projective plane we have to eliminate a fixed line and all the points contained in it. Conversely, from an affine plane, a projective plane can be obtained by adding a point for each class of parallel lines and connecting all these new points by a unique line. Affine planes can also be represented by finite vector spaces, and parallelism is an equivalence relation.

Compared to the other types of finite planes, Bolyai-Lobachevsky's ones are found in a few pieces of literature, many of them were written in the sixties or seventies of the 20th century, with few pages dedicated to planes and almost non-existent literature dedicated to spaces of this type (Korchmáros & Sonnino, 2012). A relevant result is that from a projective plane of order greater than or equal to seven, a hyperbolic one can be obtained by eliminating three lines whose intersection is empty (Sandler, 1963). Other ways generalize this method (Olgun & Gunaltili, 2007). It is also possible to define hyperbolic planes from inversive planes of even order (Crowe, 1965, 1966). Inversive planes are defined from points and circles instead of lines, which satisfy certain axioms.

An idea that has inspired this chapter is to define Mixed or Smarandache Geometry (Iseri, 2002; Mao, 2006; Smarandache, 1969, 2009), wherein a finite geometric plane some elements satisfy the projective axioms, others the affine axioms, and a third set the hyperbolic axioms. We will call this the Mixed Projective-Affine-Hyperbolic plane (MPAH plane for short). This type of geometric structure is an example of NeutroGeometry, taking into account that each of these planes can be divided into three sets of lines satisfying different axioms of parallelism.

Smarandache's idea is justified because mixed geometries contain properties more in line with mathematical problems and their applications than if one works with a single type of plane. For example, the non-existence of finite projective planes of certain orders does not imply that there are no finite mixed planes of these orders, which would be a solution to this problem.

Constructing mixed geometries is another way to define NeutroGeometries different from Smarandache's concept of neutrosophication to do so. Neutrosophication is the process to divide a concept, theory, structure, algebra, geometry, axiom, among others denoted by <A>, into a triad where <AntiA> denotes the opposite of <A>, and <NeutA> represents both non <A> and non <AntiA>.

This chapter is an approach to the problem of constructing mixed planes mainly from projective planes. An index is proposed to measure the degree to which the mixed plane satisfies the axioms of each of the three types of "pure" planes. We study the incidence matrix and its properties. Additionally, we provide a code in Python programming language, which the reader can use to obtain and test the mixed planes in the examples of finite planes.

The structure of the chapter is the following; first, the Background of this subject is developed, recalling concepts like NeutroGeometry and the main notions of finite geometries. The next section is dedicated to introducing and studying the Mixed Projective-Affine-Hyperbolic planes. The chapter finishes with a section on Discussion and another on Conclusion. One appendix offers to the reader the python code used to run the examples and another appendix summarizes some additional examples of finite MPAH planes.

BACKGROUND

Definition 1 (Smarandache, 2019, 2020, 2021): Let S be a classical geometric space, and <A> be a geometric concept (such as postulate, axiom, theorem, property, function, transformation, operator, theory, etc.). <AntiA> is the opposite of <A>, while <NeutA> (also called <NeutroA>) is the neutral (or indeterminate) part between <A> and <AntiA>.

This is the so-called *Neutrosophication* process to obtain from <A> another two related geometric concepts, where <NeutroA> contains at least an axiom or a definition that is not satisfied by 0% neither 100% of the elements of S.

Finite Geometries are geometric structures containing a finite set of points, and therefore, contain a finite set of lines, which are defined as non-singleton elements of the power set of the set of points (Dembowski, 1997). Also, there is a finite set of planes, spaces, and so on, when we deal with higher dimensional spaces. In this chapter, we restrict our study to 2-dimensional spaces or planes.

The concept of incidence structures is fundamental in finite planes; it is formally defined as follows:

Definition 2 (Dembowski, 1997): An *incidence structure* is a triple $\boldsymbol{S}=(\mathcal{P},\mathcal{L},\mathcal{I})$, where $\mathcal{P},\mathcal{L},\mathcal{I}$ are sets satisfying: $\mathcal{P}\cap\mathcal{L}=\varnothing$ and $\mathcal{I}\subseteq\mathcal{P}\times\mathcal{L}$. The elements of $\mathcal{P}$ are called *points*, those of $\mathcal{L}$ *lines*, and $\mathcal{I}$ is a symmetric relation called *incidence*.

A line is an element of the power set of $\mathcal{P}$ containing at least two points. A point P is incident in a line l and vice versa, a line is incident in the point P if it is contained in l and l passes through P. It is preferred to denote this by $P\mathcal{I}l$ and $l\mathcal{I}P$ instead of $(P,l)\in\mathcal{I}$ and $(l,P)\in\mathcal{I}$, respectively.

Definition 3 (Dembowski, 1997): The *dual structure* $\bar{\boldsymbol{S}}=(\bar{\mathcal{P}},\bar{\mathcal{L}},\bar{\mathcal{I}})$ of the incidence structure $\boldsymbol{S}=(\mathcal{P},\mathcal{L},\mathcal{I})$ is defined by $\bar{\mathcal{P}}=\mathcal{L}$, $\bar{\mathcal{L}}=\mathcal{P}$, for $P\in\mathcal{P}$ and $l\in\mathcal{L}$; $l\bar{\mathcal{I}}P$ if and only if $P\mathcal{I}l$, and $P\bar{\mathcal{I}}l$ if and only if $l\mathcal{I}P$. I.e., the points are interchanged with lines.

Definition 4 (Dembowski, 1997): If $\boldsymbol{S}=(\mathcal{P},\mathcal{L},\mathcal{I})$ is an incidence structure, $\mathcal{Q}\subset\mathcal{P}$ and $\mathcal{R}\subset\mathcal{L}$, then a *substructure* of $\boldsymbol{S}$ is the incidence structure $\boldsymbol{T}=(\mathcal{Q},\mathcal{R},\mathcal{J})$ with $\mathcal{J}=\mathcal{I}\cap(\mathcal{Q}\times\mathcal{R})$.

Definition 5 (Dembowski, 1997): Let $\boldsymbol{S}=(\mathcal{P},\mathcal{L},\mathcal{I})$ be an incidence structure, where points and lines are numbered in an arbitrary and fixed way: $P_1,P_2,\ldots,P_m$ and $l_1,l_2,\ldots,l_n$. Then the following two matrices are called *incidence matrices* $C=(c_{ij})$ and $D=(d_{ij})$, where

$$c_{ij}=\begin{cases}1, & \text{if } P_i\mathcal{I}l_j\\ 0, & \text{otherwise}\end{cases} \text{ and } d_{ij}=\begin{cases}1, & \text{if } l_i\mathcal{I}P_j\\ 0, & \text{otherwise}\end{cases}.$$

Definition 6 (Kiss & Szonyi, 2020): A *projective plane* is an incidence structure $\boldsymbol{P}=(\mathcal{P},\mathcal{L},\mathcal{I})$ with the following axioms:

(i) For each pair of two distinct points $P,\ Q\in\mathcal{P}$ there is a unique line that is incident with P and Q.

(ii) Each pair of two distinct lines $l,\ s \in \mathcal{L}$ has a unique point in common.
(iii) . There are four points, such that no three are collinear, i.e., three of them do not belong to the same line.

The last axiom is equivalent to the two following axioms:

(iii'). Every line has at least three points.
(iv'). There are at least two lines in the plane.

In a projective plane of *order* n≥2, each line contains n+1 points. Also, this type of plane satisfies the *duality principle*, where if a theorem T contains the words "point", "line" and "incidence", and the statement T′ is obtained from T replacing the word "point" by "line" and vice versa, then T′ is also a theorem.

A projective plane of order n satisfies the following properties (Zumbragel, 2016):

1. The number of points in every line is n+1.
2. The number of lines passing by only one point is n+1.
3. The total number of points in the plane is n^2+n+1.
4. The total number of lines in the plane is n^2+n+1.

Theorem 1 (Bruck-Ryser) (Bruck & Ryser, 1949; Cameron, 1994; Lam, 1991): If n≡1,2(mod4), then a necessary condition for the existence of a finite projective plane of order n is that integers x, y exist satisfying $x^2+y^2=n$.

Example 1 (Zangalho-Raposo, 2014): Fano plane is the smallest projective plane, it has order n = 2, so each line has n+1 = 3 points and each point is incident with exactly n+1 = 3 lines. Seven is the total number of points and lines, see Figure 1.

Definition 7 (Kiss & Szonyi, 2020): An *affine plane* is an incidence structure $A = (\mathcal{P}, \mathcal{L}, \mathcal{I})$ with the following axioms:

(i) For each pair of two distinct points $P,\ Q \in \mathcal{P}$ there is a unique line that is incident with *P* and *Q*.
(ii) For each point $P \in \mathcal{P}$ and each line $l \in \mathcal{L}$, there exists a unique line $s \in \mathcal{L}$ such that $P\mathcal{I}s$, and *l* and *s* do not meet. This is called *parallelism* in **A** and is denoted by *l*||*s*.
(iii) Three points are not collinear.

Similarly to projective planes, the order of the affine plane is the number of points contained in one line.

Figure 1. Graphical representation of the Fano plane

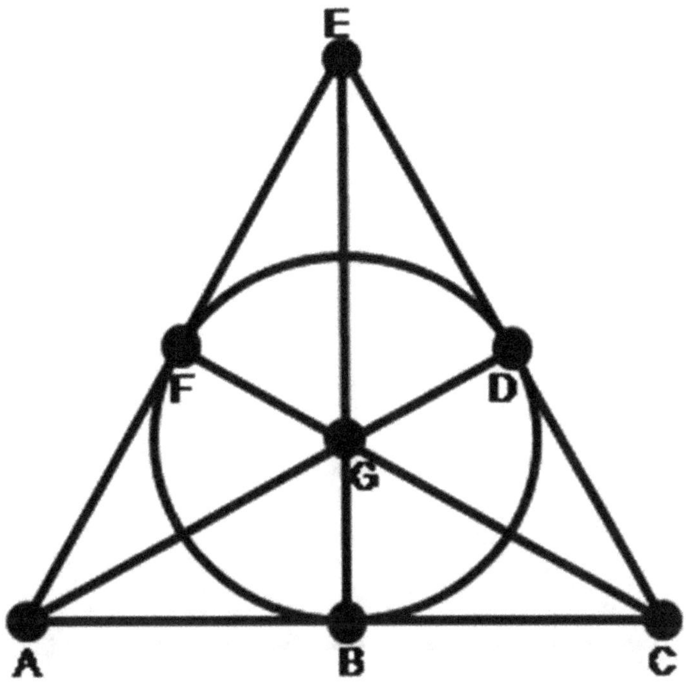

An affine plane of order n satisfies the following properties (Zumbragel, 2016):

1. Every line is incident with n points.
2. Every point is incident with n +1 lines.
3. The affine plane contains n^2 points.
4. The affine plane contains n^2+n lines.

When a fixed line and all the points contained in this line of a projective plane are removed, we obtain an affine plane.

Given the relation R over $\mathcal{L}$ in the affine plane, defined as lRs if and only if $l=s$ or $l \parallel s$ $\forall l, s \in \mathcal{L}$, then R is an equivalence relation.

Example 2: Figure 2 represents an affine plane of order 2 obtained from the Fano plane (Zangalho-Raposo, 2014) removing the line *BDF* and all its points. It contains 4 points and 6 lines.

Figure 2. Depiction of the affine plane of order 2. Parallel lines are represented by the same color.

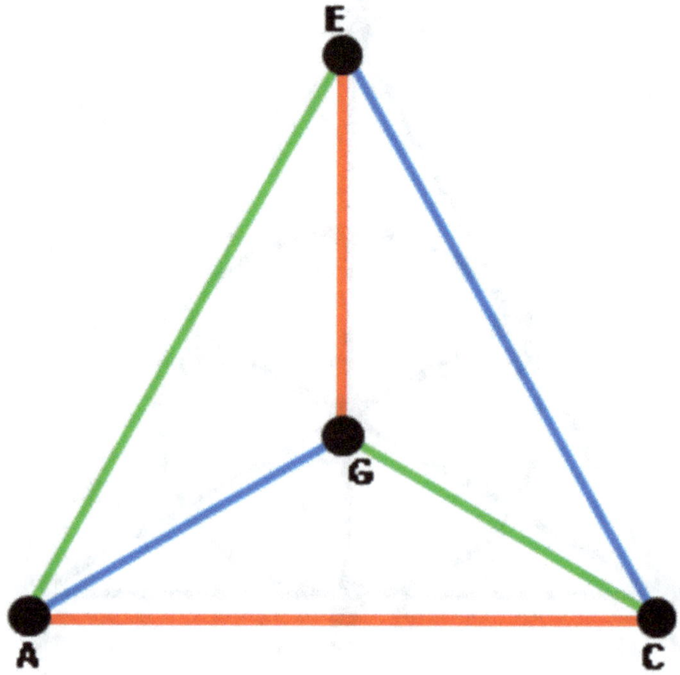

Definition 8 (Raber, 1972): A *Bolyai-Lobachevsky plane* or *Hyperbolic plane* is a geometric structure $\boldsymbol{H} = (\mathcal{P}, \mathcal{L}, \mathcal{I})$ with the following axioms:

(i) For each pair of two distinct points $P,\ Q \in \mathcal{P}$ there is a unique line that is incident with *P* and *Q*.
(ii) For each point, $P \in \mathcal{P}$ and each line $l \in \mathcal{L}$ there exists at least two lines in $\mathcal{L}$ which are incident with *P*, and these lines and *l* do not meet.
(iii) If a subset $\mathcal{S}$ of points of $\boldsymbol{H}$ contains three non-collinear points and contains all the points of the lines defined by every pair of points of $\mathcal{S}$, then $\mathcal{S}$ contains all the points of $\boldsymbol{H}$.

In this chapter we consider the third axiom dispensable in the definition of a mixed finite Geometry, see the axiomatic in (Kärteszi, 1976).

Let $\boldsymbol{P} = (\mathcal{P}, \mathcal{L}, \mathcal{I})$ be a projective plane of order n≥7, if three fixed lines $l_1, l_2, l_3 \in \mathcal{L}$ are selected such that $l_1 \cap l_2 \cap l_3 = \varnothing$, then the plane obtained by removing these lines and their points is a hyperbolic plane (Sandler, 1963). Other methods to obtain hyperbolic planes can be read in (Celik, 2001; Crowe, 1965; Olgun & Gunaltili, 2007).

Example 3: This is an example of a finite hyperbolic plane extracted from (Kärteszi, 1976). The points are denoted by capital letters from A to M in alphabetical order and the lines by a set of points as follows:

AGM, BHM, CIM, DJM, EKM, FLM, BCJ, ACK, ABL, AEF, CDF, BDE, AIJ, BIK, CGL, ADH, BFG, CEH, DGK, EJL, FHI, DIL, EGI, FJK, GHJ, HKL.

FINITE MIXED PROJECTIVE-AFFINE-HYPERBOLIC PLANES

In this section, we introduce the definitions and properties of finite Mixed Projective-Affine-Hyperbolic (MPAH) planes. The formal definition is the following:

Definition 9: A *finite Mixed Projective-Affine-Hyperbolic plane* (finite MPAH plane for short) is an incidence structure $\mathcal{MPAH} = (\mathcal{P}, \mathcal{L}, \mathcal{I})$, where $\mathcal{L} = \mathcal{L}_{\mathcal{P}} \cup \mathcal{L}_{\mathcal{A}} \cup \mathcal{L}_{\mathcal{H}}$. The elements of $\mathcal{L}_{\mathcal{P}}$ satisfy partially or totally the axiomatic of projective planes, the elements of $\mathcal{L}_{\mathcal{A}}$ satisfy partially or totally the axiomatic of affine planes, and the elements of $\mathcal{L}_{\mathcal{H}}$ satisfy partially or totally the axiomatic of hyperbolic planes. Let us call n the *order of the finite MPAH plane* if n+1 is the maximal cardinality of the elements of $\mathcal{L}$.

See that the order will be always characterized by the number of points contained in the lines of $\mathcal{L}_{\mathcal{P}}$.

Proposition 2. The finite Mixed-Projective-Affine-Hyperbolic planes are NeutroGeometries.

Proof. The main axioms that characterize every one of the three types of planes are those concerning parallelism. So, the truthfulness of the proposition is obvious. □

Remark 1.

1. See that Definition 9 includes the case where $\mathcal{L}_{\mathcal{P}}, \mathcal{L}_{\mathcal{A}}, \mathcal{L}_{\mathcal{H}}$ are pairwise disjoint.
2. The definition of MPAH planes does not include other mixed or "pure" planes; namely, projective, affine, hyperbolic, mixed projective-affine, mixed projective-hyperbolic, or mixed affine-hyperbolic. All of them are AntiPlanes in the context of this theory, e.g., mixed projective-affine planes are AntiHyperbolic geometries. Nevertheless, all those planes are related to MPAH planes.

Proposition 3. The finite Mixed-Projective-Affine-Hyperbolic planes always have an order strictly bigger than 2.

Proof. Let us begin with the Fano plane where n = 2. To obtain a line of $\mathcal{L}_{\mathcal{H}}$ we have to remove at least two points from the projective plane, but every line of the Fano plane has 3 points, so, it is not possible.□

Example 3: Now let us consider the projective plane of order n = 3, see Figure 3 (Polster, 1998).

Figure 3. Depiction of the projective plane of order 3

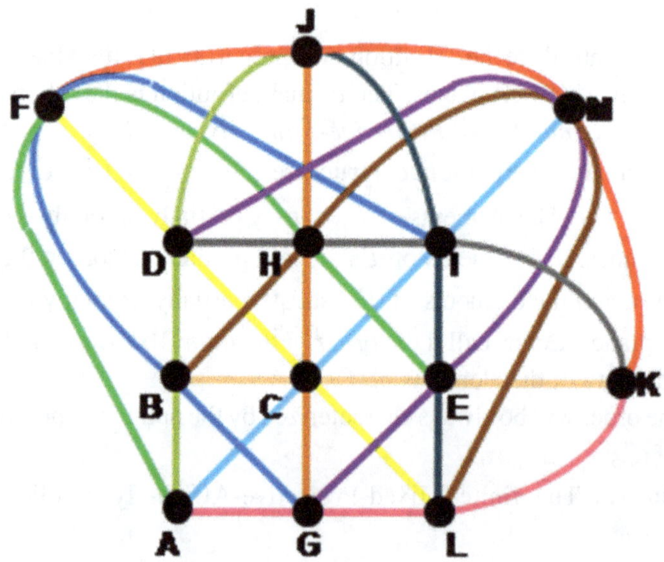

If points F and M are removed from the plane in Figure 3 we obtain the new plane in Figure 4.

It is easy to check that this is a finite MPAH plane. The line JK belongs to $\mathcal{L}_{\mathcal{H}}$, AGLK belongs to $\mathcal{L}_{\mathcal{P}}$, and AHE belongs to $\mathcal{L}_{\mathcal{A}}$. Thus, this example illustrates the existence of a finite MPAH plane of order 3. This example suggests the algorithm to obtain a finite MPAH, see Proposition 4.

Proposition 4: Given a projective plane *p* and an affine plane *a* of order n≥3, there is an MPAH that can be constructed from them.

Figure 4. A finite MPAH plane of order 3

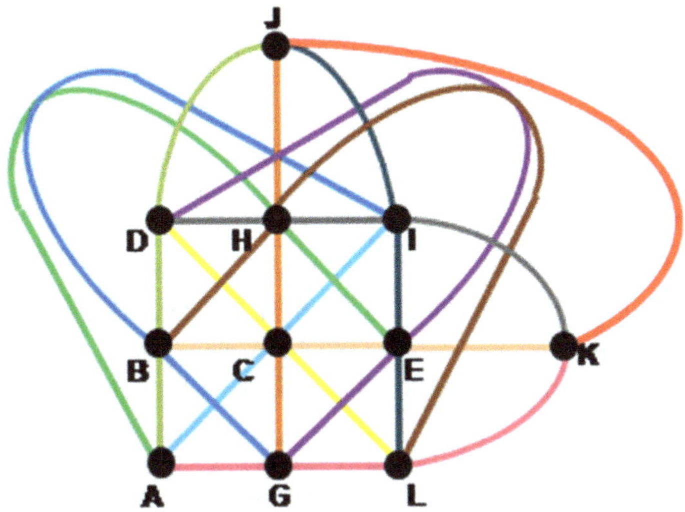

Proof. Let $p=(\mathcal{P},\mathcal{L},\mathcal{I})$ be a projective plane and $P\in\mathcal{P}$, then $\mathcal{MPAH}_p=(\mathcal{P}',\mathcal{L}',\mathcal{I}')$ such that given a fixed-line $l\in\mathcal{L}$ and two fixed points $P_1,P_2\in l$, $\mathcal{P}'=\mathcal{P}\setminus\{P_1,P_2\}$, $\mathcal{L}'=\{s\setminus\{P_1,P_2\}:s\in\mathcal{L}\}$ and $\mathcal{I}'=\mathcal{I}\cap(\mathcal{P}'\times\mathcal{L}')$, such that it is an incidence structure. In other words, we eliminate two points in the projective plane. Thus, the lines incident with P_1 (P_2) become parallels to each other when the point is removed. One of the lines obtained after removing P_1 and another one obtained after removing P_2 meet in one point Q of the plane because of the axiom of projective planes, so through this point passes two parallels of l, which corresponds to the property of hyperbolic planes. Moreover, given any point Q non-incident with l, we had the lines QP_1 and QP_2 in p, which now do not contain P_1 and P_2 so, for any external point passes two parallel lines of l. Additionally, the lines incident with P_1 (P_2) are parallels each other, so they belong to $\mathcal{L}_{\mathcal{A}}$. Given any point Q, again the lines QP_1 and QP_2 belong to p, so for Q passes only one parallel to one of the lines in $\mathcal{L}_{\mathcal{A}}$. Whereas the lines non-incident with P_1 or P_2 belong to $\mathcal{L}_{\mathcal{P}}$.

Now, let $a=(\mathcal{P},\mathcal{L},\mathcal{I})$ be an affine plane, then let us select $l_1,l_2\in\mathcal{L}$ and $l_1 \| l_2$, define $H=\{[l]:l\in\mathcal{L}\}\setminus\{[l_1],[l_2]\}$, where $[l]$ is the parallel class of l. Let us define $\mathcal{MPAH}_a=(\mathcal{P}',\mathcal{L}',\mathcal{I}')$ where for each parallel class in H we add a point in $\mathcal{P}'$ incident with them and we add a line l_3 in $\mathcal{L}'$ containing all those new points. Thus, every line with an added point satisfies the axiomatic of projective planes, the lines parallel to l_1 maintain the axiomatic of affine planes as well as those parallels to l_2. Finally, for each point external to l_3 passes both one line in $[l_1]$ and one in $[l_2]$, then l_3 fulfills the axiomatic of the hyperbolic plane. □

Remark 2.

1. This procedure can be repeated in more than one line, always that at least one line of the plane remains with n+1 points, which is the line satisfying the axiomatic of projective planes.
2. The procedure can be extended to more than two points for $\mathcal{MPAH}_p$ or more than two lines for $\mathcal{MPAH}_a$ always that at least one line contains n+1 points and every line contains at least two points. See Figure 5 where points C and E are removed from the finite MPAH plane in Figure 4.

Figure 5. A finite MPAH plane of order 3 was obtained after removing points C and E in the plane of Figure 4

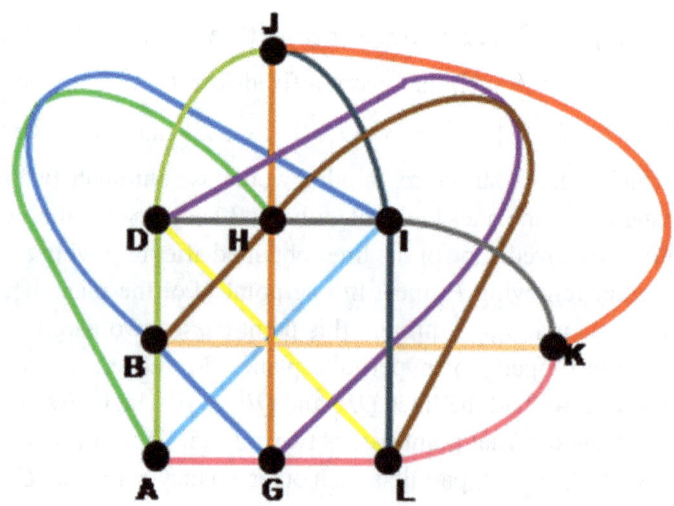

Corollary 1. A finite MPAH plane $\mathcal{MPAH}_p$ or $\mathcal{MPAH}_a$ of order n has three types of lines, namely with n+1 points corresponding to projective properties, with n points corresponding to affine properties, and with at most n-1 points corresponding to hyperbolic properties.

Proof. Obvious.

Corollary 1 implies that we can classify every line of the plane either in projective, affine, or hyperbolic according to the number of points. This is also true if the procedure is applied twice or more times according to Remark 2.

Definition 10: Let an $\mathcal{MPAH}$ ($\mathcal{MPAH}_p$ or $\mathcal{MPAH}_a$ or one obtained according to Remark 2) be a Mixed Projective-Affine-Hyperbolic plane, the *Degree of Type of Geometry* (DTG) is a triad (*p*,*a*,*h*), where

$$p=\frac{|\mathcal{PL}|}{|\mathcal{L}|}, a=\frac{|\mathcal{AL}|}{|\mathcal{L}|}, \text{ and } h=\frac{|\mathcal{HL}|}{|\mathcal{L}|},$$

where $|\mathcal{L}|$ is the number of lines in $\mathcal{MPAH}$, whereas, $|\mathcal{PL}|, |\mathcal{AL}|$, and $|\mathcal{HL}|$ are the cardinality of the lines with n+1 points, with n points, and with n-1 points or less, respectively.

Evidently, $p,a,h \in (0,1)$ and $p+a+h=1$.

Example 4: For calculating the DTG of the plane in Figure 4 we have: $\mathcal{PL}=$ {*ABDJ*, *LEIJ*, *AGLK*, *GCHJ*, *DHIK*, *BCEK*}, $\mathcal{AL}=$ {*DEG*, *BHL*, *DCL*, *ACI*, *AHE*, *GBI*}, and $\mathcal{HL}=\{JK\}$. So,

$$p=\frac{6}{13}, a=\frac{6}{13}, \text{ and } h=\frac{1}{13}.$$

Whereas, for the plane in Figure 5, $\mathcal{PL}=$ {*ABDJ*, *AGLK*, *DHIK*}, $\mathcal{AL}=$ {*BHL*, *GBI*, *LIJ*, *GHJ*}, and $\mathcal{HL}=$ {*JK*, *DG*, *DL*, *AI*, *AH*, *BK*}, therefore,

$$p=\frac{3}{13}, a=\frac{4}{13}, \text{ and } h=\frac{6}{13}.$$

Definition 11: A *Mixed Approximate Plane* is a plane whose incidence matrix $\mathrm{A}=(\mathrm{a}_{ij})_{0\le i,j\le n}$ is one of the solutions to the equation $\min_{a_{ij}\in\{0,1\}}\mathrm{AA}^{\mathrm{T}}-\mathrm{nI}-\mathrm{J}$.

Where A^T is the transpose of A, I is the identity matrix, J is the matrix of all 1's, n is the order of the plane, and $\|\bullet\|$ is a norm.

The projective plane satisfies the equation $AA^T=nI+J$ as the consequence of the following conditions:

(i) A has constant row sum n+1,
(ii) A has constant column sum n+1,
(iii) The inner product of two distinct rows is 1,
(iv) The inner product of two distinct columns is 1.

The condition $\min_{a_{ij}\in\{0,1\}} AA^{T} - nI - J$ in Definition 11 is a binary programming problem where A is the variable. When $AA^{T}-nI-J=0$ then A is the incidence matrix of a projective plane, where 0 means the null matrix.

The incidence matrix of the projective plane of order 3 is the following:

$$A_{\mathcal{P}} = \begin{bmatrix} 0 & 0 & 0 & 0 & 0 & 1 & 0 & 0 & 0 & 1 & 1 & 0 & 1 \\ 0 & 0 & 0 & 1 & 1 & 0 & 1 & 0 & 0 & 0 & 0 & 0 & 1 \\ 0 & 1 & 0 & 0 & 0 & 0 & 0 & 1 & 0 & 0 & 0 & 1 & 1 \\ 0 & 0 & 1 & 1 & 0 & 1 & 0 & 0 & 0 & 0 & 0 & 1 & 0 \\ 1 & 0 & 1 & 0 & 0 & 0 & 0 & 0 & 1 & 0 & 0 & 0 & 1 \\ 1 & 0 & 0 & 0 & 1 & 1 & 0 & 1 & 0 & 0 & 0 & 0 & 0 \\ 0 & 1 & 0 & 0 & 0 & 1 & 1 & 0 & 1 & 0 & 0 & 0 & 0 \\ 1 & 1 & 0 & 1 & 0 & 0 & 0 & 0 & 0 & 1 & 0 & 0 & 0 \\ 0 & 0 & 0 & 0 & 1 & 0 & 0 & 0 & 1 & 1 & 0 & 1 & 0 \\ 1 & 0 & 0 & 0 & 0 & 0 & 1 & 0 & 0 & 0 & 1 & 1 & 0 \\ 0 & 0 & 1 & 0 & 0 & 0 & 1 & 1 & 0 & 1 & 0 & 0 & 0 \\ 0 & 0 & 0 & 1 & 0 & 0 & 0 & 1 & 1 & 0 & 1 & 0 & 0 \\ 0 & 1 & 1 & 0 & 1 & 0 & 0 & 0 & 0 & 0 & 1 & 0 & 0 \end{bmatrix}$$

The incidence matrix of the MPAH plane represented in Figure 4 is:

$$
A_{\mathcal{PAH}_1} = \begin{bmatrix}
0 & 0 & 0 & 0 & 0 & 0 & 0 & 0 & 0 & 1 & 1 & 0 & 0 \\
0 & 0 & 0 & 1 & 1 & 0 & 1 & 0 & 0 & 0 & 0 & 0 & 0 \\
0 & 1 & 0 & 0 & 0 & 0 & 0 & 1 & 0 & 0 & 0 & 1 & 0 \\
0 & 0 & 1 & 1 & 0 & 0 & 0 & 0 & 0 & 0 & 0 & 1 & 0 \\
1 & 0 & 1 & 0 & 0 & 0 & 0 & 0 & 1 & 0 & 0 & 0 & 0 \\
1 & 0 & 0 & 0 & 1 & 0 & 0 & 1 & 0 & 0 & 0 & 0 & 0 \\
0 & 1 & 0 & 0 & 0 & 0 & 1 & 0 & 1 & 0 & 0 & 0 & 0 \\
1 & 1 & 0 & 1 & 0 & 0 & 0 & 0 & 0 & 1 & 0 & 0 & 0 \\
0 & 0 & 0 & 0 & 1 & 0 & 0 & 0 & 1 & 1 & 0 & 1 & 0 \\
1 & 0 & 0 & 0 & 0 & 0 & 1 & 0 & 0 & 0 & 1 & 1 & 0 \\
0 & 0 & 1 & 0 & 0 & 0 & 1 & 1 & 0 & 1 & 0 & 0 & 0 \\
0 & 0 & 0 & 1 & 0 & 0 & 0 & 1 & 1 & 0 & 1 & 0 & 0 \\
0 & 1 & 1 & 0 & 1 & 0 & 0 & 0 & 0 & 0 & 1 & 0 & 0
\end{bmatrix}
$$

Whereas, the incidence matrix of the MPAH plane represented in Figure 5 is:

$$
A_{\mathcal{PAH}_2} = \begin{bmatrix}
0 & 0 & 0 & 0 & 0 & 0 & 0 & 0 & 0 & 1 & 1 & 0 & 0 \\
0 & 0 & 0 & 1 & 0 & 0 & 1 & 0 & 0 & 0 & 0 & 0 & 0 \\
0 & 1 & 0 & 0 & 0 & 0 & 0 & 1 & 0 & 0 & 0 & 1 & 0 \\
0 & 0 & 0 & 1 & 0 & 0 & 0 & 0 & 0 & 0 & 0 & 1 & 0 \\
1 & 0 & 0 & 0 & 0 & 0 & 0 & 0 & 1 & 0 & 0 & 0 & 0 \\
1 & 0 & 0 & 0 & 0 & 0 & 0 & 1 & 0 & 0 & 0 & 0 & 0 \\
0 & 1 & 0 & 0 & 0 & 0 & 1 & 0 & 1 & 0 & 0 & 0 & 0 \\
1 & 1 & 0 & 1 & 0 & 0 & 0 & 0 & 0 & 1 & 0 & 0 & 0 \\
0 & 0 & 0 & 0 & 0 & 0 & 0 & 0 & 1 & 1 & 0 & 1 & 0 \\
1 & 0 & 0 & 0 & 0 & 0 & 1 & 0 & 0 & 0 & 1 & 1 & 0 \\
0 & 0 & 0 & 0 & 0 & 0 & 1 & 1 & 0 & 1 & 0 & 0 & 0 \\
0 & 0 & 0 & 1 & 0 & 0 & 0 & 1 & 1 & 0 & 1 & 0 & 0 \\
0 & 1 & 0 & 0 & 0 & 0 & 0 & 0 & 0 & 0 & 1 & 0 & 0
\end{bmatrix}
$$

Example 5: Obviously, $A_{\mathcal{P}}A_{\mathcal{P}}^{T} - nI - J_{Frob} = 0$.

$$A_{\mathcal{PAH}_1}A_{\mathcal{PAH}_1}^{T} = \begin{bmatrix} 2 & 0 & 0 & 0 & 0 & 0 & 0 & 1 & 1 & 1 & 1 & 1 & 1 \\ 0 & 3 & 0 & 1 & 0 & 1 & 1 & 1 & 1 & 1 & 1 & 1 & 1 \\ 0 & 0 & 3 & 1 & 0 & 1 & 1 & 1 & 1 & 1 & 1 & 1 & 1 \\ 0 & 1 & 1 & 3 & 1 & 0 & 0 & 1 & 1 & 1 & 1 & 1 & 1 \\ 0 & 0 & 0 & 1 & 3 & 1 & 1 & 1 & 1 & 1 & 1 & 1 & 1 \\ 0 & 1 & 1 & 0 & 1 & 3 & 0 & 1 & 1 & 1 & 1 & 1 & 1 \\ 0 & 1 & 1 & 0 & 1 & 0 & 3 & 1 & 1 & 1 & 1 & 1 & 1 \\ 1 & 1 & 1 & 1 & 1 & 1 & 1 & 4 & 1 & 1 & 1 & 1 & 1 \\ 1 & 1 & 1 & 1 & 1 & 1 & 1 & 1 & 4 & 1 & 1 & 1 & 1 \\ 1 & 1 & 1 & 1 & 1 & 1 & 1 & 1 & 1 & 4 & 1 & 1 & 1 \\ 1 & 1 & 1 & 1 & 1 & 1 & 1 & 1 & 1 & 1 & 4 & 1 & 1 \\ 1 & 1 & 1 & 1 & 1 & 1 & 1 & 1 & 1 & 1 & 1 & 4 & 1 \\ 1 & 1 & 1 & 1 & 1 & 1 & 1 & 1 & 1 & 1 & 1 & 1 & 4 \end{bmatrix}$$

$$A_{\mathcal{PAH}_2}A_{\mathcal{PAH}_2}^{T} = \begin{bmatrix} 2 & 0 & 0 & 0 & 0 & 0 & 0 & 1 & 1 & 1 & 1 & 1 & 1 \\ 0 & 2 & 0 & 1 & 0 & 0 & 1 & 1 & 0 & 1 & 1 & 1 & 0 \\ 0 & 0 & 3 & 1 & 0 & 1 & 1 & 1 & 1 & 1 & 1 & 1 & 1 \\ 0 & 1 & 1 & 2 & 0 & 0 & 0 & 1 & 1 & 1 & 0 & 1 & 0 \\ 0 & 0 & 0 & 0 & 2 & 1 & 1 & 1 & 1 & 1 & 0 & 1 & 0 \\ 0 & 0 & 1 & 0 & 1 & 2 & 0 & 1 & 0 & 1 & 1 & 1 & 0 \\ 0 & 1 & 1 & 0 & 1 & 0 & 3 & 1 & 1 & 1 & 1 & 1 & 1 \\ 1 & 1 & 1 & 1 & 1 & 1 & 1 & 4 & 1 & 1 & 1 & 1 & 1 \\ 1 & 0 & 1 & 1 & 1 & 0 & 1 & 1 & 3 & 1 & 1 & 1 & 0 \\ 1 & 1 & 1 & 1 & 1 & 1 & 1 & 1 & 1 & 4 & 1 & 1 & 1 \\ 1 & 1 & 1 & 0 & 0 & 1 & 1 & 1 & 1 & 1 & 3 & 1 & 0 \\ 1 & 1 & 1 & 1 & 1 & 1 & 1 & 1 & 1 & 1 & 1 & 4 & 1 \\ 1 & 0 & 1 & 0 & 0 & 0 & 1 & 1 & 0 & 1 & 0 & 1 & 2 \end{bmatrix}$$

Then,

$$A_{\mathcal{PAH}_1}A^{T}_{\mathcal{PAH}_1} - nI - J_{\text{Frob}} = 5.8310 \text{ and } A_{\mathcal{PAH}_2}A^{T}_{\mathcal{PAH}_2} - nI - J_{\text{Frob}} = 8.7178,$$

where $\|\bullet\|_{\text{Frob}}$ is the Frobenius norm.

Let us note that $A_{\mathcal{PAH}_1}A^{T}_{\mathcal{PAH}_1} - nI - J_{\text{Frob}} < A_{\mathcal{PAH}_2}A^{T}_{\mathcal{PAH}_2} - nI - J_{\text{Frob}}$ because $A_{\mathcal{PAH}_1}$ is nearer than $A_{\mathcal{PAH}_2}$ to represent a projective plane.

An interesting problem is to find the matrix satisfying $\min_{a_{ij}\in\{0,1\}} AA^{T} - 10I - J$, where A is a square matrix of order 111 of an MPAH plane. So, A is the most approximate MPAH plane to the projective plane of order 10.

The next proposition explains the relationship between DTG and incidence matrix of the finite MPAH plane.

Proposition 5: Let A be the incidence matrix of an MPAH plane ($\mathcal{MPAH}_p$ or $\mathcal{MPAH}_a$) of order n. If m_1 is the number of rows in AA^T having n zeros and m_2 is the number of rows in AA^T having more than n zeros, then the triad $a = \frac{m_1}{n^2+n+1}, h = \frac{m_2}{n^2+n+1}$, and $p = 1-a-h$ are the components of the DTG of the plane.

Proof. If two lines in the plane are parallel, the inner product between their rows in the incidence matrix is 0. In $\mathcal{MPAH}_p$ or $\mathcal{MPAH}_a$ there are sets of n lines in $\mathcal{L}_A$ satisfying parallelism among them. On the other hand, the line in the hyperbolic part is parallel to all these sets of n lines. Finally, the lines in the projective part have no parallels and their rows do not contain 0. Considering n^2+n+1 is the total number of lines in the projective plane then Proposition 5 is satisfied. □

Example 6: The planes in Figure 4 and Figure 5 satisfy $n^2+n+1=13$. $A_{\mathcal{PAH}_1}A^{T}_{\mathcal{PAH}_1}$ has 6 rows without zeros, $m_1=6$ rows with n = 3 zeros, and $m_2=1$ row with 6> n zeros, therefore, the DTG of the plane in Figure 4 is formed by $p = \frac{6}{13}, a = \frac{6}{13}$, and $h = \frac{1}{13}$.

Additionally, the plane in Figure 5 has 3 rows without zeros, 4 rows with 3 zeros, and 6 rows with 6 zeros, thus, $p = \frac{3}{13}, a = \frac{4}{13}$, and $h = \frac{6}{13}$.

Proposition 6 is a more practical version of Proposition 5.

Proposition 6: Let A be the incidence matrix of an MPAH plane ($\mathcal{MPAH}_p$ or $\mathcal{MPAH}_a$) of order n. Let us count the elements of the main diagonal of AA^T dividing them into three numbers, m_0 is the number of elements equal to n+1, m_1

is the number of elements equal to n, and m_2 is the number of elements less than or equal to n-1. Then $p = \frac{m_0}{n^2+n+1}$, $a = \frac{m_1}{n^2+n+1}$, and $h = \frac{m_2}{n^2+n+1}$.

Proof. The inner product of a row in A with itself corresponding to a line is the number of points contained in this line. This value is in the main diagonal of AA^T, so according to Corollary 1, these values characterize the type of axiomatic satisfied by the line, thus the number of lines characterizes the values a, h, and p.

In Example 5, the main diagonal of $A_{\mathcal{PAH}_1} A_{\mathcal{PAH}_1}^T$ has six 4, six 3, and only one 2.

DISCUSSION

The theory of finite geometries has fascinated many mathematicians around the entire world. Countless scientific papers and books have been dedicated to the study of this topic for decades. However, there are important unreached points of study within this theory. One of these points is the search and demonstration of conditions on the existence of projective planes of certain orders. One of the milestones was the computational demonstration of the non-existence of projective planes of order 10, which in itself is controversial, as it is controversial to use mathematical demonstrations with the use of programming codes. An argument against it is the possible presence of undetected bugs in the code.

Within the applications of this theory is the theory of codes and cryptography, in addition to the purely mathematical interest. The link between finite geometries and orthogonal Latin squares is also known. However, for applications, it seems that the projective planes are the only interesting for these purposes.

On the other hand, there is an evident imbalance between the amount of literature dedicated to the study of the different types of finite geometries. The vast majority of them correspond to the investigation of projective planes, projective spaces, and their applications. To a lesser degree, there is literature dedicated to finite affine planes, perhaps due to the close relationship between both types of spaces, as we have seen above, a finite projective space can be converted into an affine one only by eliminating a hyperplane and all its subplanes. For the finite Bolyai-Lobachevsky planes, the literature dedicated to them is scarce with few results.

Possibly the limitations that appear in the subjects of finite geometries are due to mathematicians having only studied pure projective, affine, or hyperbolic geometries. E.g., the problem of finding finite Mixed Projective-Affine-Hyperbolic planes of order 10 can have a solution, and it is an alternative variant for having an approximation to a projective plane of order 10.

In any case, this chapter continues a tradition from the original idea of Professor Florentin Smarandache to create mixed geometries or Smarandache Geometries, because they correspond more adequately to models of the real world (Iseri, 2002). It is very difficult to find in the universal physical space purely Euclidean, elliptical, or hyperbolic spaces, but a mixture of them.

This chapter intends only to give an introduction to this topic; still, there are issues to be developed, like the possible application of finite MPAH planes, the existence of MPAH planes of a certain order, or the generalization of these results to spaces of dimension greater than 2, but this is a debt left for future work.

A practical contribution of the author for the readers is that the first appendix provides python language codes that were used to study the examples that appear in the chapter. Another appendix is dedicated to offering the study carried out on examples of planes such as those of order n = 5 and n = 7, using the functions programmed in python code, therefore, the reader will have the opportunity to run these examples as a practice of what is proposed in this document.

CONCLUSION

This chapter introduces the finite Mixed Projective-Affine-Hyperbolic planes as an extension of the theory of finite geometries and NeutroGeometries simultaneously. Finite geometries are geometric structures with a finite number of points, while NeutroGeometries are geometric structures where at least one axiom or definition is fulfilled by a part of the elements of the space and not fulfilled by other parts. In the chapter, some properties of finite MPAH planes are demonstrated, the algorithm to obtain one of them either from a projective or an affine plane is provided, and an index is defined to quantify the satisfaction by the MPAH of the axiomatic of each of the three types of planes. This index is called the Degree of Type of Geometry. Finally, we studied the relationship between the properties of the incidence matrix of an MPAH plane and its geometric properties.

REFERENCES

Al-Seraji, N. A. M., & Jafar, Z. S. (2020). Applications of the Projective Plane in Coding Theory. *Journal of Southwest Jiaotong University*, *55*(1), 1–8. doi:10.35741/issn.0258-2724.55.1.2

Beutelspacher, A. (1990). Applications of finite geometry to cryptography. In G. Longo, M. Marchi, & A. Sgarro (Eds.), *Geometries, Codes and Cryptography* (pp. 161–186). Springer Verlag. doi:10.1007/978-3-7091-2838-1_6

Bruck, R. H., & Ryser, H. J. (1949). The Nonexistence of Certain Finite Projective Planes. *Canadian Journal of Mathematics, 1*(1), 88–93. doi:10.4153/CJM-1949-009-2

Cameron, P. J. (1994). *Combinatorics: Topics, Techniques, Algorithms*. Cambridge University Press.

Cameron, P. J. (2000). *Projective and Polar Spaces*. Queen Mary and Westfield College.

Celik, B. (2001). On Some Hyperbolic Planes from Finite Projective Planes. *International Journal of Mathematics and Mathematical Sciences, 25*(12), 757–762. doi:10.1155/S0161171201006184

Crowe, D. W. (1965). The Construction of Finite Regular Hyperbolic Planes from Inversive Planes of Even Order. *Colloquium Mathematicum, 13*(2), 247–250. doi:10.4064/cm-13-2-247-250

Crowe, D. W. (1966). Projective and Inversive Models for Finite Hyperbolic Planes. *The Michigan Mathematical Journal, 13*(2), 251–255. doi:10.1307/mmj/1028999550

Dembowski, P. (1997). Finite Geometries. Berlin: Springer.

Dissett, L. A. (2000). *Combinatorial and Computational Aspects of Finite Geometries* [Unpublished Ph.D. Thesis]. University of Toronto.

Hirschfeld, J. W. P. (1979). *Projective Geometries over Finite Fields*. Clarendon Press.

Iseri, H. (2002). Partially Paradoxist Smarandache Geometries. *Smarandache Notions Journal, 13*(1-2-3), 5-12.

Kärteszi, F. (1976). *Introduction to finite geometries*. American Elsevier Publishing Company, Inc.

Kiss, G., & Szonyi, T. (2020). *Finite Geometries*. CRC Press Taylor & Francis Group.

Klein, A., & Storme, L. (2011). Applications of finite geometry in coding theory and cryptography. In *Information Security, Coding Theory and Related Combinatorics* (pp. 38–58). IOS Press.

Korchmáros, G., & Sonnino, A. (2012). Finite Bolyai-Lobachevskii Planes. *Acta Mathematica Hungarica, 134*(4), 405–415. doi:10.100710474-011-0181-4

Lam, C. W. H. (1991). The Search for a Finite Projective Plane of Order 10. *The American Mathematical Monthly, 98*(4), 305–318. doi:10.1080/00029890.1991.12000759

Lam, C. W. H., Thiel, L., & Swiercz, S. (1989). The Non-Existence of Finite Projective Planes Of Order 10. *Canadian Journal of Mathematics, XLI*(6), 1117–1123. doi:10.4153/CJM-1989-049-4

Mao, L. (2006). *Smarandache Geometries & Map Theory with Applications (I)*. Chinese Branch Xiquan House.

Olgun, S., & Gunaltili, I. (2007). On Finite Homogeneous Bolyai-Lobachevsky (B-L) n– Spaces, n ≥ 2. *International Mathematical Forum, 2*(2), 69 - 73.

Polster, B. (1998). *A Geometrical Picture Book*. Springer. doi:10.1007/978-1-4419-8526-2

Raber, N. C. (1972). On Ostrom's Finite Hyperbolic Planes [Unpublished Ph.D.]. The Ohio State University.

Sandler, R. (1963). Finite Homogeneous Bolyai-Lobachevsky Planes. *The American Mathematical Monthly, 70*(8), 853–855. doi:10.2307/2312671

Smarandache, F. (1969). *Mixed non euclidean geometries*. Infinite Study.

Smarandache, F. (2009). *Degree of Negation of an Axiom*. Infinite Study.

Smarandache, F. (2019). Introduction to NeutroAlgebraic Structures and AntiAlgebraic Structures. In *Advances of Standard and Nonstandard Neutrosophic Theories* (pp. 240–265). Pons Publishing House.

Smarandache, F. (2020). Introduction to NeutroAlgebraic Structures and AntiAlgebraic Structures (revisited). *Neutrosophic Sets and Systems, 31*, 1–16.

Smarandache, F. (2021). NeutroGeometry & AntiGeometry are alternatives and generalizations of the Non-Euclidean Geometries. *Neutrosophic Sets and Systems, 46*(2021), 456-476.

Zangalho-Raposo, A. P. (2014). *Geometrias Finitas* [Unpublished Master Thesis]. University of Évora.

Zumbragel, J. (2016). *Designs and Codes in Affine Geometry*. arXiv e-prints arXiv -1605.03789.

ADDITIONAL READING

Akça, Z., & Altıntas, A. (2021). A Note on Fano Configurations in the Projective Space PG(5,2). *Konuralp Journal of Mathematics*, *9*, 190–192.

Al-Seraji, N. A. M., & Jafar, Z. S. (2020). Applications of the Projective Plane in Coding Theory. *Journal of Southwest Jiaotong University*, *55*(1), 1–8. doi:10.35741/issn.0258-2724.55.1.2

Ali, A. T. (2010). Special Smarandache Curves in the Euclidean Space. *International Journal of Mathematical Combinatorics*, *2*, 30–36.

Korchmáros, G., & Sonnino, A. (2014). Old and Recent Results on Finite Bolyai-Lobachevsky Planes. *Mathematica*, *56*, 59–73.

Singh, P. K. (2021). Anti-Geometry and NeutroGeometry Characterization of Non-Euclidean Data. *Journal of Neutrosophic and Fuzzy Systems*, *1*, 24–33. doi:10.54216/JNFS.010102

KEY TERMS AND DEFINITIONS

Affine Plane: Geometric structure of dimension 2 such that given a line and an external point, there is a unique line parallel to the aforementioned one, which contains the point.

Finite Field: Algebraic structure consisting of a finite set of elements and two internal operations + and ×, such that + and × are commutative groups except for 0 in ×; 0 is the neutral element of +; 1 is the neutral element of ×; + and × are distributive each other; 0 has no inverse for ×.

Finite Geometry: Geometric system with only a finite number of points.

Finite Vector Space: Vector space over a finite field.

Hyperbolic Plane: Also known as the *Bolyai-Lobachevsky plane*, it is the geometric structure of dimension 2 such that given a line and an external point; there are at least two parallel lines to the aforementioned one, which contain the point.

Incidence Matrix: A matrix whose every row represents every line of the finite plane and every column represents every point, or vice versa, such that the element of the matrix is 1 if there is an incidence relation between the corresponding pair of line and point, and it is 0 otherwise.

Incidence Structure: Symmetric relation among points and lines, where if a line l contains a point P it is said that l is incident with P and P is incident with l.

Mixed Geometry: Also called *Smarandache Geometry*, it is geometry where some elements satisfy the axiomatic of some type of geometry, and other elements satisfy other types of geometries, all in the same space. E.g., there are geometric models which are partially Euclidean, partially hyperbolic, and partially elliptic.

NeutroGeometry: Geometric structure whose definition or at least one axiom is simultaneously satisfied by some elements and not satisfied by others.

Projective Plane: Geometric structure of dimension 2 where there is non-parallelism between the lines.

APPENDIX 1

This appendix contains the python code used in this chapter.

The functions used were the following and were saved in the file named MPAHgeometry.py:

```
def rest(P,line):#Set of points not contained in the line
    return [p for p in P if p not in line]
def are_parallels(line1,line2):#Boolean function of parallelism
between two lines
    meet = [p for p in line1 if p in line2]
    return meet == []
def parallels_point(P,plane,line,extpoint):#List of parallels
to the line throughout an external point
    ans = [l for l in plane if are_parallels(l,line) and
extpoint in l]
    return ans
def parallels(P,plane,line):#List of parallels to the line for
every external point
    for point in rest(P,line):
        ans = parallels_point(P,plane,line,point)
        print(ans)
def search_parallels(plane):#Seeking for pair of parallels in
the plane
    ans = []
    for i in range(len(plane)):
        for j in range(i,len(plane)):
            if are_parallels(plane[i],plane[j]):
                ans.append([plane[i],plane[j]])
    return(ans)
def convert(plane,points):#Convert plane in a new plane with
lines no incident in the set of points
    newplane = []
    for line in plane:
        line1 = [x for x in line if x not in points]
        if(line1 != []):
            newplane.append(line1)
    return(newplane)
def plane2im(plane,points):#Output the incidence matrix of the
plane
```

```
    cardplane = len(plane)
    cardpoints = len(points)
    M = []
    for i in range(cardplane):
        A = []
        for j in range(cardpoints):
            if points[j] in plane[i]:
                A = A+[1]
            else:
               A = A+[0]
        print(A)
def DTG(plane,points):#Output the DTG coefficient of the plane
obtained from a projective plane
         l = len(points)
         order = (sqrt(4*l-3)-1)/2
         proj = [line for line in plane if len(line) ==
order+1]
         aff = [line for line in plane if len(line) == order]
         hyp = [line for line in plane if len(line) < order]
         coeff = [len(proj)/len(points),len(aff)/
len(points),len(hyp)/len(points)]
         print(coeff)
```

The code of the main program containing the examples is the following:

```
from MPAHgeometry import *
P = 'ABCDEGHIJKL' #String containing the set of points in the
MPAH plane
plane = ['JK','DEG','BHL','DCL','ACI','AHE','GBI','DBAJ','IELJ'
,'AGLK','JHCG','DHIK','BCEK'] #List containing the set of lines
in the MPAH plane
hP = 'ABCDEFGHIJKLM' #String containing the set of points in
the hyperbolic plane
hplane = ['AGM','BHM','CIM','DJM','EKM','FLM','BCJ','ACK','ABL'
,'AEF','CDF','BDE','AIJ','BIK','CGL','ADH','BFG','CEH','DGK','
EJL','FHI','DIL','EGI','FJK','GHJ','HKL'] #List containing the
set of lines in the hyperbolic plane
PP3 = 'ABCDEFGHIJKLM' #String containing the set of points in
the projective plane n = 3
pplane3 = ['FJMK','DMEG','BHML','FDCL','ACIM','AFHE','GBFI','D
```

```
BAJ','IELJ','AGLK','JHCG','DHIK','BCEK'] #List containing the
set of lines in the projective plane n = 3
PP5 = range(31) #List containing the set of points in the
projective plane n = 5
pplane5 = [[0,1,3,8,12,18],[1,2,4,9,13,19],[2,3,5,10,14,20],[3,
4,6,11,15,21],[4,5,7,12,16,22],[5,6,8,13,17,23],[6,7,9,14,18,24
],[7,8,10,15,19,25],[8,9,11,16,20,26],[9,10,12,17,21,27],[10,11
,13,18,22,28],[11,12,14,19,23,29],[12,13,15,20,24,30],[0,13,14
,16,21,25],[1,14,15,17,22,26],[2,15,16,18,23,27],[3,16,17,19,2
4,28],[4,17,18,20,25,29],[5,18,19,21,26,30],[0,6,19,20,22,27],
[1,7,20,21,23,28],[2,8,21,22,24,29],[3,9,22,23,25,30],[0,4,10,
23,24,26],[1,5,11,24,25,27],[2,6,12,25,26,28],[3,7,13,26,27,29
],[4,8,14,27,28,30],[0,5,9,15,28,29],[1,6,10,16,29,30],[0,2,7,
11,17,30]] #List containing the set of lines in the projective
plane n = 5
PP7 = range(57) #List containing the set of points in the
projective plane n = 7
pplane7 = [[0,1,3,13,32,36,43,52],[1,2,4,14,33,37,44,53],[2,3,5
,15,34,38,45,54],[3,4,6,16,35,39,46,55],[4,5,7,17,36,40,47,56],
[0,5,6,8,18,37,41,48],[1,6,7,9,19,38,42,49],[2,7,8,10,20,39,43,
50],[3,8,9,11,21,40,44,51],[4,9,10,12,22,41,45,52],[5,10,11,13,
23,42,46,53],[6,11,12,14,24,43,47,54],[7,12,13,15,25,44,48,55],
[8,13,14,16,26,45,49,56],[0,9,14,15,17,27,46,50],[1,10,15,16,18
,28,47,51],[2,11,16,17,19,29,48,52],[3,12,17,18,20,30,49,53],[4
,13,18,19,21,31,50,54],[5,14,19,20,22,32,51,55],[6,15,20,21,23,
33,52,56],[0,7,16,21,22,24,34,53],[1,8,17,22,23,25,35,54],[2,9,
18,23,24,26,36,55],[3,10,19,24,25,27,37,56],[0,4,11,20,25,26,2
8,38],[1,5,12,21,26,27,29,39],[2,6,13,22,27,28,30,40],[3,7,14,
23,28,29,31,41],[4,8,15,24,29,30,32,42],[5,9,16,25,30,31,33,43
],[6,10,17,26,31,32,34,44],[7,11,18,27,32,33,35,45],[8,12,19,28
,33,34,36,46],[9,13,20,29,34,35,37,47],[10,14,21,30,35,36,38,48
],[11,15,22,31,36,37,39,49],[12,16,23,32,37,38,40,50],[13,17,24
,33,38,39,41,51],[14,18,25,34,39,40,42,52],[15,19,26,35,40,41,4
3,53],[16,20,27,36,41,42,44,54],[17,21,28,37,42,43,45,55],[18,2
2,29,38,43,44,46,56],[0,19,23,30,39,44,45,47],[1,20,24,31,40,45
,46,48],[2,21,25,32,41,46,47,49],[3,22,26,33,42,47,48,50],[4,2
3,27,34,43,48,49,51],[5,24,28,35,44,49,50,52],[6,25,29,36,45,5
0,51,53],[7,26,30,37,46,51,52,54],[8,27,31,38,47,52,53,55],[9,
28,32,39,48,53,54,56],[0,10,29,33,40,49,54,55],[1,11,30,34,41,
```

```
50,55,56],[0,2,12,31,35,42,51,56]] #List containing the set of
lines in the projective plane n = 7
```

APPENDIX 2

In this appendix, we summarize some interesting results concerning MPAH planes of orders 5 and 7. Projective planes of orders 5 and 7 can be found in (Al-Seraji & Jafar, 2020).

Firstly, from pplane5 in Appendix 1, using the function *convert(plane,points)*, where plane = pplane5 and points is the list of numbers from 0 to 30 except [0,1,3,4,5]. 0, 1, and 3 are three points of the first line in pplane5, whereas 4, and 5 are two points in another line in the same plane, then the incidence matrix calculated with the function *plane2im(plane,points)* is the following:

```
[0, 0, 0, 0, 0, 0, 0, 0, 1, 0, 0, 0, 1, 0, 0, 0, 0, 0, 1, 0, 0,
0, 0, 0, 0, 0, 0, 0, 0, 0, 0]
[0, 0, 1, 0, 0, 0, 0, 0, 0, 1, 0, 0, 0, 1, 0, 0, 0, 0, 0, 1, 0,
0, 0, 0, 0, 0, 0, 0, 0, 0, 0]
[0, 0, 1, 0, 0, 0, 0, 0, 0, 0, 1, 0, 0, 0, 1, 0, 0, 0, 0, 0, 1,
0, 0, 0, 0, 0, 0, 0, 0, 0, 0]
[0, 0, 0, 0, 0, 0, 1, 0, 0, 0, 0, 1, 0, 0, 0, 1, 0, 0, 0, 0, 0,
1, 0, 0, 0, 0, 0, 0, 0, 0, 0]
[0, 0, 0, 0, 0, 0, 0, 1, 0, 0, 0, 0, 1, 0, 0, 0, 1, 0, 0, 0, 0,
0, 1, 0, 0, 0, 0, 0, 0, 0, 0]
[0, 0, 0, 0, 0, 0, 1, 0, 1, 0, 0, 0, 0, 1, 0, 0, 0, 1, 0, 0, 0,
0, 0, 1, 0, 0, 0, 0, 0, 0, 0]
[0, 0, 0, 0, 0, 0, 1, 1, 0, 1, 0, 0, 0, 0, 1, 0, 0, 0, 1, 0, 0,
0, 0, 0, 1, 0, 0, 0, 0, 0, 0]
[0, 0, 0, 0, 0, 0, 0, 1, 1, 0, 1, 0, 0, 0, 0, 1, 0, 0, 0, 1, 0,
0, 0, 0, 0, 1, 0, 0, 0, 0, 0]
[0, 0, 0, 0, 0, 0, 0, 0, 1, 1, 0, 1, 0, 0, 0, 0, 1, 0, 0, 0, 1,
0, 0, 0, 0, 0, 1, 0, 0, 0, 0]
[0, 0, 0, 0, 0, 0, 0, 0, 0, 1, 1, 0, 1, 0, 0, 0, 0, 1, 0, 0, 0,
1, 0, 0, 0, 0, 0, 1, 0, 0, 0]
[0, 0, 0, 0, 0, 0, 0, 0, 0, 0, 1, 1, 0, 1, 0, 0, 0, 0, 1, 0, 0,
0, 1, 0, 0, 0, 0, 0, 1, 0, 0]
[0, 0, 0, 0, 0, 0, 0, 0, 0, 0, 0, 1, 1, 0, 1, 0, 0, 0, 0, 1, 0,
0, 0, 1, 0, 0, 0, 0, 0, 1, 0]
```

```
[0, 0, 0, 0, 0, 0, 0, 0, 0, 0, 0, 0, 1, 1, 0, 1, 0, 0, 0, 0, 1,
0, 0, 0, 1, 0, 0, 0, 0, 0, 1]
[0, 0, 0, 0, 0, 0, 0, 0, 0, 0, 0, 0, 0, 1, 1, 0, 1, 0, 0, 0, 0,
1, 0, 0, 0, 1, 0, 0, 0, 0, 0]
[0, 0, 0, 0, 0, 0, 0, 0, 0, 0, 0, 0, 0, 0, 1, 1, 0, 1, 0, 0, 0,
0, 1, 0, 0, 0, 1, 0, 0, 0, 0]
[0, 0, 1, 0, 0, 0, 0, 0, 0, 0, 0, 0, 0, 0, 0, 1, 1, 0, 1, 0, 0,
0, 0, 1, 0, 0, 0, 1, 0, 0, 0]
[0, 0, 0, 0, 0, 0, 0, 0, 0, 0, 0, 0, 0, 0, 0, 0, 1, 1, 0, 1, 0,
0, 0, 0, 1, 0, 0, 0, 1, 0, 0]
[0, 0, 0, 0, 0, 0, 0, 0, 0, 0, 0, 0, 0, 0, 0, 0, 0, 1, 1, 0, 1,
0, 0, 0, 0, 1, 0, 0, 0, 1, 0]
[0, 0, 0, 0, 0, 0, 0, 0, 0, 0, 0, 0, 0, 0, 0, 0, 0, 0, 1, 1, 0,
1, 0, 0, 0, 0, 1, 0, 0, 0, 1]
[0, 0, 0, 0, 0, 0, 1, 0, 0, 0, 0, 0, 0, 0, 0, 0, 0, 0, 0, 1, 1,
0, 1, 0, 0, 0, 0, 1, 0, 0, 0]
[0, 0, 0, 0, 0, 0, 0, 1, 0, 0, 0, 0, 0, 0, 0, 0, 0, 0, 0, 0, 1,
1, 0, 1, 0, 0, 0, 0, 1, 0, 0]
[0, 0, 1, 0, 0, 0, 0, 0, 1, 0, 0, 0, 0, 0, 0, 0, 0, 0, 0, 0, 0,
1, 1, 0, 1, 0, 0, 0, 0, 1, 0]
[0, 0, 0, 0, 0, 0, 0, 0, 0, 1, 0, 0, 0, 0, 0, 0, 0, 0, 0, 0, 0,
0, 1, 1, 0, 1, 0, 0, 0, 0, 1]
[0, 0, 0, 0, 0, 0, 0, 0, 0, 0, 1, 0, 0, 0, 0, 0, 0, 0, 0, 0, 0,
0, 0, 1, 1, 0, 1, 0, 0, 0, 0]
[0, 0, 0, 0, 0, 0, 0, 0, 0, 0, 0, 1, 0, 0, 0, 0, 0, 0, 0, 0, 0,
0, 0, 0, 1, 1, 0, 1, 0, 0, 0]
[0, 0, 1, 0, 0, 0, 1, 0, 0, 0, 0, 0, 1, 0, 0, 0, 0, 0, 0, 0, 0,
0, 0, 0, 0, 1, 1, 0, 1, 0, 0]
[0, 0, 0, 0, 0, 0, 0, 1, 0, 0, 0, 0, 0, 1, 0, 0, 0, 0, 0, 0, 0,
0, 0, 0, 0, 0, 1, 1, 0, 1, 0]
[0, 0, 0, 0, 0, 0, 0, 0, 1, 0, 0, 0, 0, 0, 1, 0, 0, 0, 0, 0, 0,
0, 0, 0, 0, 0, 0, 1, 1, 0, 1]
[0, 0, 0, 0, 0, 0, 0, 0, 0, 1, 0, 0, 0, 0, 0, 1, 0, 0, 0, 0, 0,
0, 0, 0, 0, 0, 0, 0, 1, 1, 0]
[0, 0, 0, 0, 0, 0, 1, 0, 0, 0, 1, 0, 0, 0, 0, 0, 1, 0, 0, 0, 0,
0, 0, 0, 0, 0, 0, 0, 0, 1, 1]
[0, 0, 1, 0, 0, 0, 0, 1, 0, 0, 0, 1, 0, 0, 0, 0, 0, 1, 0, 0, 0,
0, 0, 0, 0, 0, 0, 0, 0, 0, 1]
```

The new plane contains lines without parallels, others with only one parallel passing by an external point, others having two parallels per external point, and finally, other lines have three parallels passing by every external point.

The list of parallels can be obtained by running the python function *search_parallels(plane)*.

The number of lines corresponding to the axiomatic of projective planes is 10, thirteen lines correspond to the affine plane axioms, seven lines have two parallels by exterior point and one line has three parallels by external point. So, the DTG is $\left(\frac{10}{31}, \frac{13}{31}, \frac{8}{31}\right)$.

Now, let us obtain an MPAH plane of order 7 from the projective plane of this order, where some lines have either 4, 3, 2, or 1 parallels. We removed from the projective plane the points numbered 0, 1, 3, 4, 5, 6, 7, 11, and 13. The incidence matrix is the following:

```
[0,0,0,0,0,0,0,0,0,0,0,0,0,0,0,0,0,0,0,0,0,0,0,0,0,0,0,0,0,0,0,
0,1,0,0,0,1,0,0,0,0,0,0,1,0,0,0,0,0,0,0,0,1,0,0,0,0]
[0,0,1,0,0,0,0,0,0,0,0,0,0,0,1,0,0,0,0,0,0,0,0,0,0,0,0,0,0,0,0,
0,0,1,0,0,0,1,0,0,0,0,0,0,1,0,0,0,0,0,0,0,0,1,0,0,0]
[0,0,1,0,0,0,0,0,0,0,0,0,0,0,0,1,0,0,0,0,0,0,0,0,0,0,0,0,0,0,0,
0,0,0,1,0,0,0,1,0,0,0,0,0,0,1,0,0,0,0,0,0,0,0,1,0,0]
[0,0,0,0,0,0,0,0,0,0,0,0,0,0,0,0,1,0,0,0,0,0,0,0,0,0,0,0,0,0,0,
0,0,0,0,1,0,0,0,1,0,0,0,0,0,0,1,0,0,0,0,0,0,0,0,1,0]
[0,0,0,0,0,0,0,0,0,0,0,0,0,0,0,0,0,1,0,0,0,0,0,0,0,0,0,0,0,0,0,
0,0,0,0,0,1,0,0,0,1,0,0,0,0,0,0,1,0,0,0,0,0,0,0,0,1]
[0,0,0,0,0,0,0,0,1,0,0,0,0,0,0,0,0,0,1,0,0,0,0,0,0,0,0,0,0,0,0,
0,0,0,0,0,0,1,0,0,0,1,0,0,0,0,0,0,1,0,0,0,0,0,0,0,0]
[0,0,0,0,0,0,0,0,0,1,0,0,0,0,0,0,0,0,0,1,0,0,0,0,0,0,0,0,0,0,0,
0,0,0,0,0,0,0,1,0,0,0,1,0,0,0,0,0,0,1,0,0,0,0,0,0,0]
[0,0,1,0,0,0,0,0,1,0,1,0,0,0,0,0,0,0,0,0,1,0,0,0,0,0,0,0,0,0,0,
0,0,0,0,0,0,0,0,1,0,0,0,1,0,0,0,0,0,0,1,0,0,0,0,0,0]
[0,0,0,0,0,0,0,0,1,1,0,0,0,0,0,0,0,0,0,0,0,1,0,0,0,0,0,0,0,0,0,
0,0,0,0,0,0,0,0,0,1,0,0,0,1,0,0,0,0,0,0,1,0,0,0,0,0]
[0,0,0,0,0,0,0,0,0,1,1,0,1,0,0,0,0,0,0,0,0,0,1,0,0,0,0,0,0,0,0,
0,0,0,0,0,0,0,0,0,0,1,0,0,0,1,0,0,0,0,0,0,1,0,0,0,0]
[0,0,0,0,0,0,0,0,0,0,1,0,0,0,0,0,0,0,0,0,0,0,0,1,0,0,0,0,0,0,0,
0,0,0,0,0,0,0,0,0,0,0,1,0,0,0,1,0,0,0,0,0,0,1,0,0,0]
[0,0,0,0,0,0,0,0,0,0,0,0,1,0,1,0,0,0,0,0,0,0,0,0,1,0,0,0,0,0,0,
0,0,0,0,0,0,0,0,0,0,0,0,1,0,0,0,1,0,0,0,0,0,0,1,0,0]
```

```
[0,0,0,0,0,0,0,0,0,0,0,0,1,0,0,1,0,0,0,0,0,0,0,0,0,1,0,0,0,0,0,
0,0,0,0,0,0,0,0,0,0,0,0,0,1,0,0,0,1,0,0,0,0,0,0,1,0]
[0,0,0,0,0,0,0,0,1,0,0,0,0,0,1,0,1,0,0,0,0,0,0,0,0,0,1,0,0,0,0,
0,0,0,0,0,0,0,0,0,0,0,0,0,0,1,0,0,0,1,0,0,0,0,0,0,1]
[0,0,0,0,0,0,0,0,0,1,0,0,0,0,1,1,0,1,0,0,0,0,0,0,0,0,0,1,0,0,0,
0,0,0,0,0,0,0,0,0,0,0,0,0,0,0,1,0,0,0,1,0,0,0,0,0,0]
[0,0,0,0,0,0,0,0,0,0,1,0,0,0,0,1,1,0,1,0,0,0,0,0,0,0,0,0,1,0,0,
0,0,0,0,0,0,0,0,0,0,0,0,0,0,0,0,1,0,0,0,1,0,0,0,0,0]
[0,0,1,0,0,0,0,0,0,0,0,0,0,0,0,0,1,1,0,1,0,0,0,0,0,0,0,0,0,1,0,
0,0,0,0,0,0,0,0,0,0,0,0,0,0,0,0,0,1,0,0,0,1,0,0,0,0]
[0,0,0,0,0,0,0,0,0,0,0,0,1,0,0,0,0,1,1,0,1,0,0,0,0,0,0,0,0,0,1,
0,0,0,0,0,0,0,0,0,0,0,0,0,0,0,0,0,0,1,0,0,0,1,0,0,0]
[0,0,0,0,0,0,0,0,0,0,0,0,0,0,0,0,0,0,1,1,0,1,0,0,0,0,0,0,0,0,0,
1,0,0,0,0,0,0,0,0,0,0,0,0,0,0,0,0,0,0,1,0,0,0,1,0,0]
[0,0,0,0,0,0,0,0,0,0,0,0,0,0,1,0,0,0,0,1,1,0,1,0,0,0,0,0,0,0,0,
0,1,0,0,0,0,0,0,0,0,0,0,0,0,0,0,0,0,0,0,1,0,0,0,1,0]
[0,0,0,0,0,0,0,0,0,0,0,0,0,0,0,1,0,0,0,0,1,1,0,1,0,0,0,0,0,0,0,
0,0,1,0,0,0,0,0,0,0,0,0,0,0,0,0,0,0,0,0,0,1,0,0,0,1]
[0,0,0,0,0,0,0,0,0,0,0,0,0,0,0,0,1,0,0,0,0,1,1,0,1,0,0,0,0,0,0,
0,0,0,1,0,0,0,0,0,0,0,0,0,0,0,0,0,0,0,0,0,0,1,0,0,0]
[0,0,0,0,0,0,0,0,1,0,0,0,0,0,0,0,0,1,0,0,0,0,1,1,0,1,0,0,0,0,0,
0,0,0,0,1,0,0,0,0,0,0,0,0,0,0,0,0,0,0,0,0,0,0,1,0,0]
[0,0,1,0,0,0,0,0,0,1,0,0,0,0,0,0,0,0,1,0,0,0,0,1,1,0,1,0,0,0,0,
0,0,0,0,0,1,0,0,0,0,0,0,0,0,0,0,0,0,0,0,0,0,0,0,1,0]
[0,0,0,0,0,0,0,0,0,0,1,0,0,0,0,0,0,0,0,1,0,0,0,0,1,1,0,1,0,0,0,
0,0,0,0,0,0,1,0,0,0,0,0,0,0,0,0,0,0,0,0,0,0,0,0,0,1]
[0,0,0,0,0,0,0,0,0,0,0,0,0,0,0,0,0,0,0,0,1,0,0,0,0,1,1,0,1,0,0,
0,0,0,0,0,0,0,1,0,0,0,0,0,0,0,0,0,0,0,0,0,0,0,0,0,0]
[0,0,0,0,0,0,0,0,0,0,0,0,1,0,0,0,0,0,0,0,0,1,0,0,0,0,1,1,0,1,0,
0,0,0,0,0,0,0,0,1,0,0,0,0,0,0,0,0,0,0,0,0,0,0,0,0,0]
[0,0,1,0,0,0,0,0,0,0,0,0,0,0,0,0,0,0,0,0,0,0,1,0,0,0,0,1,1,0,1,
0,0,0,0,0,0,0,0,0,1,0,0,0,0,0,0,0,0,0,0,0,0,0,0,0,0]
[0,0,0,0,0,0,0,0,0,0,0,0,0,0,1,0,0,0,0,0,0,0,0,1,0,0,0,0,1,1,0,
1,0,0,0,0,0,0,0,0,0,1,0,0,0,0,0,0,0,0,0,0,0,0,0,0,0]
[0,0,0,0,0,0,0,0,1,0,0,0,0,0,0,1,0,0,0,0,0,0,0,0,1,0,0,0,0,1,1,
0,1,0,0,0,0,0,0,0,0,0,1,0,0,0,0,0,0,0,0,0,0,0,0,0,0]
[0,0,0,0,0,0,0,0,0,1,0,0,0,0,0,0,1,0,0,0,0,0,0,0,0,1,0,0,0,0,1,
1,0,1,0,0,0,0,0,0,0,0,0,1,0,0,0,0,0,0,0,0,0,0,0,0,0]
[0,0,0,0,0,0,0,0,0,0,1,0,0,0,0,0,0,1,0,0,0,0,0,0,0,0,1,0,0,0,0,
1,1,0,1,0,0,0,0,0,0,0,0,0,1,0,0,0,0,0,0,0,0,0,0,0,0]
```

[0,0,0,0,0,0,0,0,0,0,0,0,0,0,0,0,0,0,1,0,0,0,0,0,0,0,0,1,0,0,0,
0,1,1,0,1,0,0,0,0,0,0,0,0,0,1,0,0,0,0,0,0,0,0,0,0,0]
[0,0,0,0,0,0,0,0,1,0,0,0,1,0,0,0,0,0,0,1,0,0,0,0,0,0,0,0,1,0,0,
0,0,1,1,0,1,0,0,0,0,0,0,0,0,0,1,0,0,0,0,0,0,0,0,0,0]
[0,0,0,0,0,0,0,0,0,1,0,0,0,0,0,0,0,0,0,0,1,0,0,0,0,0,0,0,0,1,0,
0,0,0,1,1,0,1,0,0,0,0,0,0,0,0,0,1,0,0,0,0,0,0,0,0,0]
[0,0,0,0,0,0,0,0,0,0,1,0,0,0,1,0,0,0,0,0,0,1,0,0,0,0,0,0,0,0,1,
0,0,0,0,1,1,0,1,0,0,0,0,0,0,0,0,0,1,0,0,0,0,0,0,0,0]
[0,0,0,0,0,0,0,0,0,0,0,0,0,0,0,1,0,0,0,0,0,0,1,0,0,0,0,0,0,0,0,
1,0,0,0,0,1,1,0,1,0,0,0,0,0,0,0,0,0,1,0,0,0,0,0,0,0]
[0,0,0,0,0,0,0,0,0,0,0,0,1,0,0,0,1,0,0,0,0,0,0,1,0,0,0,0,0,0,0,
0,1,0,0,0,0,1,1,0,1,0,0,0,0,0,0,0,0,0,1,0,0,0,0,0,0]
[0,0,0,0,0,0,0,0,0,0,0,0,0,0,0,0,0,1,0,0,0,0,0,0,1,0,0,0,0,0,0,
0,0,1,0,0,0,0,1,1,0,1,0,0,0,0,0,0,0,0,0,1,0,0,0,0,0]
[0,0,0,0,0,0,0,0,0,0,0,0,0,0,1,0,0,0,1,0,0,0,0,0,0,1,0,0,0,0,0,
0,0,0,1,0,0,0,0,1,1,0,1,0,0,0,0,0,0,0,0,0,1,0,0,0,0]
[0,0,0,0,0,0,0,0,0,0,0,0,0,0,0,1,0,0,0,1,0,0,0,0,0,0,1,0,0,0,0,
0,0,0,0,1,0,0,0,0,1,1,0,1,0,0,0,0,0,0,0,0,0,1,0,0,0]
[0,0,0,0,0,0,0,0,0,0,0,0,0,0,0,0,1,0,0,0,1,0,0,0,0,0,0,1,0,0,0,
0,0,0,0,0,1,0,0,0,0,1,1,0,1,0,0,0,0,0,0,0,0,0,1,0,0]
[0,0,0,0,0,0,0,0,0,0,0,0,0,0,0,0,0,1,0,0,0,1,0,0,0,0,0,0,1,0,0,
0,0,0,0,0,0,1,0,0,0,0,1,1,0,1,0,0,0,0,0,0,0,0,0,1,0]
[0,0,0,0,0,0,0,0,0,0,0,0,0,0,0,0,0,0,1,0,0,0,1,0,0,0,0,0,0,1,0,
0,0,0,0,0,0,0,1,0,0,0,0,1,1,0,1,0,0,0,0,0,0,0,0,0,1]
[0,0,0,0,0,0,0,0,0,0,0,0,0,0,0,0,0,0,0,1,0,0,0,1,0,0,0,0,0,0,1,
0,0,0,0,0,0,0,0,1,0,0,0,0,1,1,0,1,0,0,0,0,0,0,0,0,0]
[0,0,0,0,0,0,0,0,0,0,0,0,0,0,0,0,0,0,0,0,1,0,0,0,1,0,0,0,0,0,0,
1,0,0,0,0,0,0,0,0,1,0,0,0,0,1,1,0,1,0,0,0,0,0,0,0,0]
[0,0,1,0,0,0,0,0,0,0,0,0,0,0,0,0,0,0,0,0,0,1,0,0,0,1,0,0,0,0,0,
0,1,0,0,0,0,0,0,0,0,1,0,0,0,0,1,1,0,1,0,0,0,0,0,0,0]
[0,0,0,0,0,0,0,0,0,0,0,0,0,0,0,0,0,0,0,0,0,0,1,0,0,0,1,0,0,0,0,
0,0,1,0,0,0,0,0,0,0,0,1,0,0,0,0,1,1,0,1,0,0,0,0,0,0]
[0,0,0,0,0,0,0,0,0,0,0,0,0,0,0,0,0,0,0,0,0,0,0,1,0,0,0,1,0,0,0,
0,0,0,1,0,0,0,0,0,0,0,0,1,0,0,0,0,1,1,0,1,0,0,0,0,0]
[0,0,0,0,0,0,0,0,0,0,0,0,0,0,0,0,0,0,0,0,0,0,0,0,1,0,0,0,1,0,0,
0,0,0,0,1,0,0,0,0,0,0,0,0,1,0,0,0,0,1,1,0,1,0,0,0,0]
[0,0,0,0,0,0,0,0,0,0,0,0,0,0,0,0,0,0,0,0,0,0,0,0,0,1,0,0,0,1,0,
0,0,0,0,0,1,0,0,0,0,0,0,0,0,1,0,0,0,0,1,1,0,1,0,0,0]
[0,0,0,0,0,0,0,0,0,0,0,0,0,0,0,0,0,0,0,0,0,0,0,0,0,0,1,0,0,0,1,
0,0,0,0,0,0,1,0,0,0,0,0,0,0,0,1,0,0,0,0,1,1,0,1,0,0]

```
[0,0,0,0,0,0,0,0,1,0,0,0,0,0,0,0,0,0,0,0,0,0,0,0,0,0,0,1,0,0,0,
1,0,0,0,0,0,0,1,0,0,0,0,0,0,0,0,1,0,0,0,0,1,1,0,1,0]
[0,0,0,0,0,0,0,0,0,1,0,0,0,0,0,0,0,0,0,0,0,0,0,0,0,0,0,0,1,0,0,
0,1,0,0,0,0,0,0,1,0,0,0,0,0,0,0,0,1,0,0,0,0,1,1,0,1]
[0,0,0,0,0,0,0,0,0,0,1,0,0,0,0,0,0,0,0,0,0,0,0,0,0,0,0,0,0,1,0,
0,0,1,0,0,0,0,0,0,1,0,0,0,0,0,0,0,0,1,0,0,0,0,1,1,0]
[0,0,0,0,0,0,0,0,0,0,0,0,0,0,0,0,0,0,0,0,0,0,0,0,0,0,0,0,0,0,1,
0,0,0,1,0,0,0,0,0,0,1,0,0,0,0,0,0,0,0,1,0,0,0,0,1,1]
[0,0,1,0,0,0,0,0,0,0,0,0,1,0,0,0,0,0,0,0,0,0,0,0,0,0,0,0,0,0,0,
1,0,0,0,1,0,0,0,0,0,0,1,0,0,0,0,0,0,0,0,1,0,0,0,0,1]
```

Its DTG is $\left(\frac{12}{57},\frac{26}{57},\frac{19}{57}\right)$.

Chapter 9
A Study on Neutro-Topological Neighbourhood and Neutro-Topological Base

Bhimraj Basumatary
Bodoland University, India

Alympica Talukdar
Bodoland University, India

ABSTRACT

Recently, Smarandache introduced NeutroAlgebra and AntiAlgebras. NeutroAlgebras and AntiAlgebras represent a new research subject that is based on real-world examples. He studied the concept of neutro-structures and anti-structures. From the concepts of NeutroAlgebra, he showed that if a statement (theorem, lemma, consequence, property, etc.) is totally true in a classical Algebra, it does not mean that it is also totally true in a NeutroAlgebra or an AntiAlgebra. It depends on the operations and axioms (if they are totally true, partially true, totally false, or partially or totally indeterminate) it is based upon. In this study, the concept of Neutro-Topological neighbourhood and Neutro-Topological base are studied. Some examples of Neutro-Neighbourhood and Neutro-Base are given and compare the results of classical topological neighbourhood and Neutro-Topological Neighbourhood as well as classical topological base and Neutro-Topological Base.

DOI: 10.4018/978-1-6684-4740-6.ch009

INTRODUCTION

Set theory enabled us to study various concepts of mathematics. But we have many uncertainities in our lives. So many times classical mathematical logic is not sufficient to overcome these uncertainities. Then the concept of fuzzy sets is introduced by Zadeh (1965). After three years Chang (1968) introduced and developed the theory of fuzzy topological spaces. Then another new concept called intuitionistic fuzzy set was established by Atanassov (1983). After that Coker (1997) introduced the concept of intuitionistic fuzzy topological space. The idea of neutrosophic set was first given by Smarandache (1998). In these concept there is a degree of membership T, degree of uncertainity I and degree of non-membership F. In other words, in neutrosophy, a situation is handled according to its trueness, its falsity, and its uncertainty. Therefore, neutrosophic logic and neutrosophic sets help us explain many uncertainties in our lives. Therefore many researchers have made studies on this subject. Sahin *et al* (2021) obtain some operations for interval valued neutrosophic sets. Smarandache (2019, 2020a) defined neutro-structures and anti-structures. Similar to neutrosophic logic, an algebraic structure divides into three regions: A, the set of elements that satisfy the conditions of algebraic structure, the truth region; Neutro-A, the set of elements that do not meet the conditions of the algebraic structure, the uncertainty region and the anti-A, the set of elements that do not meet the conditions of the algebraic structure, the inaccuracy region. Thus, the structure of neutrosophic logic has been transferred to the structure of classical algebras, without using neutrosophic sets and neutrosophic numbers. In recent years, the academic community has witnessed growing research interests in neutrosophic set theory. Therefore, neutro-algebraic structures, which have a more general structure than classical algebras, can be obtained. In addition, the region of the elements that do not satisfy any of the classical algebras are also taken as anti-algebraic structures. Recently, Smarandache (2020b) studied neutro-algebra; Razaei *et al* (2020) obtained the neutrosophic triplet of BI-algebras; Smarandache and Hamidi (2020) introduced neutro-BCK-algebra; Ibrahim and Agboola (2020a, 2020b) studied neutro-hypergroups.

In this chapter, Neutro-Topology, neighbourhood and base are studied. The definition of Neutro-Neighbourhood and comparison table of neighbourhood and Neutro-Neighbourhood with respect to the result of classical topological neighbourhood is studied. Also, the definition of Neutro-Base, Neutro-subBase and the comparison table of base and Neutro-Base as well as sub-base and Neutro-sub-Base are given.

PRELIMINARIES

Definition 1. (Smarandache, 2020c) The NeutroSophication of the Law

(i) Let X be a non-empty set and $*$ be a binary operation. For some elements $(a,b)\in(X,X)$, $(a*b)\in X$ (degree of well defined (T)) and for other elements $(x,y),(p,q)\in(X,X)$; [$x*y$ is indeterminate (degree of indeterminacy (I)), or $p*q\notin X$ (degree of outer-defined (F)], where (T, I, F) is different from (1,0,0) that represents the Classical Law, and from (0,0,1) that represents the AntiLaw.

(ii) In Neutro Algebra, the classical well-defined for binary operation $*$ is divided into three regions: degree of well-defined (T), degree of indeterminacy (I) and degree of outer-defined (F) similar to neutrosophic set and neutrosophic logic.

Definition 2. (Şahin *et al.*, 2021) Let X be the non-empty set and τ be a collection of subsets of X. Then τ is said to be a Neutro Topology on X *and* the pair (X,τ) is said to be a Neutro Topological space, if at least one of the following conditions hold good:

(i) $\left[\left(\varnothing_N \in \tau, X_N \notin \tau\right) \text{ or } \left(X_N \in \tau, \varnothing_N \notin \tau\right)\right]$ or $\left[\varnothing_N, X_N \in_{\sim} \tau\right]$.

(ii) For some n elements $a_1,a_2,\ldots,a_n\in\tau$, $\bigcup_{i=1}^{n} a_i \in \tau$ [degree of truth T] and for other n elements $b_1,b_2,\ldots,b_n\in\tau, p_1,p_2,\ldots,p_n\in\tau$; [$\left(\bigcup_{i=1}^{n} b_i \notin \tau\right)$ [degree of falsehood F] or ($\bigcup_{i=1}^{n} p_i$ is indeterminate (degree of indeterminacy I)], where n is finite; [where (T, I, F) is different from (1,0,0) that represents the Classical Axiom, and from (0,0,1) that represents the AntiAxiom].

(iii) For some n elements $a_1,a_2,\ldots,a_n\in\tau$, $\bigcap_{i=1}^{n} a_i \in \tau$ [degree of truth T] and for other n elements $b_1,b_2,\ldots,b_n\in\tau$, $p_1,p_2,\ldots,p_n\in\tau$; [$\left(\bigcap_{i=I}^{n} b_i \notin \tau\right)$ [degree of falsehood F] or ($\bigcap_{i=I}^{n} p_i$ is indeterminate (degree of indeterminacy I)], where n is finite; [where (T, I, F) is different from (1,0,0) that represents the Classical Axiom, and from (0,0,1) that represents the AntiAxiom].

Remark 1. (Şahin *et al.*, 2021) The symbol "$\in_{\sim}$" will be used for situations where it is an unclear appurtenance (not sure if an element belongs or not to a set). For example, if it is not certain whether "a" is a member of the set P, then it is denoted by a $\in_{\sim} P$.

NEUTRO-TOPOLOGICAL NEIGHBOURHOOD

Definition 1. Let (X,τ) be a Neutro-Topological space and let $x \in X$. A subset N of X *is* said to be a τ-Neutro-Neighbourhood of x *if* and only if there exists a τ-Neutro-Open set G *such* that $x \in G \subset N$. Similarly, N *is* called a τ-Neutro-Neighbourhood of $A \subset X$ if and only if there exists a τ-Neutro- Open set G *such* that $A \subset G \subset N$.

The collection of all τ-Neutro-Neighbourhood of $x \subset X$ is called the Neutro-Neighbourhood system at x.

Example 1. Let, $X=\{a,b,c,d\}$ be a set and $\tau= \{\phi\{a\}, \{a,b\}, \{b,c\}, \{c,d\}\}$ be a collection of subsets of X. Then,

i. It is clear that $X \notin \tau$ and $\phi \in \tau$.
ii. Let $p_1=\{a\}, p_2=\{a,b\}, q_1=\{c,d\}, q_2=\{b,c\}$. We obtain that $\left(\bigcap_{i=1}^{2} p_i \in \tau\right)$ and $\left(\bigcap_{i=1}^{2} q_i \notin \tau\right)$.
iii. Let $p_1=\{a,b\}, p_2=\{c,d\}, q_1=\{a\}, q_2=\{b,c\}$. We obtain that $\bigcup_{i=1}^{2} p_i \in \tau)$ and $\left(\bigcup_{i=1}^{2} q_i \in \tau\right)$

Thus (X,τ) satisfies the conditions of Neutro-Topology.
Hence, (X,τ) is a Neutro-Topological space.

τ- Neutro-Neighbourhoods of a are, $\{a\}, \{a,b\}, \{a,c\}, \{a,d\}, \{a,b,c\},\{a,c,d\},\{a,b,d\}$, X.
τ- Neutro-Neighbourhoods of b are, $\{a,b\}, \{b,c\}, \{a,b,c\}, \{a,b,d\}, \{a,b,c\}$, X.
τ- Neutro-Neighbourhoods of c are, $\{c,d\}, \{b,c\}, \{a,c,d\},\{b,c,d\},\{a,b,c\}$, X.
τ- Neutro-Neighbourhoods of d are, $\{c,d\}, \{a,c,d\}, \{b,c,d\}$, X.

Now we compare the General Topology and Neutro Topology in terms of neighbourhood.

Table 1 is the comparison table.

Table 1. Neighbourhood and neutro-neighbourhood

General Topology	Neutro Topology
1. A subset of a topological space is open if and only if it is neighbourhood of each of its points. Example: Let $X=\{1,2,3,4,5\}$ and $\tau=\{\phi, \{1\}, \{1,2\}, \{1,2,5\}, \{1,3,4\}, \{1,2,3,4\}, X\}$ Then, τ-neighbourhoods of 1 are: {1}, {1,2}, {1,3}, {1,4}, {1,5}, {1,2,3}, {1,2,3,4}, {1,2,3,5}, {1,2,4,5}, {1,3,4,5}, and X. τ-neighbourhoods of 2 are: {1,2}, {1,2,3}, {1,2,4}, {1,2,5}, {1,2,3,4}, {1,2,3,5}, {1,2,4,5}, and X τ-neighbourhoods of 3 are: {1,3,4}, {1,2,3,4}, {1,3,4,5}, and X τ-neighbourhoods of 4 are: {1,3,4}, {1,2,3,4}, {1,3,4,5}, and X. τ-neighbourhoods of 5 are: {1,2,5}, {1,2,3,5}, {1,2,4,5}, and X Here *X* is neighbourhood of each of its points then we get that X is open since $X^0=X$. Conversely, let X is open then it is seen from the above example that X is neighbourhood of each of its points since for all $x\in X$ there exists a $A\in\tau$ such that $x \in A \subseteq X.$	1. A subset of a Neutro-Topological space is Neutro-Open if and only if it is Neutro-Neighbourhood of each of its points. Let $X=\{a,b,c,d\}$ and $\tau=\{\phi, \{a\}, \{a,b\}, \{b,c\}, \{c,d\}\}$ τ- Neutro-Neighbourhoods of a are, {*a*}, {*a,b*}, {*a,c*}, {*a,d*}, {*a,b,c*}, {*a,c,d*}, {*a,b,d*}, *X*. *T*- Neutro-Neighbourhoods of b are: {*a,b*}, {*b,c*}, {*a,b,c*}, {*a,b,d*}, {*a,b,c*}, *X*. τ- Neutro-Neighbourhoods of c are: {*c,d*}, {*b,c*}, {*a,c,d*}, {*b,c,d*}, {*a,b,c*}, *X*. τ- Neutro-Neighbourhoods of d are: {*c,d*}, {*a,c,d*}, {*b,c,d*}, *X*. Now, it is cleared that that A= {*a,c,b*} is τ- Neutro-Neighbourhood of each of its points. Consequently *A* is a Neutro-Open set since Neutro-Interior of *A* is *A* ($A^0=A$). a is Neutro-Interior point of *A*, since $a\in A$ and there exists $\{a\}\in\tau$ such that $a \in \{a\} \subseteq A.$ b is Neutro-Interior point of *A*, since $b\in A$ and there exists $\{a,b\}\in\tau$ such that $b \in \{a,b\} \subseteq A.$ c is Neutro-Interior point of *A*, since $c\in A$ and there exists $\{b,c\}\in\tau$ such that $c \in \{b,c\} \subseteq A$. d is not an Neutro-Interior point of *A*, since $d \notin A$ and there does not exists $G\in\tau$ such that $d \in G \subseteq A$. Therefore, $A^0=A$. Conversely, $A=\{a,b,c\}$ is Neutro-Open as $A^0=A$. Clearly, it is τ- Neutro-Neighbourhood of each of its points.
2. Let *X*be a topological space, and for each $x\in X$, let $N(x)$ be the collection of all neighbourhoods of *x*. Then [*N*0]: $\forall x\in X, N(x)\neq\phi$ i.e every point *x* has at least on neighbourhood. [*N*(1)]: $N\in N(x) \Rightarrow x\in N$. i,e every neighbourhood of x contains x. $[N(2)]: N \in N(x), M \supset N \Rightarrow M \in N(x).$ i.e every set containing a neighbourhood of *x* is a neighbourhood of *x*. [*N*(3)]: $N\in N(x)\ M\in N(x) \Rightarrow N\cap M\in N(x)$. [*N*(4)]: $N\in N(x) \Rightarrow \exists M\in N(x)$ such that, $M \subset N$ and $M\in N(y)\ \forall y\in M$	2. Let *X* be a Neutro-Topological space, and for each $x\in X$, let $N(x)$ be the collection of all τ- Neutro-neighbourhoods of *x*. Then [*N*0]: $\forall x\in X, N(x)\neq\phi$ i.e every point x has atleast one τ-Neutro-Neighbourhood. [*N*(1)]: $N\in N(x) \Rightarrow x\in N$. i.e every τ- Neutro-Neighbourhood of x contains x. $[N(2)]: N \in N(x), M \supset N \Rightarrow M \in N(x).$ i.e every set containing a τ-Neutro-Neighbourhood of *x* is a τ-Neutro-Neighbourhood of *x*. $N[3]: N \in N(x), M \in N(x) \not\Rightarrow N \cap M \in N(x)$. To show this, an example is cited below: Example: Let $X=\{0,1,2,3\}$ and $\tau=\{\phi, X, \{1\}, \{2\}, \{2,3\}, \{1,3\}\}$. τ- Neutro-Neighbourhoods of 0 is X. τ- Neutro-Neighbourhoods of 1 are: {1}, {0,1}, {1,2}, {1,3}, {0,1,2}, {0,1,3}, {1,2,3}, and X. τ- Neutro-Neighbourhoods of 2 are: {2}, {0,2}, {1,2}, {2,3}, {0,1,2}, {1,2,3}, and X. τ- Neutro-Neighbourhoods of 3 are: {1,3}, {2.3}, {0,2,3}, {0,1,3}, {1,2,3}, and X. e.g $\{0,1,3\}\in N(3)$, $\{0,2,3\}\in N(3)$ but $\{0,1,3\}\cap\{0,2,3\}=\{0,3\}\notin N(3)$. $[N(4)]: N \in N(x) \not\Rightarrow M \in N(x)$ such that $M \subset N$ and $M\in N(y)\forall y\in M$. e.g $\{0,1,2\}\in N(1)$ then $\{0,1\}\in N(1)$ such that $\{0,1\} \subset \{0,1,2\}$ but $\{0,1\}\notin N(y)\forall y\notin M$.

continues on following page

Table 1. Continued

General Topology	Neutro Topology
3. Let X be a non-empty set and with each $x \in X$, let there be associated a family $N(x)$ of subsets of X, called neighbourhoods, satisfying the following conditions: $[N0]$: $N(x) \neq \phi \forall x \in X$. $[N1]$: $N \in N(x) \Rightarrow x \in N$. $[N2]: N \in N(x), M \supset N \Rightarrow M \in N(x)$. $[N3]$: $N \in N(x)$, $M \in N(x) \Rightarrow N \cap M \in N(x)$ $[N4]$: $N \in N(x) \Rightarrow \exists M \in N(x)$ such that $M \subset N$ and $M \in N(y) \forall y \in M$. Then there exists a unique topology τ on X in such a way that if $N^*(x)$ is the collection of neighbourhoods of x, defined by the topology τ, then $N^*(x)=N(x)$.	3. Let X be a non-empty set and with each $x \in X$, let there be associated a family $N(x)$ of subsets of X, called τ- Neutro-Neighbourhoods. $[N0]$: $N(x) \neq \phi \forall x \in X$. $[N1]$: $N \in N(x) \Rightarrow x \in N$. $[N2]: N \in N(x), M \supset N \Rightarrow M \in N(x)$. $[N3]: N \in N(x), M \in N(x) \nRightarrow N \cap M \in N(x)$. To show this, an example is cited below: Let $X=\{0,1,2,3\}$ and $\tau= \{\phi, X \{1\}, \{2\}, \{2,3\}, \{1,3\}\}$. τ- Neutro-Neighbourhoods of 0 is X. τ- Neutro-Neighbourhoods of 1 are: $\{1\}, \{0,1\}, \{1,2\}, \{1,3\}, \{0,1,2\}, \{0,1,3\}, \{1,2,3\}$, and X. τ- Neutro-Neighbourhoods of 2 are: $\{2\},\{0,2\},\{1,2\},\{2,3\},\{0,1,2\}, \{1,2,3\}$, and X. τ- Neutro-Neighbourhoods of 3 are: $\{1,3\}, \{2,3\}, \{0,2,3\}, \{0,1,3\}, \{1,2,3\}$, and X. $N[4]: N \in N(x) \nRightarrow M \in N(x)$ such that $M \subset N$ and $M \in N(y) \forall y \in M$. *e.g* $\{0,1,2\} \in N$, $\{0,1\} \in N(1)$ such that $\{0,1\} \subset \{0,1,2\}$ but $\{0,1\} \notin N(y)/\{0,1\} \notin N(y) \forall y \in M$. Then there doesnot exists *a* unique Neutro-Topology τ on X *in* such a way that if $N^*(x)$ is the collection of τ- Neutro-Neighbourhoods of x, *defined* by the Neutro-Topology τ since all the properties are not satisfied by τ- Neutro-Neighbourhood.
4. Let X be a non-empty set, and for each $x \in X$, let $N(x)$ be a nonempty collection of subsets of X satisfying the following conditions: $[M1]$: $N \in N(x) \Rightarrow c \in N$ $[M2]$: $N \in N(x)$, $M \in N(x) \Rightarrow N \cap M \in N(x)$ Let τ consists of the empty and all those non-empty subsets G of X having the property that $x \in G$ implies that there exists a $N \in N(x)$ such that $x \in N \subset G$. Then τ is a topology for X.	4. Let X be a non-empty set, and for each $x \in X$, let $N(x)$ be a non-empty collection of subsets of X. $[M1]$: $N \in N(x) \Rightarrow x \in N$ $[M2]: N \in N(X), M \in N(x) \nRightarrow N \cap M \in N(x)$. To show this, an example is cited below: Let $X=\{a,b,c,d\}$ $\tau= \{\phi, \{a\}, \{a,b\}, \{b,c\}, \{c,d\}\}$ τ- Neutro-Neighbourhoods of a are: $\{a\}, \{a,b\}, \{a,c\}, \{a,d\}, \{a,b,c\}, \{a,c,d\}, \{a,b,d\}, X$. τ- Neutro-Neighbourhoods of b are: $\{a,b\}, \{b,c\}, \{a,b,c\}, \{a,b,d\}, \{a,b,c\}, X$. τ- Neutro-Neighbourhoods of c are: $\{c,d\}, \{b,c\}, \{a,c,d\}, \{b,c,d\}, \{a,b,c\}, X$. τ- Neutro-Neighbourhoods of d are: $\{c,d\}, \{a,c,d\}, \{b,c,d\}, X$. Now $\{a,b\} \in N(b)$ and $\{b,c\} \in N(b)$ but $\{a,b\} \cap \{b,c\}=\{b\} \notin N(b)$. The second condition is not satisfied by Neutro-Neighbourhood.

NEUTRO-TOPOLOGICAL BASE

Definition 1. Let (X, τ) be a Neutro-Topological space. Then a non-empty subcollection B of subsets of X is said to be a Neutro-Base for some Neutro-Topology on X if the following conditions satisfied:

i. For all $x \in X$ there exist $A \in B$ such that $x \in A$
ii. For some $A_1, A_2 \in B$ and for $x \in A_1 \cap A_2$ there exist $A_3 \in B$ such that $x \in A_3 \subseteq A_1 \cap A_2$ and for some $A_1, A_2 \in B$ and for all $x \in A_1 \cap A_2$ there may not exist $A_3 \in B$ such that $x \in A_3 \subseteq A_1 \cap A_2$

Example 1. Let $X=\{a,b,c,d\}$ and $\tau= \{\phi,\{a\}, \{a,b\}, \{c,d\}, \{b,c\}, \{a,c,d\}\}$ be a Neutro-Topological space. Then $B= \{\{a\}, \{a,b\}, \{b,c\}, \{c,d\}\}$ is a Neutro-Topological base since

i. For all $x \in X$ there exist $A \in B$ such that $x \in A$.
ii. For $\{a\}, \{a,b\} \in B$ and for $x \in \{a\} \cap \{a,b\}$ there exist $\{a\} \in B$ such that $x \in \{a\} \subseteq \{a\} \cap \{a,b\}$ and for $\{b,c\},\{c,d\} \in B$ and $x \in \{b,c\} \cap \{c,d\}$, there does not exist any $A_3 \in B$ such that $x \in A_3 \subseteq A_1 \cap A_2$

We compare the Topological base and Neutro-Topological-Base.
Table 2 is the comparison table.

Table 2. Base and Neutro-Base

General Topology	Neutro Topology
1. Let (X,τ) be a topological space. A sub-collection B of τ is a base for τ *if* and only if every τ-open set can be expressed as the union of members of B.	1. Let (X,τ) be a Neutro-Topological space. A sub-collection B *of* τ is a Neutro-Base for τ if every Neutro-Open set can be expressed as the union of members of B *b*ut the converse is not true. To show that the converse part is not true an example is cited below. Example: Let X= {a,*b,c,d}* τ= {ϕ, {a}, *{*a,b}, *{*b,c}, *{*c,d}*}* *B*= {*{*a}, *{*a,b}, *{*c,d}*}* *H*ere not every τ-Neutro-Open set can be expressed as the union of members of B. It is seen that {b,c} *can* not be expressed as the union of members of B.

continues on following page

Table 2. Continued

General Topology	Neutro Topology
2. Let (X,τ) be a topological space and B be a base for τ. Then B has the following properties: (1) For every $x\in X$, there exists a $A\in B$ such that $x \in B$ i.e $X= \cup\{A: A\in B\}$ (2) For every $A_1, A_2 \in B$ and every point $x\in A_1 \cap A_2$ there exists a $A\in B$ such that $x \in A \subset A_1 \cap A_2$ that is the intersection of any two members of B is a union of members of B.	2. Let (X,τ) be a Neutro-Topological space and B be a Neutro-Base for τ. Then B has the following properties: (1) For every $x\in X$ there exists a $A\in B$ such that $x\in B$ that is $X= \cup\{A: A\in B\}$ Therefore we get $X= \cup\{A: A\in B\}$ (2) For some $A_1 A_2 \in B$ and every point $x\in A_1 \cap A_2$ there exists $A\in B$ such that $x \in A \subset A_1 \cap A_2$ and for some $A_1, A_2 \in B$ for any $x\in A_1 \cap A_2$ there does not exist A such that $x \in A \subset A_1 \cap A_2$. Example: Let $X= \{a,b,c,d\}$ $\tau= \{\{a\}, \{a,b\}, \{c,d\}\}$ (1) For all $x\in X$ there exists $A\in B$ such that $X= \cup\{A: A\in B\}$ (2) $\{a\},\{a,b\} \in B$, then there exists $\{a\}\in B$ such that a $\in\{a\}\subset\{a\}\subseteq\{a,b\}$. Again $\{a,b\}, \{c,d\}\in B$ then $\{a,b\}\cap\{c,d\}=\phi$ but there does not exist any $A\in B$ such that $x \in A \subset \{a,b\} \cap \{c,d\}$. Therefore the second condition of the theorem is not satisfied by all $A\in B$.
3. Let X be a non-empty set and let B be a collection of subsets of X satisfying the following conditions: (1) For every $x\in X$, there exists a $A\in B$ such that $x\in B$ i.e. $X=\cup\{A: A\in B\}$ (2) For every $A_1\in B$, $A_2\in B$ and every point $x\in A_1\cap A_2$ there exists a $A\in B$ such that $x \in A \subset A_1 \cap A_2$ that is the intersection of any two members of B is a union of members of B. Then there exists a unique topology τ for X such that B is a base for τ.	3. Let X be a non-empty set and let B be a collection of subsets of X. Then (1) For every $x\in X$, there exists a $A\in B$ such that $x\in B$ i.e $X= \cup\{A: A\in B\}$ (2) For some $A_1 A_2 \in B$ and every point $x\in A_1\cap A_2$ there exists $A\in B$ such that $x \in A \subset A_1 \cap A_2$ and for some $A_1 A_2 \in B$ for any $x\in A_1\cap A_2$ there may not exist A such that $x \in A \subset A_1 \cap A_2$. Let $X= \{a,b,c,d,e\}$ $\tau= \{\phi, \{a\}, \{a,b\}, \{b,c\}, \{b,d\}, \{c,d\}, \{c,d,e\}, \{a,b,c,d\}\}$ $B= \{\phi, \{a\}, \{a,b\}, \{b,d\}, \{c,d\}, \{c,d,e\}, \{a,b,c,d\}\}$ $\{a\}, \{a,b\}\in B$ then $\{a\}\cap\{a,b\}=\{a\}$. For $a\in\{a\}$, $\exists\{a\}\in B$ such that $x \in \{a\} \subset \{a\} \cap \{a,b\}$. But $\{a,b\}, \{b,d\}\in B$, $\{a,b\}\cap\{b,d\}=\{b\}$ But there does not exists any $A \in B \subset \{a,b\} \cap \{b,d\}$. Then there exists a unique Neutro -Topology τ for X such that B is a Neutro-Base for τ if the above conditions satisfied by B. So, it is seen that the second condition is not satisfied.

Definition 2. Let (X,τ) be a Neutro-Topological space. A collection B' of subsets of X is called a Neutro-sub-Base for the Neutro-Topology τ iff $B' \subset \tau$ and finite intersections of members of B' form a Neutro-Base for τ.

Example 2. Let $X= \{a,b,c,d,e,f\}$

$\tau= \{\phi, \{a,b\}, \{b,c\}, \{d,e\}, \{c,d,e\}, \{a,b,c,d\}, \{e,d\}\}$

Let $B' = \{\{a,b\},\{b,c\},\{d,e\},\{e,f\},\{a,b,c,d\}\}$

The finite intersection of members of B' are

$B_0= \{\{b\}, \{d\}, \{e\}, \{a,b\}, \{b,c\}, \{d,e\}, \{e,f\}, \{a,b,c,d\}\}$

Then B_0 is a base since

(1) For every $x \in X$, $\exists$ a $A \in B_0$ such that $x \in B_0$.
(2) $\{a,b\} \in B_0$, $\{a,b,c,d\} \in B_0$, $\{a,b\} \cap \{a,b,c,d\}=\{a,b\}$, then for every $x \in \{a,b\} \exists$ $A \in B_0$ such that $x \in B_0 \subset \{a,b\} \cap \{a,b,c,d\}$.

Again $\{b,c\},\{c,d\} \in B_0$, $\{b,c\} \cap \{d,c\}=\{c\}$.

But there does not exists any A such that for every $x \in B_0 \subset \{b,c\} \cap \{c,d\}$.

We compare the Topological sub-base and Neutro -Topological-sub-Base. Table 3 is the comparison table.

CONCLUSION

In this study, we introduced the notion of Neutro-Topological Neighbourhood and Neutro-Topological Base via Neutro-Topology. We proved some results of neighbourhood and base. Similarities and differences between neighbourhood and Neutro-Neighbourhood as well as base and Neutro-Base, sub-base and Neutro-sub-Base are discussed.

Table 3. Sub-base and Neutro-sub-Base

General Topology	Neutro-Topology
Let B^* be a collection of subsets of a non-empty set X. Then B_0 is a sub-base for a unique topology τ for X, that is finite intersections of members of B^* form a base for τ.	Let B^* be a collection of subsets of a non-empty set X. Then B_0 is a Neutro-sub-Base for a unique Neutro Topology τ for X, that is finite intersections of members of B^* form a base for τ if the following conditions are satisfied by B_0: (1) For every $x \in X$, $\exists A \in B_0$ such that $x \in B0$ such that $X = \cup\{A: A \in B0\}$ (2) For every $A_1 \in B_0$, $A_2 \in B_0$, every point $x \in A_1 \cap A_2 \exists A \in B_0$ such that $x \in A \subset A_1 \cap A_2$ and for some $A_1, A_2 \in B_0$ there may not exist any $A_3 \in B$ such that $x \in A_3 \subseteq A_1 \cap A_2$ Let $X = \{a,b,c,d,e,f\}$ $\tau = \{\phi,\{a,b\},\{b,c\},\{d,e\},\{c,d,e\},\{a,b,c,d\},\{e,f\}\}$ Let $B^* = \{\{a,b\}, \{b,c\}, \{d,e\}, \{e,f\}, \{c,d,e\}, \{a,b,c,d\}\}$ The finite intersection of members of B^* are $B_0 = \{\{b\}, \{d\}, \{e\}, \{a,b\}, \{b,c\}, \{c,d\}, \{d,e\}, \{e,f\}, \{a,b,c,d\}\}$ Then B_0 is a Neutro-Base as (1) For every $x \in X, \exists$a $A \in B_0$ such that $x \in B_0$. (2) $\{a,b\} \in B_0$, $\{a,b,c,d\} \in B_0$ $\{a,b\} \cap \{a,b,c,d\} = \{a,b\}$, then for every $x \in \{a,b\} \exists$ a $A \in B_0$ such that $x \in B_0 \subset \{a,b\} \cap \{a,b,c,d\}$. Again, $\{b,c\},\{c,d\} \in B_0$, $\{b,c\} \cap \{c,d\} = \{c\}$ but there does not exist any $A_3 \in B_0$ such that $x \in A_3 \subseteq A_1 \cap A_2$ Therefore, the second condition of Neutro-sub-Base is not satisfied here.

REFERENCES

Atanassov, K. T. (1986). Intuitionistic fuzzy sets. *Fuzzy Sets and Systems*, *20*(1), 87–96. doi:10.1016/S0165-0114(86)80034-3

Chang, C. L. (1968). Fuzzy Topological Space. *Journal of Mathematical Analysis and Applications*, *24*(1), 182–190. doi:10.1016/0022-247X(68)90057-7

Coker, D. (1997). An introduction to intuitionistic fuzzy topological spaces. *Fuzzy Sets and Systems*, *88*(1), 81–89. doi:10.1016/S0165-0114(96)00076-0

Ibrahim, M. A., & Agboola, A. A. A. (2020a). NeutroVectorSpaces I. *Neutrosophic Sets and Systems*, *36*, 328–351.

Ibrahim, M. A., & Agboola, A. A. A. (2020b). Introduction to NeutroHyperGroups. *Neutrosophic Sets and Systems*, *38*(1), 2.

Şahin, M., Kargın, A., & Yücel, M. (2021). Neutro-Topological space and Anti-Topological space. *NeutroAlgebra Theory*, *I*, 16.

Salama, A. A., Smarandache, F., & Alblowi, S. A. (2014). New neutrosophic crisp topological concepts. *Neutrosophic Sets and Systems, 4*, 50-54.

Smarandache, F. (1998). Neutrosophy: Neutrosophic Probability, Set and Logic. Amer. Research Press.

Smarandache, F. (2019). Introduction to NeutroAlgebraic Structures and AntiAlgebraic Structures. In *Advances of Standard and Nonstandard Neutrosophic Theories*. Pons Publishing House Brussels.

Smarandache, F. (2020a). Introduction to NeutroAlgebraic Structures and AntiAlgebraic Structures (revisited). *Neutrosophic Sets and Systems*, *31*, 1–16. doi:10.5281/zenodo.3638232

Smarandache, F. (2020b). NeutroAlgebra is a Generalization of Partial Algebra. *International Journal of Neutrosophic Science, 2*(1), 8-17.

Smarandache, F. (2020c). NeutroAlgebra is a generalization of partial algebra. *International Journal of Neutrosophic Science*, *2*, 8–17. doi:10.54216/IJNS.020103

Smarandache, F., & Hamidi, M. (2020). Neutro-BCK-algebra. *International Journal of Neutrosophic Science*, *8*(2), 110.

ADDITIONAL READING

Agboola, A. A. A. (2020b). Introduction to NeutroRings. *International Journal of Neutrosophic Science*, *7*(2), 62–73.

Agboola, A. A. A., Ibrahim, M. A., & Adeleke, E. O. (2020a). Elementary Examination of NeutroAlgebras and AntiAlgebras viz-a-viz the Classical Number Systems. *International Journal of Neutrosophic Science*, *4*(1), 16–19. doi:10.54216/IJNS.040102

Ahin, S. M., Kargın, A., & Kılı¸c, A. (2020). Generalized neutrosophic quadruple sets and numbers. *Quadruple Neutrosophic Theory and Applications*, *1*, 11–22.

Bakbak, D., & Uluc¸ay, V. (2020b). A Theoretic Approach to Decision Making Problems in Architecture with Neutrosophic Soft Set. *Quadruple Neutrosophic Theory and Applications*, *01*, 113–126.

Broumi, S., Bakali, A., Talea, M., Smarandache, F., & Ulu¸cay, V. (2017, December). Minimum spanning tree in trapezoidal fuzzy neutrosophic environment. In *International Conference on Innovations in BioInspired Computing and Applications* (pp. 25-35). Springer.

Chandran, K., Sundaramoorthy, S. S., Smarandache, F., & Jafari, S. (2020). On Product of Smooth Neutrosophic Topological Spaces. *Symmetry, 12*(9), 1557. doi:10.3390ym12091557

Dhavaseelan, R., Jafari, S., & Smarandache, F. (2017). Compact open topology and evaluation map via neutrosophic sets. *Neutrosophic Set and Systems, 16*, 35–38.

Ecemi¸s, O., S¸ahin, M., & Kargın, A. (2018). Single valued neutrosophic number valued generalized neutrosophic triplet groups and its applications for decision making applications. *Asian Journal of Mathematics and Computer Research, 24*(5), 205–218.

Hassan, N., Uluc¸ay, V., & S¸ahin, M. (2018). Q-neutrosophic soft expert set and its application in decision making. *International Journal of Fuzzy System Applications, 7*(4), 37–61. doi:10.4018/IJFSA.2018100103

Kargın, A., Dayan, A., & S¸ahin, N. M. (2021). Generalized Hamming Similarity Measure Based on Neutrosophic Quadruple Numbers and Its Applications to Law Sciences. *Neutrosophic Sets and Systems, 40*(1), 4.

Kargın, A., Dayan, A., Yıldız, I., & Kılıc¸, A. (2020). Neutrosophic Triplet m-Banach Spaces. *Neutrosophic Set and Systems, 38*, 383–398.

Khalifa, N. E. M., Smarandache, F., Manogaran, G., & Loney, M. (2021). A study of the neutrosophic set significance on deep transfer learning models: An experimental case on a limited covid-19 chest x-ray dataset. *Cognitive Computation*, 1–10. doi:10.100712559-020-09802-9 PMID:33425043

Mohammed, F. M., & Wadei, A. O. (2020). Continuity and contra continuity via preopen sets in new construction fuzzy neutrosophic topology. In *Optimization Theory Based on Neutrosophic and Plithogenic Sets* (pp. 215–233). Academic Press. doi:10.1016/B978-0-12-819670-0.00010-X

Rezaei, A., & Smarandache, F. (2020a). On Neutro-BE-algebras and Anti-BE-algebras (revisited). *International Journal of Neutrosophic Science, 4*(1), 8–15.

Rezaei, A., & Smarandache, F. (2020b). The Neutrosophic Triplet of BI-algebras. *Neutrosophic Sets and Systems, 33*, 313–321.

Sahin, M., Alkhazaleh, S., & Ulucay, V. (2015). Neutrosophic soft expert sets. *Applied Mathematics*, *6*(1), 116–127. doi:10.4236/am.2015.61012

Sahin, M., & Kargın, A. (2017a). Neutrosophic triplet normed space. *Open Physics*, *15*(1), 697–704. doi:10.1515/phys-2017-0082

Sahin, M., & Kargın, A. (2018a). Neutrosophic Triplet v-Generalized Metric Space. *Axioms*, *7*(3), 67. doi:10.3390/axioms7030067

Sahin, M., & Kargın, A. (2018b). Neutrosophic triplet normed ring space. *Neutrosophic Set and Systems*, *21*, 20–27.

Sahin, M. ¸., & Kargın, A. (2019b). Neutrosophic triplet Lie algebras. *Neutrosophic Triplet Research*, *1*(6), 68–78.

Sahin, M., & Kargın, A. (2019c). *Neutrosophic Triplet Partial v-Generalized Metric Space. Quadruple* (Vol. I). Neutrosophic Theory And Applications.

Sahin, M., & Kargın, A. (2019d). Neutrosophic triplet metric topology. *Neutrosophic Set and Systems*, *27*, 154–162.

Sahin M., Kargın A. (2019e). Single valued neutrosophic quadruple graphs. *Asian Journal of Mathematics and Computer Research,* 243-250.

Sahin, M., & Kargın, A. (2019f). Neutrosophic Triplet b-Metric Space. *Neutrosophic Triplet Structures*, *1*(7), 79–89.

Sahin, M., & Kargın, A. (2019g). Neutrosophic Triplet Partial Inner Product Spaces. *Neutrosophic Triplet Stuructures*, *1*(1), 10–21.

Sahin, M., & Kargın, A. (2019h). Neutrosophic triplet group based on set valued neutrosophic quadruple numbers. *Neutrosophic Sets and Systems*, *30*, 122–131.

Sahin, M., Kargın, A., & Smarandache, F. (2019a). Neutrosophic triplet topology. *Neutrosophic Triplet Research*, *1*(4), 43–54.

Sahin, M., Kargın, A., & Uz, M. S. (2020e). Neutrosophic Triplet Partial Bipolar Metric Spaces. *Neutrosophic Sets and Systems*, *33*, 297–312.

Sahin, M., Kargın, A., & Yıldız, I. (2020c). Neutrosophic Triplet Field and Neutrosophic Triplet Vector˙ Space Based on Set Valued Neutrosophic Quadruple Number. *Quadruple Neutrosophic Theory And Applications*, *I*, 52.

Sahin, M., Kargın, A., & Yu¨cel, M. (2020a). Neutrosophic triplet g - metric space. *Neutrosophic Quadruple Research*, *1*(13), 181–202.

Sahin, M., Kargın, A., & Yu¨cel, M. (2020d). Neutrosophic Triplet Partial g-Metric Spaces. *Neutrosophic Sets and Systems*, *33*, 116–133.

Sahin, M., Olgun, N., Ulu¸cay, V., Kargın, A., & Smarandache, F. (2017b). A new similarity measure on falsity value between single valued neutrosophic sets based on the centroid points of transformed single valued neutrosophic numbers with applications to pattern recognition. *Neutrosophic Sets and Systems*, *15*, 31–48.

Sahin, M., Ulucay, V., & Menekse, M. (2018). Some new operations of (α,β,γ) interval cut set of interval valued neutrosophic sets. Int*rosophic Journal of Mathematical and Fundamental Sciences, 50(2)*, 103–120.

Sahin, S., Kargın, A., & Yu¨cel, M. (2021a). Hausdorff Measures on Generalized Set Valued Neutrosophic Quadruple Numbers and Decision Making Applications for Adequacy of Online Education. *Neutrosophic Sets and Systems*, *40*(1), 6.

Thivagar, M. L., Jafari, S., & Devi, V. S. (2018b). The ingenuity of neutrosophic topology N – Topology. *Neutrosophic Set and Systems*, *19*, 91–100.

Thivagar, M. L., Jafari, S., Devi, V. S., & Antonysamy, V. (2018a). A novel approach to nano topology via neutrosophic sets. *Neutrosophic Set and Systems*, *20*, 86–94.

Uluc¸ay, V. (2020b). *Some concepts on interval-valued refined neutrosophic sets and their applications*. J Ambient Intell Human Comput., doi:10.100712652-020-02512-y

Ulucay, V., Deli, I., & S¸ahin, M. (2018b). Similarity measures of bipolar neutrosophic sets and their application to multiple criteria decision making. *Neural Computing & Applications*, *29*(3), 739–748. doi:10.100700521-016-2479-1

Ulucay, V., Kılıc¸, A., Sahin, M., & Deniz, H. (2019b). A new hybrid distance-based similarity measure for refined neutrosophic sets and its application in medical diagnosis. *Matematika*, *35*(1), 83–96. doi:10.11113/matematika.v35.n1.1063

Ulucay, V., & Sahin, M. (2019a). Neutrosophic Multigroups and Applications. *Mathematics*, *7*(1), 95. doi:10.3390/math7010095

Uluc¸ay, V., & S¸ahin, M. (2020a). Decision-making method based on neutrosophic soft expert graphs. In *Neutrosophic Graph Theory and Algorithms* (pp. 33–76). IGI Global. doi:10.4018/978-1-7998-1313-2.ch002

Uluc¸ay, V., S¸ahin, M., & Hassan, N. (2018a). Generalized neutrosophic soft expert set for multiple-criteria decision-making. *Symmetry*, *10*(10), 437. doi:10.3390ym10100437

Ulucay, V., Sahin, M., & Olgun, N. (2018e). Time-neutrosophic soft expert sets and its decision making problem. *Matematika*, *34*(2), 246–260. doi:10.11113/matematika.v34.n2.934

Uluc¸ay, V., S¸ahin, M., Olgun, N., & Kilicman, A. (2017). On neutrosophic soft lattices. *Afrika Matematika*, *28*(3-4), 379–388. doi:10.100713370-016-0447-7

Zadeh, L. A. (1965). Fuzzy sets. *Information and Control*, *8*(3), 338–353. doi:10.1016/S0019-9958(65)90241-X

Chapter 10
Formulae for Finding the Number of NeutroTopological Spaces

Bhimraj Basumatary
Bodoland University, India

Jili Basumatary
Bodoland University, India

Nijwm Wary
Bodoland University, India

ABSTRACT

There is a relation between classical topological space and NeutroTopological space. Every classical topological space generates a NeutroTopological space, and every anti-topological space generates a NeutroTopological space. According to NeutroTopological spaces, NeutroTopological spaces have a broader structure. Thus, the neutrosophic theory's (T, I, F) components are added to traditional topological space, yielding a new structure. Topology is a branch of mathematics that works with particular definitions for spatial structure notions, compares them, and analyses the relationships between the structure and the set's qualities. In topology, the initial step is to define a general definition of the fit, followed by an investigation of the connections between topological structures derived from various methodologies. Finding the number of topologies in a set is challenging work. In this chapter, the formulas for finding the number of NeutroTopological spaces having two and three open sets are defined. Further, formulae for non-NeutroTopological space are defined.

DOI: 10.4018/978-1-6684-4740-6.ch010

INTRODUCTION

Smarandache (1998) defined the idea of neutrosophic logic and the concept of a neutrosophic set. After defining the neutrosophic set many authors applied the idea of the neutrosophic set in many branches of science and technology. Many authors extended the idea of the neutrosophic set to various fields of pure and applied mathematics. Topology has a wide range of applications, from analysis to geometry. Because of these features, many researchers have worked on topology. Salama and Alblowi (2012) proposed the concept of neutrosophic topological space. Sumathi and Arockiarani (2016) introduced the idea of the topological group structure of the neutrosophic set after defining the neutrosophic group, while Sumathi and Arockiarani (2015) researched the fuzzy neutrosophic group. Separation axioms in ordered neutrosophic bitopological space were studied by Devi *et al.* (2017). Mwchahary and Basumatary (2020) studied neutrosophic bitopological space.

The centroid single valued triangular neutrosophic numbers and their applications were studied by Şahin *et al.* (2017), a generalized neutrosophic soft expert set for multiple-criteria decision-making was obtained by Ulu∩ay *et al.* (2018), soft neutrosophic modules were recently introduced by Bal *et al.* (2018), and neutrosophic soft expert multiset and their applications were recently studied by Bakbak *et al.* (2019a, 2019b).

An emerging subject of study called NeutroAlgebras & AntiAlgebras is motivated by the reality around us. All of the axioms and all of the operations are fully defined in classical algebraic structures, but in real life, in many cases, these restrictions are too harsh since in our world we have things that only partially verify some laws or some operations. The concept of neutro-structures and anti-structures was first defined by Smarandache (2019, 2020a). Smarandache (2020c) studied the NeutroAlgebra as a generalization of partial algebra. Şahin *et al.* (2021) discussed the idea of NeutroTopological space and anti-topological space.

The concepts of neutro-groups, neutro-sub-groups, neutro-rings, neutro-subrings, neutro-ideal, neutro-quotient-rings, and anti-groups were explicitly introduced by Agboola (2020a, 2020b, 2020c, 2020d) who also proved some properties of these structures and their sub-structures. Neutro-BE-algebras and anti-BE-algebras are concepts that were presented by Rezaei and Smarandache (2020a). Anti-rings were introduced and some of their features were established by Agboola and Ibrahim (2020). The concept of neutro-vector spaces was first presented by Ibrahim and Agboola (2020b). Rezaei and Smarandache (2020b) worked on the neutrosophic triplet of BIalgebras. Smarandache and Hamidi (2020) introduced neutro-bck-algebra. Ibrahim and Agboola (2020a) studied neutro-hypergroups. Smarandache *et al.* (2020), proposed a new concept of neutro-CI-algebras and anti-CI-algebras. Jiménez *et al.* (2021) studied neutroalgebra for the evaluation of barriers to migrants' access.

Al-Tahan *et al.* (2021a, 2021b) obtained NeutroOrderedAlgebra. Applications to Semigroups. Basumatary *et al.* (2021) studied the properties of NeutroTopological spaces.

The concept of a neutrosophic filter was discussed and neutrosophic set theory was applied to BE-algebras by Rezaei *et al.* (2015). Some mistakes and deficiencies in the original definition of neutrosophic filter are pointed out by some examples by Zhang *et al.* (2020). Moreover, a new definition of the neutrosophic filter is established, some basic properties are presented, and the relationship between fuzzy filters and neutrosophic filters is discussed.

IMPORTANCE OF THE STUDY

The topology of a set describes how its components are spatially related to one another, and it is known that several topologies can exist for the same set. If X is a finite set having n elements. Then the efficient computation of the number of finite topologies $T(n)$ on X in the classical case is an outstanding and open problem. There is still no formula for calculating $T(n)$ and many researchers are working in this particular area. Finding the number of topological spaces is not an easy task. Studying this particular area is also very important in the field of topology and is one of the fascinating and challenging research areas. It suggests that each calculation is valuable on its own. In this study, we find the formulae of finding numbers of NeutroTopological spaces. In a finite set, it can be seen that there are more NeutroTopological spaces than topological spaces. By observing this remarkable fact, the present work has forwarded to find the number of NeutroTopological spaces.

BACKGROUND

Definition 1. (Smarandache, 2020c) The NeutroSophication of the Law

(i) Let X be a non-empty set and $*$ be a binary operation. For some elements $(a,b)\in(X,X)$, $(a*b)\in X$ (degree of well defined (T)) and for other elements (x,y), $(p,q)\in(X,X)$; $[x*y$ is indeterminate (degree of indeterminacy (I)), or $p*q\notin X$ (degree of outer-defined $(F)]$, where (T, I, F) is different from (1,0,0) that represents the Classical Law, and from (0,0,1) that represents the AntiLaw.

(ii) In NeutroAlgebra, the classical well-defined for binary operation $*$ is divided into three regions: degree of well-defined (T), degree of indeterminacy (I) and degree of outer-defined (F) similar to neutrosophic set and neutrosophic logic.

Definition 2. (Şahin *et al.*, 2021) Let *X* be the non-empty set and τ be a collection of subsets of X. Then τ is said to be a NeutroTopology on X *a*nd the pair (X,τ) is said to be a NeutroTopological space, if at least one of the following conditions hold good:

(i) $\left[\left(\varnothing_N \in \tau, X_N \notin \tau\right) \text{or} \left(X_N \in \tau, \varnothing_N \notin \tau\right)\right]$ or $\left[\varnothing_N, X_N \in_{\sim} \tau\right]$.

(ii) For some *n* elements $a_1, a_2, \ldots, a_n \in \tau$, $\cap_{i=1}^{n} a_i \in \tau$ [degree of truth T] and for other *n* elements $b_1, b_2, \ldots, b_n \in \tau, p_1, p_2, \ldots, p_n \in \tau$; [$\left(\cap_{i=1}^{n} b_i \notin \tau\right)$ [degree of falsehood F] or ($\cap_{i=1}^{n} p_i$ is indeterminate (degree of indeterminacy I)], where *n* is finite; [where (T, I, F) is different from (1,0,0) that represents the Classical Axiom, and from (0,0,1) that represents the AntiAxiom].

(iii) For some *n* elements $a_1, a_2, \ldots, a_n \in \tau$, $\cap_{i=1} a_i \in \tau$ [degree of truth T] and for other *n* elements $b_1, b_2, \ldots, b_n \in \tau, p_1, p_2, \ldots, p_n \in \tau$; [$\left(\cap_{i=1} b_i \notin \tau\right)$ [degree of falsehood F] or ($\cap_{i=1} p_i$ is indeterminate (degree of indeterminacy I)], where *n* is finite; [where (T, I, F) is different from (1,0,0) that represents the Classical Axiom, and from (0,0,1) that represents the AntiAxiom].

Remark 1. (Şahin *et al.*, 2021) The symbol "$\in_{\sim}$" will be used for situations where it is an unclear appurtenance (not sure if an element belongs or not to a set). For example, if it is not certain whether "a" is a member of the set *P*, then it is denoted by a $\in_{\sim} P$.

Theorem 1. (Şahin *et al.*, 2021) Let (*X*,τ) be a classical topological space. Then (X,$\tau - \varnothing$) is a NeutroTopological space.

Theorem 2. (Şahin *et al.*, 2021) Let (X,τ) be a classical topological space. Then (X,τ – X) *is* a NeutroTopological space.

MAIN FOCUS OF THE CHAPTER

In this section, some results related to the number of NeutroTopologies and non-NeutroTopologies are obtained.

Result 1. Let *X* be a non-empty finite set such that $|X|=n\geq1$. Then, the number of NeutroTopologies having one open set is two.

Proof. For any non-empty finite set having *n* elements, $|P(X)|=2^n$. It is known that among these 2^n subsets, union and intersection of singleton set containing proper subsets of *X* is itself and so it satisfies none of the three conditions of NeutroTopology. The singleton sets containing the remaining subsets i.e., $\varnothing$ and *X* form NeutroTopology on *X*. Hence, the number of NeutroTopologies having one open set is two.

Example 1. Let $X=\{a,b\}$, then $|P(X)|=2^2=4$ The NeutroTopology having one open set are $\tau_1=\{\emptyset\}$ and $\tau_2=\{X\}$. Hence the number of NeutroTopologies having one open set is two.

Result 2. Let X be a non-empty finite set such that $|X|=n\geq 1$. Then the number of NeutroTopologies having two open set is $4(2^{n-2}-1)$.

Proof. For $|X|=n$, the cardinality of the power set of X i.e., $|P(X)|=2^n$. In $P(X)$, clearly the number of proper subsets of X is 2^n-2. Since all the combinations of $\emptyset$ with one proper subset of X satisfies the first condition of NeutroTopology and hence a NeutroTopology having 2-open sets. Therefore, the number of such combinations equals 2^n-2.

Similarly, all the combinations of X with one proper subset of X satisfies the first condition of NeutroTopology and hence a NeutroTopology having 2-open sets. Therefore, the number of such combinations in this case also equals 2^n-2.

On the other hand, all other combinations of two subsets of X out of 2^n subsets of X satisfy none of the three conditions of NeutroTopology.

Hence the total number of NeutroTopologies having two open sets is $(2^n-2)+(2^n-2)=4(2^{n-1}-1)$.

Result 3. The number of NeutroTopologies having two open sets is twice the number of classical topologies having three open sets.

Example 2. Let $X=\{a,b\}$ then the proper subsets of X are $\{a\},\{b\}$. The NeutroTopologies having two open sets are $\tau_1=\{\emptyset,\{a\}\}$, $\tau_2=\{\emptyset,\{b\}\}$, $\tau_3=\{\{a\},X\}$, $\tau_4=\{\{b\},X\}$. Hence the number of NeutroTopologies having two open set is $4=4.1=4(2^{2-1}-1)$.

Result 4. For a non-empty finite set X, $|X|=n, n\leq 5$, the number of NeutroTopologies having three open set is obtained by $\binom{2^n}{3}-\left\{A+\left(2^n-2\right)\right\}$, where A *is* obtained by sum of the results 5-15.

Result 5. Let X be a non-empty finite set such that $|X|n\geq 1$. The number of non-NeutroTopologies having three subsets of X say $\{A,B,C\}$ such that $|A|=|B|=|C|=k$, $1\leq k<n$ is obtained by $\sum_{i=1}^{n-1}\binom{\binom{n}{i}}{3}$.

Proof. In $P(X)$, the number of proper subsets of cardinality i, $1\leq i\leq n-1$ is equal to the number of i combinations out of n-elements which is done in $\binom{n}{i}$ ways.

Now, the number of non-NeutroTopologies having three subsets of X of cardinality i is equal to number of the combinations of 3 subsets out of $\binom{n}{i}$ subsets, which is done in $\left(\begin{pmatrix} n \\ i \end{pmatrix} \atop 3\right)$ ways. Such combinations do not satisfy conditions of the NeutroTopology. Hence, the total number of such non-NeutroTopologies is $\sum_{i=1}^{n-1}\left(\begin{pmatrix} n \\ i \end{pmatrix} \atop 3\right)$.

Example 3. Let $X=\{a,b,c,d\}$, then $|X|=n=4$. The collection of combination of three subsets of X say $\{A,B,C\}$ such that $|A|=|B|=|C|=k$, $1\leq k<n$ are

$$\neg_1=\{\{a\},\{b\},\{c\}\}, \neg_2=\{\{a\},\{b\},\{d\}\}, \neg_3=\{\{a\},\{c\},\{d\}\}, \neg_4=\{\{b\},\{c\},\{d\}\},$$

$$\neg_5=\{\{a,b\},\{a,c\},\{a,d\}\}, \neg_6=\{\{a,b\},\{a,c\},\{b,c\}\}, \neg_7=\{\{a,b\},\{a,c\},\{b,d\}\},$$

$$\neg_8=\{\{a,b\},\{a,c\},\{c,d\}\}, \neg_9=\{\{a,b\},\{a,d\},\{b,c\}\}, \neg_{10}=\{\{a,b\},\{a,d\},\{b,d\}\},$$

$$\neg_{11}=\{\{a,b\},\{a,d\},\{c,d\}\}, \neg_{12}=\{\{a,b\},\{b,c\},\{b,d\}\}, \neg_{13}=\{\{a,b\},\{b,c\},\{c,d\}\},$$

$$\neg_{14}=\{\{a,b\},\{b,d\},\{c,d\}\}, \neg_{15}=\{\{a,c\},\{a,d\},\{b,c\}\}, \neg_{16}=\{\{a,c\},\{a,d\},\{b,d\}\},$$

$$\neg_{17}=\{\{a,c\},\{a,d\},\{c,d\}\}, \neg_{18}=\{\{a,c\},\{b,c\},\{b,d\}\}, \neg_{19}=\{\{a,c\},\{b,c\},\{c,d\}\},$$

$$\neg_{20}=\{\{a,c\},\{b,d\},\{c,d\}\}, \neg_{21}=\{\{a,d\},\{b,c\},\{b,d\}\}, \neg_{22}=\{\{a,d\},\{b,c\},\{c,d\}\},$$

$$\neg_{23}=\{\{a,d\},\{b,d\},\{c,d\}\}, \neg_{24}=\{\{b,c\},\{b,d\},\{c,d\}\}, \neg_{25}=\{\{a,b,c\},\{a,b,d\},\{a,c,d\}\},$$

$$\neg_{26}=\{\{a,b,c\},\{a,b,d\},\{b,c,d\}\}, \neg_{27}=\{\{a,b,c\},\{a,c,d\},\{b,c,d\}\}, \neg_{28}=\{\{a,b,d\},\{a,c,d\},\{b,c,d\}\}.$$

Clearly, these collections do not form NeutroTopology.
There are 28 collection and

$$28 = 4 + 20 + 4 = \left(\begin{pmatrix}4\\1\end{pmatrix} \atop 3\right) + \left(\begin{pmatrix}4\\2\end{pmatrix} \atop 3\right) + \left(\begin{pmatrix}4\\3\end{pmatrix} \atop 3\right) = \sum_{i=1}^{3} \left(\begin{pmatrix}4\\i\end{pmatrix} \atop 3\right) = \sum_{i=1}^{4-1} \left(\begin{pmatrix}4\\i\end{pmatrix} \atop 3\right).$$

Result 6. Let X be a non-empty finite set such that $|X|=n\geq 1$. The number of non-NeutroTopologies having three subsets of X say $\{A,B,C\}$ such that $A\subseteq B\subseteq C$ is obtained by $4!S(n,4)$.

Example 4. The collection of combinations of three subsets of $X=\{a,b,c,d\}$, say $\{A,B,C\}$ such that $A\subseteq B\subseteq C$ are

$A_1=\{\{a\},\{a,b\},\{a,b,c\}\}$, $A_2=\{\{a\},\{a,b\},\{a,b,d\}\}$, $A_3=\{\{a\},\{a,c\},\{a,b,c\}\}$, $A_4=\{\{a\},\{a,c\},\{a,c,d\}\}$, $A_5=\{\{a\},\{a,d\},\{a,b,d\}\}$, $A_6=\{\{a\},\{a,d\},\{a,c,d\}\}$, $A_7=\{\{b\},\{a,b\},\{a,b,c\}\}$, $A_8=\{\{b\},\{a,b\},\{a,b,d\}\}$, $A_9=\{\{b\},\{b,c\},\{a,b,c\}\}$, $A_{10}=\{\{b\},\{b,c\},\{b,c,d\}\}$, $A_{11}=\{\{b\},\{b,d\},\{a,b,d\}\}$, $A_{12}=\{\{b\},\{b,d\},\{b,c,d\}\}$, $A_{13}=\{\{c\},\{a,c\},\{a,b,c\}\}$, $A_{14}=\{\{c\},\{a,c\},\{a,c,d\}\}$, $A_{15}=\{\{c\},\{b,c\},\{a,b,c\}\}$, $A_{16}=\{\{c\},\{b,c\},\{b,c,d\}\}$, $A_{17}=\{\{c\},\{c,d\},\{a,c,d\}\}$, $A_{18}=\{\{c\},\{c,d\},\{b,c,d\}\}$, $A_{19}=\{\{d\},\{a,d\},\{a,b,d\}\}$, $A_{20}=\{\{d\},\{a,d\},\{a,c,d\}\}$, $A_{21}=\{\{d\},\{b,d\},\{a,b,d\}\}$, $A_{22}=\{\{d\},\{b,d\},\{b,c,d\}\}$, $A_{23}=\{\{d\},\{c,d\},\{a,c,d\}\}$, $A_{24}=\{\{d\},\{c,d\},\{b,c,d\}\}$.

Clearly these collection does not form NeutroTopology. Hence the number of non-NeutroTopologies of the form $A\subseteq B\subseteq C$ is $24=4!S(4.4)$.

Result 7. The number of non-NeutroTopologies having three subsets of a non-empty finite set X, $|X|=n$, say $\{A,B,C\}$ such that $|A|=|B|=1$, $2\leq|C|\leq n-2$ is obtained by,

$$\binom{n}{2}\times\sum_{i=2}^{n-2}\binom{n-2}{i}.$$

Example 5. The collection of combinations of three subsets of $X=\{a,b,c,d\}$, say $\{A,B,C\}$ such that $|A|=|B|=1$, $2\leq|C|\leq n-2$ and which does not form NeutroTopology are

$$\urcorner_1=\{\{a\},\{b\},\{c,d\}\},\urcorner_2=\{\{a\},\{c\},\{b,d\}\},\urcorner_3=\{\{a\},\{d\},\{b,c\}\},$$
$$\urcorner_4=\{\{b\},\{c\},\{a,d\}\},\urcorner_5=\{\{b\},\{d\},\{a,c\}\},\urcorner_6=\{\{c\},\{d\},\{a,b\}\}.$$

Hence, in this case the number of non-NeutroTopologies having three subsets of X is 6 and the number of such non NeutroTopologies is obtained by result 7.

Result 8. The number of non-NeutroTopologies having three subsets of a non-empty finite set X, $|X|=n$, say $\{A,B,C\}$ such that $|A|=1$, $|B|=|C|=k$, $2\le k\le n-2$, $A\cap B=A\cap C=\varnothing$ is obtained by

$$n\sum_{i=2}^{n-2}\begin{pmatrix}\begin{pmatrix}n-1\\ i\end{pmatrix}\\ 2\end{pmatrix}.$$

Example 6. The collection of combinations of three subsets of X say $\{A,B,C\}$ such that $|A|=1$, $|B|=|C|=k$, $2\le k\le n-2$, $A\cap B=A\cap C=\varnothing$ for $X=\{a,b,c,d\}$ and which does not form NeutroTopologies are

$$\urcorner_1=\{\{a\},\{b,c\},\{b,d\}\},\urcorner_2=\{\{a\},\{b,c\},\{c,d\}\},\urcorner_3=\{\{a\},\{b,d\},\{c,d\}\},$$

$$\urcorner_4=\{\{b\},\{a,c\},\{a,d\}\},\urcorner_5=\{\{b\},\{a,c\},\{c,d\}\},\urcorner_6=\{\{b\},\{a,d\},\{c,d\}\},$$

$$\urcorner_7=\{\{c\},\{a,b\},\{a,d\}\},\urcorner_8=\{\{c\},\{a,b\},\{b,d\}\},\urcorner_9=\{\{c\},\{a,d\},\{b,d\}\},$$

$$\urcorner_{10}=\{\{d\},\{a,b\},\{a,c\}\},\urcorner_{11}=\{\{d\},\{a,b\},\{b,c\}\},\urcorner_{12}=\{\{d\},\{a,c\},\{b,c\}\}.$$

Hence, in this case the number of non-NeutroTopologies having three subsets of X is 12 and the number of such non NeutroTopologies is obtained by result 8.

Result 9. The number of non-NeutroTopologies having three subsets of a non-empty finite set X, $|X|=n$, say $\{A,B,C\}$ such that $|A|=1$, $|A|<|B|<|C|\le n-2$, $A\cap B=A\cap C=\varnothing$, $|B\cap C|=0$ or 1 is obtained by

$$\binom{n}{1}\times\binom{n-1}{2}\times\sum_{i=1}^{n-4}\left\{\binom{n-1}{i+2}-\binom{n-3}{i}\right\}.$$

Example 7. The collection of combinations of three subsets of $X=\{a,b,c,d\}$, say $\{A,B,C\}$ such that $|A|=1$, $|A|<|B|<|C|\le n-2$, $A\cap B=A\cap C=\varnothing$, $|B\cap C|=0$ or 1 and which does not form NeutroTopologies are

$$\urcorner_1=\{\{a\},\{b,c\},\{b,d,e\}\},\urcorner_2=\{\{a\},\{b,c\},\{c,d,e\}\},\urcorner_3=\{\{a\},\{b,d\},\{b,c,e\}\},$$

$$\urcorner_{4}=\{\{a\},\{b,d\},\{c,d,e\}\},\urcorner_{5}=\{\{a\},\{b,e\},\{b,c,d\}\},\urcorner_{6}=\{\{a\},\{b,e\},\{c,d,e\}\},$$

$$\urcorner_{7}=\{\{a\},\{c,d\},\{b,c,e\}\},\urcorner_{8}=\{\{a\},\{c,d\},\{b,d,e\}\},\urcorner_{9}=\{\{a\},\{c,e\},\{b,c,d\}\},$$

$$\urcorner_{10}=\{\{a\},\{c,e\},\{b,d,e\}\},\urcorner_{11}=\{\{a\},\{d,e\},\{b,c,d\}\},\urcorner_{12}=\{\{a\},\{d,e\},\{b,c,e\}\},$$

$$\urcorner_{13}=\{\{b\},\{a,c\},\{a,d,e\}\},\urcorner_{14}=\{\{b\},\{a,c\},\{c,d,e\}\},\urcorner_{15}=\{\{b\},\{a,d\},\{a,c,e\}\},$$

$$\urcorner_{16}=\{\{b\},\{a,d\},\{c,d,e\}\},\urcorner_{17}=\{\{b\},\{a,e\},\{a,c,d\}\},\urcorner_{18}=\{\{b\},\{a,e\},\{c,d,e\}\},$$

$$\urcorner_{19}=\{\{b\},\{c,d\},\{a,c,e\}\},\urcorner_{20}=\{\{b\},\{c,d\},\{a,d,e\}\},\urcorner_{21}=\{\{b\},\{c,e\},\{a,d,e\}\},$$

$$\urcorner_{22}=\{\{b\},\{c,e\},\{a,c,d\}\},\urcorner_{23}=\{\{b\},\{d,e\},\{a,c,d\}\},\urcorner_{24}=\{\{b\},\{d,e\},\{a,c,e\}\},$$

$$\urcorner_{25}=\{\{c\},\{a,b\},\{a,d,e\}\},\urcorner_{26}=\{\{c\},\{a,b\},\{b,d,e\}\}.\urcorner_{27}=\{\{c\},\{a,d\},\{a,b,e\}\},$$

$$\urcorner_{28}=\{\{c\},\{a,d\},\{b,d,e\}\},\urcorner_{29}=\{\{c\},\{a,e\},\{a,b,d\}\},\urcorner_{30}=\{\{c\},\{a,e\},\{b,d,e\}\},$$

$$\urcorner_{31}=\{\{c\},\{b,d\},\{a,b,e\}\},\urcorner_{32}=\{\{c\},\{b,d\},\{a,d,e\}\},\urcorner_{33}=\{\{c\},\{b,e\},\{a,b,d\}\},$$

$$\urcorner_{34}=\{\{c\},\{b,e\},\{a,d,e\}\},\urcorner_{35}=\{\{c\},\{d,e\},\{a,b,d\}\},\urcorner_{36}=\{\{c\},\{d,e\},\{a,b,e\}\},$$

$$\urcorner_{37}=\{\{d\},\{a,b\},\{a,c,e\}\},\urcorner_{38}=\{\{d\},\{a,b\},\{b,c,e\}\},\urcorner_{39}=\{\{d\},\{a,c\},\{a,b,e\}\},$$

$$\urcorner_{40}=\{\{d\},\{a,c\},\{b,c,e\}\},\urcorner_{41}=\{\{d\},\{a,e\},\{a,b,c\}\},\urcorner_{42}=\{\{d\},\{a,e\},\{b,c,e\}\},$$

$$\urcorner_{43}=\{\{d\},\{b,c\},\{a,b,e\}\},\urcorner_{44}=\{\{d\},\{b,c\},\{a,c,e\}\},\urcorner_{45}=\{\{d\},\{b,e\},\{a,b,c\}\},$$

$\urcorner_{46}=\{\{d\},\{b,e\},\{a,c,e\}\},\urcorner_{47}=\{\{d\},\{c,e\},\{a,b,c\}\},\urcorner_{48}=\{\{d\},\{c,e\},\{a,b,e\}\}$,

$\urcorner_{49}=\{\{e\},\{a,b\},\{a,c,d\}\},\urcorner_{50}=\{\{e\},\{a,b\},\{b,c,d\}\},\urcorner_{51}=\{\{e\},\{a,c\},\{a,b,d\}\}$,

$\urcorner_{52}=\{\{e\},\{a,c\},\{b,c,d\}\},\urcorner_{53}=\{\{e\},\{a,d\},\{a,b,c\}\},\urcorner_{54}=\{\{e\},\{a,d\},\{b,c,d\}\}$,

$\urcorner_{55}=\{\{e\},\{b,c\},\{a,b,d\}\},\urcorner_{56}=\{\{e\},\{b,c\},\{a,c,d\}\},\urcorner_{57}=\{\{e\},\{b,d\},\{a,b,c\}\}$,

$\urcorner_{58}=\{\{e\},\{b,d\},\{a,c,d\}\},\urcorner_{59}=\{\{e\},\{c,d\},\{a,b,c\}\},\urcorner_{60}=\{\{e\},\{c,d\},\{a,b,d\}\}$.

Hence, in this case the number of non-NeutroTopologies having three subsets of X is 60 and the number of such non NeutroTopologies is obtained by result 9.

Result 10. The number of non-NeutroTopologies having three subsets of a non-empty finite set X,$|X|=n$, say $\{A,B,C\}$ such that $|A|=2$, $|B|=|C|=3$, $B\cap C\neq A$, $|A\cap B|=|A\cap C|=0$ or 1 is obtained by

$$\binom{n}{2}\times\sum_{i=1}^{\binom{n}{3}-\{(n-2)+1\}} i .$$

Example 8. Let $X=\{a,b,c,d\}$, the collection of combinations of three subsets of X say $\{A,B,C\}$ such that $|A|=2$, $|B|=|C|=3$, $B\cap C\neq A$, $|A\cap B|=|A\cap C|=0$ or 1 and which does not form NeutroTopologies are

$\urcorner_{1}=\{\{a,b\},\{a,c,d\},\{b,c,d\}\},\urcorner_{2}=\{\{a,c\},\{a,b,d\},\{b,c,d\}\},\urcorner_{3}=\{\{a,d\},\{a,b,c\},\{b,c,d\}\}$,
$\urcorner_{4}=\{\{b,c\},\{a,b,d\},\{a,c,d\}\},\urcorner_{5}=\{\{b,d\},\{a,b,c\},\{a,c,d\}\},\urcorner_{6}=\{\{c,d\},\{a,b,c\},\{a,b,d\}\}$.

Hence, in this case the number of non-NeutroTopologies having three subsets of X is 6 and the number of such non-NeutroTopologies is obtained by result 10.

Result 11. The number of non-NeutroTopologies having three subsets of a non-empty finite set X, $|X|=n$, say $\{A,B,C\}$ such that $|A|=|B|=2$, $2<|C|\leq n-1$ is obtained by

$$\left[\sum_{i=1}^{n-3}\binom{n}{i+2}-\left\{\binom{n-2}{i}+\binom{n-3}{i}\right\}\right]\times\sum_{t=1}^{n-2}a_t+\left[\sum_{i=1}^{n-4}\binom{n}{i+2}-2\binom{n-2}{i}\right]\times\sum_{p=1}^{n-3}a_p,$$

where $a_t = \sum_{j=1}^{t} 3j$ and $a_p = \frac{p(p+1)}{2} \times (p+2)$.

Result 12. The number of non-NeutroTopologies having three subsets of a non-empty finite set X, $|X|=n\leq 5$, say $\{A,B,C\}$ such that $|A|=2$, $|A|<|B|<|C|\leq n-2$, $A\bigcap B\neq A$, $A\bigcap C\neq A$, $B\bigcap C\neq B$, is obtained by

$$2\times\binom{n}{2}\times\binom{n-2}{2}\times\left[\binom{n}{4}-\left\{\binom{n-2}{2}+\binom{n-4}{1}\right\}\right]+\binom{n}{2}\times\binom{n-2}{3}\times\left[\binom{n}{4}-\left\{\binom{n-2}{2}+\binom{n-3}{1}\right\}\right].$$

Result 13. The number of non-NeutroTopologies having three subsets of a non-empty finite set X, $|X|_n\leq 5$, say $\{A,B,C\}$ such that $|A|=2$, $|B|=|C|=4$, $A\bigcap B\neq A$, $A\bigcap C\neq A$, $B\bigcap C\neq A$ is obtained by

$$\binom{n}{2}\times\binom{2\times\binom{n-2}{3}}{2}.$$

Example 9. The non-NeutroTopology having three subsets of a non-empty finite set $X=\{a,b,c,d,e\}$, $|X|=5$ say $\{A,B,C\}$, such that that $|A|=2$, $|B|=|C|=4$, $A\bigcap B\neq A$, $A\bigcap C\neq A$, $B\bigcap C\neq A$ are

$\{\{a,b\},\{a,c,d,e\},\{b,c,d,e\}\}, \{\{a,c\},\{a,b,d,e\},\{b,c,d,e\}\}, \{\{a,d\},\{a,b,c,e\},\{b,c,d,e\}\}, \{\{a,e\},\{a,b,c,d\},\{b,c,d,e\}\}, \{\{b,c\},\{a,b,d,e\},\{a,c,d,e\}\}, \{\{b,d\},\{a,b,c,e\},\{a,c,d,e\}\}, \{\{b,e\},\{a,b,c,d\},\{a,c,d,e\}\}, \{\{c,d\},\{a,b,c,e\},\{a,b,d,e\}\}, \{\{c,e\},\{a,b,c,d\},\{a,b,d,e\}\}, \{\{d,e\},\{a,b,c,d\},\{a,b,c,e\}\}$.

Result 14. The number of non-NeutroTopologies having three subsets of a non-empty finite set X, $|X|=n\leq 5$, say $\{A,B,C\}$ such that $|A|=3$, $3<|B|=|C|\leq n-1$, $A\bigcap B\neq A$, $A\bigcap C\neq A$, $B\bigcap C\neq A$ is obtained by

$$\binom{n}{3}\times\binom{\binom{n}{4}-(n-3)}{2}.$$

Result 15. The number of non-NeutroTopologies having three subsets of a non-empty finite set X, $|X|=n\leq 5$ say $\{A,B,C\}$ such that $|A|=|B|=3$, $3<|C|\leq n-1$ is obtained by

$$\binom{n}{2}\times\binom{n-2}{2}\times\left\{\binom{n}{4}-\binom{n-2}{2}\right\}+n(n-2)\times\left\{\binom{n}{4}-\binom{n-1}{3}\right\}.$$

Result 16. For a non-empty finite set X, $|X|=n\geq 3$, the number of NeutroTopologies having 2^n-1 open set is 2^n.

Proof. Let X be a non-empty finite set such that $|X|=n\geq 3$. The collection of 2^n-1 subsets of X does not contain only one subset of X and $(X,P(X))$ is a topological space. Theorem 1 and Theorem 2, implies that $(X,P(X)-\varnothing)$ and $(X,P(X)-X)$ is a NeutroTopology having 2^n-1 open sets. This gives two NeutroTopologies having 2^n-1 open sets.

Also, if any one of proper subsets of X is absent then it satisfies condition 2 or 3 of NeutroTopology. The number of such collection is equal to the number of proper subsets of X i.e., 2^n-2. Hence, the total number of NeutroTopologies having 2^n-1 open set is $2+2^n-2=2^n$.

Example 10. Let $X=\{a,b,c\}$, then, $|X|=n=3$. Therefore, the NeutroTopologies of 2^n-1 open sets are

$\tau_1=\{\varnothing,\{a\},\{b\},\{c\},\{a,b\},\{a,c\},\{b,c\}\}$, $\tau_2=\{X,\{a\},\{b\},\{c\},\{a,b\},\{a,c\},\{b,c\}\}$, $\tau_3=\{\varnothing,\{b\},\{c\},\{a,b\},\{a,c\},\{b,c\},X\}$, $\tau_4=\{\varnothing,\{a\},\{c\},\{a,b\},\{a,c\},\{b,c\},X\}$, $\tau_5=\{\varnothing,\{a\},\{b\},\{a,b\},\{a,c\},\{b,c\},X\}$, $\tau_6=\{\varnothing,\{a\},\{b\},\{c\},\{a,c\},\{b,c\},X\}$, $\tau_7=\{\varnothing,\{a\},\{b\},\{c\},\{a,b\},\{a,c\},X\}$, $\tau_8=\{\varnothing,\{a\},\{b\},\{c\},\{a,b\},\{b,c\},X\}$.

Result 17. For a non-empty finite set X, $|X|=n\geq 1$, the number of NeutroTopologies having 2^n open set is zero.

Proof. For any non-empty finite set X, $|X|=n\geq 1$, the collection of 2^n subsets of X is the power set of X, the classical discrete topology and clearly this collection does not satisfy conditions of NeutroTopology. Hence not a NeutroTopology.

Remark 2. It is observed that for any non-empty finite set X, $|X|=n\geq 1$

(a) the minimum number of open sets in a NeutroTopology on X is one.

Pro*of. P*roof is straightforward by result 1.

(b) the maximum number of open sets in a NeutroTopology on X is 2^n-1.

P*roof.* Result 16 provides the number of NeutroTopologies having 2n-1 open sets and result 17 implies that there is no NeutroTopology having 2n open sets. Also, by definition of NeutroTopology, it is clear that NeutroTopology having more than 2n open sets are not possible. Hence the maximum number of open sets in a NeutroTopology on X is 2n-1.

FUTURE RESEARCH DIRECTIONS

The results obtained in this chapter will help researchers to compute the number of NeutroTopological spaces containing 4-open, 5-open sets, and 6-open sets.

CONCLUSION

In this chapter, the formulae of the number of NeutroTopological spaces are determined with examples. Further, formulae of non-NeutroTopological space are determined. We hope, this work can help in further study of finding the number of NeutroTopological spaces.

FUNDING

This research received no specific grant from any funding agency in the public, commercial, or not-for-profit sectors.

ACKNOWLEDGMENT

The authors are grateful to the referees for their valuable comments and suggestions.

REFERENCES

Agboola, A. A. A. (2020a). Introduction to anti groups. *International Journal of Neutrosophic Science*, *12*(2), 71–80.

Agboola, A. A. A. (2020b). Introduction to NeutroGroups. *International Journal of Neutrosophic Science*, *6*, 41–47.

Agboola, A. A. A. (2020c). Introduction to NeutroRings. *International Journal of Neutrosophic Science*, *7*(2), 62–73.

Agboola, A. A. A. (2020d). On Finite NeutroGroups of Type-NG. *International Journal of Neutrosophic Science, 10*(2), 84–95.

Agboola, A. A. A., & Ibrahim, M. A. (2020). Introduction to antirings. *Neutrosophic Sets and Systems, 36*, 293–307.

Al-Tahan, M., Davvaz, B., Smarandache, F., & Anis, O. (2021a). On Some NeutroHyperstructures. *Symmetry, 13*(4), 535. doi:10.3390ym13040535

Al-Tahan, M., Smarandache, F., & Davvaz, B. (2021b). NeutroOrderedAlgebra: Applications to Semigroups. *Neutrosophic Sets and Systems, 39*, 133–147.

Bakbak, D., Ulu∩ay, V., & Sahin, M. (2019a). Neutrosophic multigroups and applications. *Mathematics, 7*(1), 95. doi:10.3390/math7010095

Bakbak, D., Ulu∩ay, V., & Şahin, M. (2019b). Neutrosophic soft expert multiset and their application to multiple criteria decision making. *Mathematics, 7*(1), 50. doi:10.3390/math7010050

Bal, M., & Olgun, N. (2018). Soft neutrosophic modules. *Mathematics, 6*(12), 323. doi:10.3390/math6120323

Basumatary, B., & Broumi, S. (2020). Interval-valued triangular neutrosophic linear programming problem. *International Journal of Neutrosophic Science, 10*(2), 105–115.

Basumatary, B., Wary, N., Mwchahary, D. D., Brahma, A. K., Moshahary, J., Basumatary, U. R., & Basumatary, J. (2021). A study on some properties of neutrosophic multi topological group. *Symmetry, 13*(9), 1689. doi:10.3390ym13091689

Devi, R., Dhavaseelan, R., & Jafari, S. (2017). On separation axioms in an ordered neutrosophic bitopological space. *Neutrosophic Sets and Systems, 18*, 27–36.

Ibrahim, M. A., & Agboola, A. A. A. (2020a). Introduction to NeutroHyperGroups. *Neutrosophic Sets and Systems, 38*, 15–32.

Ibrahim, M. A., & Agboola, A. A. A. (2020b). NeutroVectorSpaces I. *Neutrosophic Sets and Systems, 36*, 328–351.

Jiménez, D. S., Mayorga, J. A. V., Ubilla, M. E. R., & Hernández, N. B. (2021). NeutroAlgebra for the evaluation of barriers to migrants' access in Primary Health Care in Chile based on PROSPECTOR function. *Neutrosophic Sets and Systems, 39*, 1–9.

Mwchahary, D. D., & Basumatary, B. (2020). A note on neutrosophic bitopological space. *Neutrosophic Sets and Systems, 33*, 134–144.

Rezaei, A., Borumand Saeid, A., & Smarandache, F. (2015). Neutrosophic filters in BE-algebras. *Ratio Mathematica, 29*, 65–79.

Rezaei, A. & Smarandache, F. (2020a). On Neutro-BE-algebras and Anti-BE-algebras, *International Journal of Neutrosophic Science, 4*(1), 8-15.

Rezaei, A., & Smarandache, F. (2020b). The neutrosophic triplet of BI-algebras. *Neutrosophic Sets and Systems*, *33*, 313–321.

Şahin, M., Ecemiş, O., Uluɳay, V., & Kargın, A. (2017). Some new generalized aggregation operators based on centroid single valued triangular neutrosophic numbers and their applications in multi-attribute decision making. *Asian Journal of Mathematics and Computer Research*, *16*(2), 63–84.

Şahin, M., Kargın, A., & Yücel, M. (2021). Neutro-Topological space and Anti-Topological space. *NeutroAlgebra Theory*, *I*, 16.

Salama, A. A., & Alblowi, S. A. (2012). Neutrosophic set and neutrosophic topological spaces. *IOSR Journal of Mathematics*, *3*(4), 31–35. doi:10.9790/5728-0343135

Smarandache, F. (1998). Neutrosophy. Neutrosophic Probability, Set, and Logic. ProQuest Information & Learning.

Smarandache, F. (2019). *Introduction to NeutroAlgebraic Structures and AntiAlgebraic Structures. In Advances of Standard and Nonstandard Neutrosophic Theories*. Pons Publishing House Brussels.

Smarandache, F. (2020a). Introduction to NeutroAlgebraic Structures and AntiAlgebraic Structures (revisited). *Neutrosophic Sets and Systems*, *31*, 1–16.

Smarandache, F. (2020b). Generalizations and Alternatives of Classical Algebraic Structures to NeutroAlgebraic Structures and AntiAlgebraic Structures. *Journal of Fuzzy Extension and Applications*, *1*(2), 85–87.

Smarandache, F. (2020c). NeutroAlgebra is a generalization of partial algebra. *International Journal of Neutrosophic Science*, *2*, 8–17. doi:10.54216/IJNS.020103

Smarandache, F. (2020d). Structure, NeutroStructure, and AntiStructure in science. *International Journal of Neutrosophic Science*, *13*(1), 28–33.

Smarandache, F., Rezaei, A., & Kim, H. S. (2020). A new trend to extensions of CI-algebras. *International Journal of Neutrosophic Science*, *5*(1), 8. doi:10.54216/IJNS.050101

Sumathi, I. R., & Arockiarani, I. (2015). Fuzzy neutrosophic groups. *Advanced in Fuzzy Mathematics*, *10*(2), 117–122.

Sumathi, I. R., & Arockiarani, I. (2016). Topological group structure of neutrosophic set. *Journal of Advanced Studies in Topology*, *7*(1), 12–20. doi:10.20454/jast.2016.1003

Uluḁay, V., Şahin, M., & Hassan, N. (2018). Generalized neutrosophic soft expert set for multiple-criteria decision-making. *Symmetry*, *10*(10), 437. doi:10.3390ym10100437

Zhang, X., Yu, P., Smarandache, F., & Park, C. (2020). Redefined neutrosophic filters in BE-algebras. *International Journal of Pure and Applied Mathematics*, *43*, 158–176.

ADDITIONAL READING

Atanassov, K. T. (1986). Intuitionistic fuzzy sets. *Fuzzy Sets and Systems*, *20*(1), 87–96. doi:10.1016/S0165-0114(86)80034-3

Basumatary, B., Khaklary, J., Wary, N., & Smarandache, F. (2022). *On Neutro-Topological space and their properties, Theory and Applications of NeutroAlgebras as Generalizations of Classical Algebras*. IGI Global Publisher. doi:10.4018/978-1-6684-3495-6

Blizard, W. (1989). Multiset theory. *Notre Dame Journal of Formal Logic*, *30*, 36–66.

Chang, C. L. (1968). Fuzzy topological space. *Journal of Mathematical Analysis and Applications*, *24*(1), 182–190. doi:10.1016/0022-247X(68)90057-7

Coker, D. (1997). An introduction to intuitionistic fuzzy topological spaces. *Fuzzy Sets and Systems*, *88*(1), 81–89. doi:10.1016/S0165-0114(96)00076-0

Kandil, A., Nouth, A. A., & El-Sheikh, S. A. (1995). On fuzzy bitopological spaces. *Fuzzy Sets and Systems*, *74*(3), 353–363. doi:10.1016/0165-0114(94)00333-3

Onasanya, B. O., & Hoskova-Mayerova, S. (2019). Multi-fuzzy group induced by multisets. *International Journal of Pure and Applied Mathematics*, *41*, 597–604.

Salama, A. A., Smarandache, F., & Kroumov, V. (2014). Closed sets and neutrosophic continuous functions. *Neutrosophic Sets and Systems*, *4*, 4–8.

Smarandache, F., & Hamidi, M. (2020). Neutro-bck-algebra. *International Journal of Neutrosophic Science*, *8*(2), 110.

Zadeh, L. A. (1965). Fuzzy sets. *Information and Control*, *8*(3), 338–353. doi:10.1016/S0019-9958(65)90241-X

APPENDIX

Example for Result 11: Let $X=\{a,b,c,d,e\}$, then non NeutroTopologies are

Table 1.

<table>
<tr>
<td>{{a,b},{a,c},{a,d,e}},
{{a,b},{a,c},{b,c,d}},
{{a,b},{a,c},{b,c,e}},
{{a,b},{a,c},{b,d,e}},
{{a,b},{a,c},{c,d,e}},
{{a,b},{a,d},{a,c,e}},
{{a,b},{a,d},{b,c,d}},
{{a,b},{a,d},{b,c,e}},
{{a,b},{a,d},{b,d,e}},
{{a,b},{a,d},{c,d,e}},
{{a,b},{a,e},{a,c,d}},
{{a,b},{a,e},{b,c,d}},
{{a,b},{a,e},{b,c,e}},
{{a,b},{a,e},{b,d,e}},
{{a,b},{a,e},{c,d,e}},
{{a,b},{b,c},{a,c,d}},
{{a,b},{b,c},{a,c,e}},
{{a,b},{b,c},{a,d,e}},
{{a,b},{b,c},{b,d,e}},
{{a,b},{b,c},{c,d,e}},
{{a,b},{b,d},{a,c,d}},
{{a,b},{b,d},{a,c,e}},
{{a,b},{b,d},{a,d,e}},
{{a,b},{b,d},{b,c,e}},
{{a,b},{b,d},{c,d,e}},
{{a,b},{b,e},{a,c,d}},
{{a,b},{b,e},{a,c,e}},
{{a,b},{b,e},{a,d,e}},
{{a,b},{b,e},{b,c,d}},
{{a,b},{b,e},{c,d,e}},
{{a,b},{c,d},{a,c,e}},
{{a,b},{c,d},{a,d,e}},
{{a,b},{c,d},{b,c,e}},
{{a,b},{c,d},{b,d,e}},
{{a,b},{c,e},{a,c,d}},
{{a,b},{c,e},{a,d,e}},
{{a,b},{c,e},{b,c,d}},
{{a,b},{c,e},{b,d,e}},
{{a,b},{d,e},{a,c,d}},
{{a,b},{d,e},{a,c,e}},
{{a,b},{d,e},{b,c,d}},
{{a,b},{d,e},{b,c,e}},
{{a,c},{a,d},{a,b,e}},
{{a,c},{a,d},{b,c,d}},
{{a,c},{a,d},{b,c,e}},
{{a,c},{a,d},{b,d,e}},
{{a,c},{a,d},{c,d,e}},
{{a,c},{a,e},{a,b,d}},
{{a,c},{a,e},{b,c,d}},
{{a,c},{a,e},{b,c,e}},
{{a,c},{a,e},{b,d,e}},
{{a,c},{a,e},{c,d,e}},
{{a,c},{b,c},{a,b,d}},
{{a,c},{b,c},{a,b,e}},
{{a,c},{b,c},{a,d,e}},
{{a,c},{b,c},{b,d,e}},
{{a,c},{b,c},{c,d,e}},
{{a,c},{b,d},{a,b,e}},
{{a,c},{b,d},{a,d,e}},
{{a,c},{b,d},{b,c,e}},</td>
<td>{{a,c},{b,d},{c,d,e}},
{{a,c},{b,e},{a,b,d}},
{{a,c},{b,e},{a,d,e}},
{{a,c},{b,e},{b,c,d}},
{{a,c},{b,e},{c,d,e}},
{{a,c},{c,d},{a,b,d}},
{{a,c},{c,d},{a,b,e}},
{{a,c},{c,d},{a,d,e}},
{{a,c},{c,d},{b,c,e}},
{{a,c},{c,d},{d,e}},
{{a,c},{c,e},{a,b,c}},
{{a,c},{c,e},{a,b,e}},
{{a,c},{c,e},{a,d,e}},
{{a,c},{c,e},{b,c,d}},
{{a,c},{c,e},{b,d,e}},
{{a,c},{d,e},{a,b,d}},
{{a,c},{d,e},{a,b,e}},
{{a,c},{d,e},{b,c,d}},
{{a,c},{d,e},{b,c,e}},
{{a,d},{a,e},{a,b,c}},
{{a,d},{a,e},{b,c,d}},
{{a,d},{a,e},{b,c,e}},
{{a,d},{a,e},{b,d,e}},
{{a,d},{a,e},{c,d,e}},
{{a,d},{b,c},{a,b,e}},
{{a,d},{b,c},{a,c,e}},
{..},
{{a,d},{b,c},{c,d,e}},
{{a,d},{b,d},{a,b,c}},
{{a,d},{b,d},{a,b,e}},
{{a,d},{b,d},{a,c,e}},
{{a,d},{b,d},{b,c,e}},
{{a,d},{b,d},{c,d,e}},
{{a,d},{b,e},{a,b,c}},
{{a,d},{b,e},{a,c,e}},
{{a,d},{b,e},{b,c,d}},
{{a,d},{b,e},{c,d,e}},
{{a,d},{c,d},{a,b,c}},
{{a,d},{c,d},{a,b,e}},
{{a,d},{c,d},{a,c,e}},
{{a,d},{c,d},{b,c,e}},
{{a,d},{c,d},{b,d,e}},
{{a,d},{c,e},{a,b,c}},
{{a,d},{c,e},{a,b,e}},
{{a,d},{c,e},{b,c,d}},
{{a,d},{c,e},{b,d,e}},
{{a,d},{d,e},{a,b,e}},
{{a,d},{d,e},{a,c,e}},
{{a,d},{d,e},{b,c,d}},
{{a,d},{d,e},{b,c,e}},
{{a,d},{d,e},{c,d,e}},
{{a,e},{b,c},{a,b,d}},
{{a,e},{b,c},{a,c,d}},
{{a,e},{b,c},{b,d,e}},
{{a,e},{b,c},{c,d,e}},
{{a,e},{b,d},{a,b,c}},
{{a,e},{b,d},{a,c,d}},
{{a,e},{b,d},{b,c,e}},
{{a,e},{b,d},{c,d,e}},
{{a,e},{b,e},{a,b,c}},</td>
<td>{{a,e},{b,e},{a,b,d}},
{{a,e},{b,e},{a,c,d}},
{{a,e},{b,e},{b,c,d}},
{{a,e},{b,e},{c,d,e}},
{{a,e},{c,d},{a,b,c}},
{{a,e},{c,d},{a,b,d}},
{{a,e},{c,d},{b,c,e}},
{{a,e},{c,d},{b,d,e}},
{{a,e},{c,e},{a,b,d}}
{{a,e},{c,e},{a,c,d}},
{{a,e},{c,e},{b,c,d}},
{{a,e},{c,e},{b,d,e}},
{{a,e},{c,e},{a,b.c}},
{{a,e},{d,e},{a,b,d}},
{{a,e},{d,e},{b,c,d}},
{{a,e},{d,e},{b,c,e}},
{{b,c},{b,d},{a,b,e}},
{{b,c},{b,d},{a,c,d}},
{{b,c},{b,d},{a,c,e}},
{{b,c},{b,d},{a,d,e}},
{{b,c},{b,d},{c,d,e}},
{{b,c},{b,e},{a,b,d}},
{{b,c},{b,e},{a,c,d}},
{{b,c},{b,e},{a,c,e}},
{{b,c},{b,e},{a,d,e}},
{{b,c},{b,e},{c,d,e}},
{{b,c},{c,d},{a,b,d}},
{{b,c},{c,d},{a,b,e}},
{{b,c},{c,d},{a,c,e}},
{{b,c},{c,d},{a,d,e}},
{{b,c},{c,d},{b,d,e}},
{{b,c},{c,e},{a,b,d}},
{{b,c},{c,e},{a,b,e}},
{{b,c},{c,e},{a,c,d}},
{{b,c},{c,e},{a,d,e}},
{{b,c},{c,e},{b,d,e}},
{{b,c},{d,e},{a,b,d}},
{{b,c},{d,e},{a,b,e}},
{{b,c},{d,e},{a,c,d}},
{{b,c},{d,e},{a,c,e}},
{{b,d},{b,e},{a,b,c}},
{{b,d},{b,e},{a,c,d}},
{{b,d},{b,e},{a,c,e}},
{{b,d},{b,e},{a,d,e}},
{{b,d},{b,e},{c,d,e}},
{{b,d},{c,d},{a,b,c}},
{{b,d},{c,d},{a,b,e}},
{{b,d},{c,d},{a,c,e}},
{{b,d},{c,d},{a,d,e}},
{{b,d},{c,d},{b,c,e}},
{{b,d},{c,e},{a,b,c}},
{{b,d},{c,e},{a,b,e}},
{{b,d},{c,e},{a,c,d}},
{{b,d},{c,e},{a,d,e}},
{{b,d},{d,e},{a,b,c}},
{{b,d},{d,e},{a,b,e}},
{{b,d},{d,e},{a,c,d}},
{{b,d},{d,e},{a,c,e}},</td>
<td>{{b,e},{c,d},{a,b,c}},
{{b,e},{c,d},{a,b,d}},
{{b,e},{c,d},{a,c,e}},
{{b,e},{c,d},{a,d,e}},
{{b,e},{c,e},{a,b,c}},
{{b,e},{c,e},{a,b,d}},
{{b,e},{c,e},{a,c,d}},
{{b,e},{c,e},{a,d,e}},
{{b,e},{c,e},{b,c,d}},
{{b,e},{d,e},{a,b,c}},
{{b,e},{d,e},{a,b,d}},
{{b,e},{d,e},{a,c,d}},
{{b,e},{d,e},{a,c,d}},
{{b,e},{d,e},{b,c,d}},
{{c,d},{c,e},{a,b,c}},
{{c,d},{c,e},{a,b,d}},
{{c,d},{c,e},{a,b,e}},
{{c,d},{c,e},{a,d,e}},
{{c,d},{c,e},{b,d,e}},
{{c,d},{d,e},{a,b,c}},
{{c,d},{d,e},{a,b,d}},
{{c,d},{d,e},{a,b,e}},
{{c,d},{d,e},{a,c,e}},
{{c,d},{d,e},{b,c,e}},
{{c,e},{d,e},{a,b,c}},
{{c,e},{d,e},{a,b,d}},
{{c,e},{d,e},{a,b,e}},
{{c,e},{d,e},{a,b,d}},
{{c,e},{d,e},{b,c,d}},
{{a,b},{a,c},{b,c,d,e}},
{{a,b},{a,d},{b,c,d,e}},
{{a,b},{a,e},{b,c,d,e}},
{{a,b},{b,c},{a,c,d,e}},
{{a,b},{b,d},{a,c,d,e}},
{{a,b},{b,e},{a,c,d,e}},
{{a,c},{a,d},{b,c,d,e}},
{{a,c},{a,e},{b,c,d,e}},
{{a,c},{b,c},{a,b,d,e}},
{{a,c},{c,d},{a,b,d,e}},
{{a,c},{c,e},{a,b,d,e}},
{{a,d},{a,e},{b,c,d,e}},
{{a,d},{b,d},{a,b,c,e}},
{{a,d},{c,d},{a,b,c,e}},
{{a,d},{d,e},{a,b,c,e}},
{{a,e},{b,e},{a,b,c,d}},
{{a,e},{c,e},{a,b,c,d}},
{{a,e},{d,e},{a,b,c,d}},
{{b,c},{b,d},{a,c,d,e}},
{{b,c},{b,e},{a,c,d,e}},
{{b,c},{c,d},{a,b,d,e}},
{{b,c},{c,e},{a,b,d,e}},
{{b,d},{b,e},{a,c,d,e}},
{{b,d},{c,d},{a,b,c,e}},
{{b,d},{d,e},{a,b,c,e}},
{{b,e},{c,e},{a,b,c,d}},
{{b,e},{d,e},{a,b,c,d}},
{{c,d},{c,e},{a,b,d,e}},
{{c,d},{d,e},{a,b,c,e}},
{{c,e},{d,e},{a,b,c,d}},</td>
</tr>
</table>

Example for Result 12: Let $X = \{a,b,c,d,e\}$, then non NeutroTopologies are

Table 2.

{{a,b},{a,c,d},{b,c,d,e}},	{{a,e},{a,c,d},{b,c,d,e}},	{{b,e},{b,c,d},{a,c,d,e}},
{{a,b},{a,c,e},{b,c,d,e}},	{{a,e},{b,c,e},{a,b,c,d}},	{{b,e},{c,d,e},{a,b,c,d}},
{{a,b},{a,d,e},{b,c,d,e}},	{{a,e},{b,d,e},{a,b,c,d}},	{{c,d},{a,b,c},{a,b,d,e}},
{{a,b},{b,c,d},{a,c,d,e}},	{{a,e},{c,d,e},{a,b,c,d}},	{{c,d},{a,b,d},{a,b,c,e}},
{{a,b},{b,c,e},{a,c,d,e}},	{{b,c},{a,b,d},{a,c,d,e}},	{{c,d},{a,c,e},{a,b,d,e}},
{{a,b},{b,d,e},{a,c,d,e}},	{{b,c},{a,b,d},{a,c,d,e}},	{{c,d},{a,d,e},{a,b,c,e}},
{{a,c},{a,b,d},{b,c,d,e}},	{{b,c},{a,b,e},{a,c,d,e}},	{{c,d},{b,c,e},{a,b,d,e}},
{{a,c},{a,b,e},{b,c,d,e}},	{{b,c},{a,c,d},{a,b,d,e}},	{{c,d},{b,d,e},{a,b,c,e}},
{{a,c},{a,d,e},{b,c,d,e}},	{{b,c},{a,c,e},{a,b,d,e}},	{{c,e},{a,b,c},{a,b,d,e}},
{{a,c},{b,c,d},{a,b,d,e}},	{{b,c},{b,d,e},{a,c,d,e}},	{{c,e},{a,b,e},{a,b,c,d}},
{{a,c},{b,c,e},{a,b,d,e}},	{{b,c},{c,d,e},{a,b,d,e}},	{{c,e},{a,c,d},{a,b,d,e}},
{{a,c},{c,d,e},{a,b,d,e}},	{{b,d},{a,b,c},{a,c,d,e}},	{{c,e},{a,d,e},{a,b,c,d}},
{{a,d},{a,b,c},{b,c,d,e}},	{{b,d},{a,b,e},{a,c,d,e}},	{{c,e},{b,c,d},{a,b,d,e}},
{{a,d},{a,b,e},{b,c,d,e}},	{{b,d},{a,c,d},{a,b,c,e}},	{{c,e},{b,d,e},{a,b,c,d}},
{{a,d},{a,c,e},{b,c,d,e}},	{{b,d},{a,d,e},{a,b,c,e}},	{{d,e},{a,b,d},{a,b,c,e}},
{{a,d},{b,c,d},{a,b,c,e}},	{{b,d},{b,c,e},{a,c,d,e}},	{{d,e},{a,b,e},{a,b,c,d}},
{{a,d},{b,d,e},{a,b,c,e}},	{{b,d},{c,d,e},{a,b,c,e}},	{{d,e},{a,c,d},{a,b,c,e}},
{{a,d},{c,d,e},{a,b,c,e}},	{{b,e},{a,b,c},{a,c,d,e}},	{{d,e},{a,c,e},{a,b,c,d}},
{{a,e},{a,b,c},{b,c,d,e}},	{{b,e},{a,b,d},{a,c,d,e}},	{{d,e},{b,c,d},{a,b,c,e}},
{{a,e},{a,b,d},{b,c,d,e}},	{{b,e},{a,c,e},{a,b,c,d}},	{{d,e},{b,c,e},{a,b,c,d}},
	{{b,e},{a,d,e},{a,b,c,d}},	

Example for Result 14: Let $X = \{a,b,c,d,e\}$, then non NeutroTopologies are

Table 3.

{{a,b,c},{a,b,d,e},{a,c,d,e}},	{{a,c,d},{a,b,c,e},{b,c,d,e}},	{{b,c,d},{a,b,d,e},{a,c,d,e}},
{{a,b,c},{a,b,d,e},{b,c,d,e}},	{{a,c,d},{a,b,d,e},{b,c,d,e}},	{{b,c,e},{a,b,c,d},{a,b,d,e}},
{{a,b,c},{a,c,d,e},{b,c,d,e}},	{{a,c,e},{a,b,c,d},{a,b,d,e}},	{{b,c,e},{a,b,c,d},{a,c,d,e}},
{{a,b,d},{a,b,c,e},{a,c,d,e}},	{{a,c,e},{a,b,c,d},{b,c,d,e}},	{{b,c,e},{a,b,d,e},{a,c,d,e}},
{{a,b,d},{a,b,c,e},{b,c,d,e}},	{{a,c,e},{a,b,d,e},{b,c,d,e}},	{{b,d,e},{a,b,c,d},{a,b,c,e}},
{{a,b,d},{a,c,d,e},{b,c,d,e}},	{{a,d,e},{a,b,c,d},{a,b,c,e}},	{{b,d,e},{a,b,c,d},{a,c,d,e}},
{{a,b,e},{a,b,c,d},{a,c,d,e}},	{{a,d,e},{a,b,c,d},{b,c,d,e}},	{{b,d,e},{a,b,c,e},{a,c,d,e}},
{{a,b,e},{a,b,c,d},{b,c,d,e}},	{{a,d,e},{a,b,c,e},{b,c,d,e}},	{{c,d,e},{a,b,c,d},{a,b,c,e}},
{{a,b,e},{a,c,d,e},{b,c,d,e}},	{{b,c,d},{a,b,c,e},{a,b,d,e}},	{{c,d,e},{a,b,c,d},{a,b,d,e}},
{{a,c,d},{a,b,c,e},{a,b,d,e}},	{{b,c,d},{a,b,c,e},{a,c,d,e}},	{{c,d,e},{a,b,c,e},{a,b,d,e}},

Example for Result 15: Let $X= \{a,b,c,d,e\}$, then non NeutroTopologies are

Table 4.

{{a,b,c},{a,b,d},{a,c,d,e}},	{{a,b,d},{a,d,e},{a,b,c,e}},	{{a,c,d},{a,c,e},{b,c,d,e}},	{{a,d,e},{b,d,e},{a,b,c,e}},
{{a,b,c},{a,b,d},{b,c,d,e}},	{{a,b,d},{a,d,e},{b,c,d,e}},	{{a,c,d},{a,d,e},{a,b,c,e}},	{{a,d,e},{c,d,e},{a,b,c,d}},
{{a,b,c},{a,b,e},{a,c,d,e}},	{{a,b,d},{b,c,d},{a,b,c,e}},	{{a,c,d},{a,d,e},{b,c,d,e}},	{{a,d,e},{c,d,e},{a,b,c,e}},
{{a,b,c},{a,b,e},{b,c,d,e}},	{{a,b,d},{b,c,d},{a,c,d,e}},	{{a,c,d},{b,c,d},{a,b,c,e}},	{{b,c,d},{b,c,e},{a,b,d,e}},
{{a,b,c},{a,c,d},{a,b,d,e}},	{{a,b,d},{b,c,e},{a,c,d,e}},	{{a,c,d},{b,c,d},{a,b,d,e}},	{{b,c,d},{b,c,e},{a,c,d,e}},
{{a,b,c},{a,c,d},{b,c,d,e}},	{{a,b,d},{b,d,e},{a,b,c,e}},	{{a,c,d},{b,c,e},{a,b,d,e}},	{{b,c,d},{b,d,e},{a,b,c,e}},
{{a,b,c},{a,c,e},{a,b,d,e}},	{{a,b,d},{b,d,e},{a,c,d,e}},	{{a,c,d},{b,d,e},{a,b,c,e}},	{{b,c,d},{b,d,e},{a,c,d,e}},
{{a,b,c},{a,c,e},{b,c,d,e}},	{{a,b,d},{c,d,e},{a,b,c,e}},	{{a,c,d},{c,d,e},{a,b,c,e}},	{{b,c,d},{c,d,e},{a,b,c,e}},
{{a,b,c},{a,d,e},{b,c,d,e}},	{{a,b,e},{a,c,d},{b,c,d,e}},	{{a,c,d},{c,d,e},{a,b,d,e}},	{{b,c,d},{c,d,e},{a,b,d,e}},
{{a,b,c},{b,c,d},{a,b,d,e}},	{{a,b,e},{a,c,e},{a,b,c,d}},	{{a,c,e},{a,d,e},{a,b,c,d}},	{{b,c,e},{b,d,e},{a,b,c,d}},
{{a,b,c},{b,c,d},{a,c,d,e}},	{{a,b,e},{a,c,e},{b,c,d,e}},	{{a,c,e},{a,d,e},{b,c,d,e}},	{{b,c,e},{b,d,e},{a,c,d,e}},
{{a,b,c},{b,c,e},{a,b,d,e}},	{{a,b,e},{a,d,e},{a,b,c,d}},	{{a,c,e},{b,c,d},{a,b,d,e}},	{{b,c,e},{c,d,e},{a,b,c,d}},
{{a,b,c},{b,c,e},{a,c,d,e}},	{{a,b,e},{a,d,e},{b,c,d,e}},	{{a,c,e},{b,c,e},{a,b,c,d}},	{{b,c,e},{c,d,e},{a,b,d,e}},
{{a,b,c},{b,d,e},{a,d,d,e}},	{{a,b,e},{b,c,d},{a,c,d,e}},	{{a,c,e},{b,c,e},{a,b,d,e}},	{{b,d,e},{c,d,e},{a,b,c,d}},
{{a,b,c},{c,d,e},{a,b,d,e}},	{{a,b,e},{b,c,e},{a,b,c,d}},	{{a,c,e},{b,d,e},{a,b,c,d}},	{{b,d,e},{c,d,e},{a,b,c,e}}.
{{a,b,d},{a,b,e},{a,c,d,e}},	{{a,b,e},{b,c,e},{a,c,d,e}},	{{a,c,e},{c,d,e},{a,b,c,d}},	
{{a,b,d},{a,b,e},{b,c,d,e}},	{{a,b,e},{b,d,e},{a,b,c,d}},	{{a,c,e},{c,d,e},{a,b,d,e}},	
{{a,b,d},{a,c,d},{a,b,c,e}},	{{a,b,e},{b,d,e},{a,c,d,e}},	{{a,d,e},{b,c,d},{a,b,c,e}},	
{{a,b,d},{a,c,d},{b,c,d,e}},	{{a,b,e},{c,d,e},{a,b,c,d}},	{{a,d,e},{b,c,e},{a,b,c,d}},	
{{a,b,d},{a,c,e},{b,c,d,e}},	{{a,c,d},{a,c,e},{a,b,d,e}},	{{a,d,e},{b,d,e},{a,b,c,d}},	

Compilation of References

Adhikari, M. R. (2016). *Basic algebraic topology and its applications*. Springer. doi:10.1007/978-81-322-2843-1

Agboola, A. A. A. (2020a). Introduction to anti groups. *International Journal of Neutrosophic Science*, *12*(2), 71–80.

Agboola, A. A. A. (2020b). Introduction to NeutroGroups. *International Journal of Neutrosophic Science*, *6*, 41–47.

Agboola, A. A. A. (2020c). Introduction to NeutroRings. *International Journal of Neutrosophic Science*, *7*(2), 62–73.

Agboola, A. A. A. (2020d). On Finite NeutroGroups of Type-NG. *International Journal of Neutrosophic Science*, *10*(2), 84–95.

Agboola, A. A. A., & Ibrahim, M. A. (2020). Introduction to antirings. *Neutrosophic Sets and Systems*, *36*, 293–307.

Al-Seraji, N. A. M., & Jafar, Z. S. (2020). Applications of the Projective Plane in Coding Theory. *Journal of Southwest Jiaotong University*, *55*(1), 1–8. doi:10.35741/issn.0258-2724.55.1.2

Al-Tahan, M., & Davvaz, B. (2017). Algebraic Hyperstructures Associated to Biological Inheritance. *Mathematical Biosciences*, *285*, 112–118. doi:10.1016/j.mbs.2017.01.002 PMID:28108295

Al-Tahan, M., & Davvaz, B. (2018). Weak Chemical Hyperstructures Associated to Electrochemical Cells. *Iranian Journal of Mathematical Chemistry*, *9*(1), 65–75.

Al-Tahan, M., & Davvaz, B. (2019). Chemical hyperstructures for Astatine, Tellurium and for Bismuth. *Bull. Comput. Appl. Math*, *7*(1), 9–25.

Al-Tahan, M., & Davvaz, B. (2022). Chemical Hyperstructures for Element with Four Oxidation States. *Iranian Journal of Mathematical Chemistry*, *13*(2), 85–97.

Al-Tahan, M., Davvaz, B., Smarandache, F., & Anis, O. (2021). On some NeutroHyperstructures. *Symmetry*, *13*(4), 535. doi:10.3390ym13040535

Al-Tahan, M., Smarandache, F., & Davvaz, B. (2021). NeutroOrderedAlgebra: Applications to semigroups. *Neutrosophic Sets and System*, *39*, 133–147.

Al-Tahan, M., Smarandache, F., & Davvaz, B. (2021b). NeutroOrderedAlgebra: Applications to Semigroups. *Neutrosophic Sets and Systems*, *39*, 133–147.

Aristotle. (1908). Nicomachean Ethics (W. D. Ross, Trans.). Clarendon Press.

Atanassov, K. (1986). Intuitionistic. *Fuzzy Sets and Systems*, *20*(1), 87–96. doi:10.1016/S0165-0114(86)80034-3

Athanassopoulos, E., & Voskoglou, M. (2020). A Philosophical Treatise on the Connection of Scientific Reasoning with Fuzzy Logic. *Mathematics*, *8*(6), 875. doi:10.3390/math8060875

Bakbak, D., Uluçay, V., & Sahin, M. (2019a). Neutrosophic multigroups and applications. *Mathematics*, *7*(1), 95. doi:10.3390/math7010095

Bakbak, D., Uluçay, V., & Şahin, M. (2019b). Neutrosophic soft expert multiset and their application to multiple criteria decision making. *Mathematics*, *7*(1), 50. doi:10.3390/math7010050

Balachandran, A. P. (1991). Classical topology and quantum statistics.*International Journal of Modern Physics B, 5*(16-17), 2585-2623.

Bal, M., & Olgun, N. (2018). Soft neutrosophic modules. *Mathematics*, *6*(12), 323. doi:10.3390/math6120323

Bal, M., Shalla, M., & Olgun, N. (2018). Neutrosophic Triplet Cosets and Quotient Groups. *Symmetry*, *10*(4), 126. doi:10.3390ym10040126

Bal, M., Singh, P. K., & Ahmad, K. D. (2022). A short introduction to the concept of symbolic Turiyam matrix. *Journal of Neutrosophic and Fuzzy Systems*, *2*, 88–99. doi:10.54216/JNFS.020108

Bal, M., Singh, P. K., & Ahmad, K. D. (2022). A Short Introduction To The Symbolic Turiyam Vector Spaces and Complex Numbers. *Journal of Neutrosophic and Fuzzy Systems*, *2*, 76–87. doi:10.54216/JNFS.020107

Bal, M., Singh, P. K., & Ahmad, K. D. (2022). An introduction to the symbolic Turiyam R-modules and Turiyam modulo integers. *Journal of Neutrosophic and Fuzzy Systems*, *2*, 8–19. doi:10.54216/JNFS.020201

Basumatary, B., & Broumi, S. (2020). Interval-valued triangular neutrosophic linear programming problem. *International Journal of Neutrosophic Science*, *10*(2), 105–115.

Basumatary, B., Wary, N., Mwchahary, D. D., Brahma, A. K., Moshahary, J., Basumatary, U. R., & Basumatary, J. (2021). A study on some properties of neutrosophic multi topological group. *Symmetry*, *13*(9), 1689. doi:10.3390ym13091689

Belnap, N.D. (1977). A useful four valued logic. *Modern Uses of Multiple Valued Logic*, 9–37.

Bentham, J. (1789). An Introduction to the Principles of Morals and Legislation. Dover Publications Inc.

Beutelspacher, A. (1990). Applications of finite geometry to cryptography. In G. Longo, M. Marchi, & A. Sgarro (Eds.), *Geometries, Codes and Cryptography* (pp. 161–186). Springer Verlag. doi:10.1007/978-3-7091-2838-1_6

Bhattacharya, S. (2004). A model to the Smarandache geometries. *Journal of Recreational Mathematics, 33*, 166.

Bishnoi, A., Gijsiwijt, D., D'haesleer, J., & Potukuchi, A. (2022). *Trifferent codes and affine blocking sets*. Paper presented at the Applications of Computer Algebra-ACA 2022.

Boyer, C. (1989). *Historia de la geometría*. Editorial Coimoff.

Bresar, M. (2014). *Introduction to noncommutative algebra*. Springer. doi:10.1007/978-3-319-08693-4

Bröcker, T., & Jänich, K. (1982). *Introduction to differential topology*. Cambridge University Press.

Bruck, R. H., & Ryser, H. J. (1949). The Nonexistence of Certain Finite Projective Planes. *Canadian Journal of Mathematics, 1*(1), 88–93. doi:10.4153/CJM-1949-009-2

Brython. (n.d.). http://www.brython.info/

Cameron, P. J. (1994). *Combinatorics: Topics, Techniques, Algorithms*. Cambridge University Press.

Cameron, P. J. (2000). *Projective and Polar Spaces*. Queen Mary and Westfield College.

Celik, B. (2001). On Some Hyperbolic Planes from Finite Projective Planes. *International Journal of Mathematics and Mathematical Sciences, 25*(12), 757–762. doi:10.1155/S0161171201006184

Chang, C. L. (1968). Fuzzy Topological Space. *Journal of Mathematical Analysis and Applications, 24*(1), 182–190. doi:10.1016/0022-247X(68)90057-7

Chapouthier, G. (2004). To what extent is moral judgment natural? *European Review (Chichester, England), 12*(2), 179–183. doi:10.1017/S1062798704000171

Chatterjee, R., Majumdar, P., & Samanta, S. K. (2016). On some similarity measures and entropy on quadripartitioned single valued neutrosophic sets. *Journal of Intelligent & Fuzzy Systems, 30*(4), 2475–2485. doi:10.3233/IFS-152017

Chen, B. C., & Kikuchi, N. (2001). Topology optimization with design-dependent loads. *Finite Elements in Analysis and Design, 37*(1), 57–70. doi:10.1016/S0168-874X(00)00021-4

Cheng-Ming, H. (1985). Fuzzy topological spaces. *Journal of Mathematical Analysis and Applications, 110*(1), 141–178. doi:10.1016/0022-247X(85)90340-3

Chimienti, S. P., & Bencze, M. (1998a). Smarandache paradoxist geometry. *Smarandache Notions Journal, 9*(1-3), 42.

Chimienti, S. P., & Bencze, M. (1998b). Smarandache anti-geometry. *Smarandache Notions Journal, 9*(1-3), 48.

Chung, S. C. (2014). Chemical Hyperstructures for Vanadium. *Journal of Chungcheong Mathematical Society*, *27*(2), 309–317. doi:10.14403/jcms.2014.27.2.309

Chung, S. C. (2019). Chemical Hyperstructures For Ozone Depletion. *Journal of Chungcheong Mathematical Society*, *32*(4), 491–508.

Coker, D. (1997). An introduction to intuitionistic fuzzy topological spaces. *Fuzzy Sets and Systems*, *88*(1), 81–89. doi:10.1016/S0165-0114(96)00076-0

Corsini, P. (1993). Prolegomena of Hypergroup Theory. Udine Aviani Editore.

Coxeter, H.S.M. (1942). *Non-Euclidean geometry*. University of Toronto Press.

Crowe, D. W. (1965). The Construction of Finite Regular Hyperbolic Planes from Inversive Planes of Even Order. *Colloquium Mathematicum*, *13*(2), 247–250. doi:10.4064/cm-13-2-247-250

Crowe, D. W. (1966). Projective and Inversive Models for Finite Hyperbolic Planes. *The Michigan Mathematical Journal*, *13*(2), 251–255. doi:10.1307/mmj/1028999550

CullinaneS. H. (2017). finite-geometry/galois: Release with Jan. 9, 2016, compressed folder of finitegeometry.org (v1.0). Zenodo. doi:10.5281/zenodo.1038121

Das, S., Das, R., & Granados, C. (2021). Topology on quadripartitioned neutrosophic sets. *Neutrosophic Sets and Systems*, *45*, 54–61.

Davvaz, B. & Leoreanu-Fotea. (2022). *Hypergroup Theory*. World Scientific.

Davvaz, B. (1998). On H*υ*–rings and fuzzy H*υ*–ideal. J*ournal of Fuzzy Mathematics,* 6(*1*), 33–42.

Davvaz, B., Nezhad, A. D., & Benvidi, A. (2012). Chemical Hyperalgebra: Dismutation Reactions. M*ATCH Commun.Math.Comput.Chem,* 67*(*1), 55–63.

Davvaz, B., Nezhad, A. D., & Mazloum-Ardakani, M. (2014). Chemical Hyperalgebra: Redox Reactions. *MATCH Commun.Math.Comput.Chem*, *71*(2), 323–331.

Dembowski, P. (1997). Finite Geometries. Berlin: Springer.

Deng, J. (1982). Control Problems of Grey Systems. *Systems & Control Letters*, 288–294.

Devi, R., Dhavaseelan, R., & Jafari, S. (2017). On separation axioms in an ordered neutrosophic bitopological space. *Neutrosophic Sets and Systems*, *18*, 27–36.

Dillon, M. I. (2018). Neutral geometry. In *Geometry Through History*. Springer. doi:10.1007/978-3-319-74135-2_2

Dissett, L. A. (2000). *Combinatorial and Computational Aspects of Finite Geometries* [Unpublished Ph.D. Thesis]. University of Toronto.

Driver, J. (2022). Moral Theory. The Stanford Encyclopedia of Philosophy.

Dubois, D., & Prade, H. (2005), Interval-Valued Fuzzy Sets, Possibility Theory and Imprecise Probability. *Proceedings EUSFLAT-LFA,* 314-319.

Encyclopedia Britannica. (1971). *History of Ethics - Ethics Comparative*. William Benton.

Farrow, C. L., Juhas, P., Liu, J., Bryndin, D., Božin, E., Bloch, T., Proffen, T., & Billinge, S. (2007). PDFfit2 and PDFgui: Computer programs for studying nanostructure in crystals. *Journal of Physics Condensed Matter, 19*(33), 335219. doi:10.1088/0953-8984/19/33/335219 PMID:21694142

García, E. (2018). El medio ambiente sano: La consolidación de un derecho. *Iuris Tantum Revista Boliviana de Derecho, 25*, 550–569.

Ginzburg, V. L. (1979). *Theoretical Physics and Astrophysics*. Pergamon Press.

Goleman, D. (2007). Emotional Intelligence: Why It Can Matter More Than IQ (10th ed.). Random House USA Inc.

Granados, C. (2022). A note on AntiGeometry and NeutroGeometry and their application to real life. *Neutrosophic Sets and Systems, 49*(1), 579–592.

Granados, C. (2022). A note on AntiGeometry and NeutroGeometry and their application to real life. *NeutrosophicSets and Systems, 49*, 579–593. doi:10.5281/zenodo.6466520

Hájek, P. (1998). Metamathematics of Fuzzy Logic. Dordrecht: Springer Science+Business Media. doi:10.1007/978-94-011-5300-3

Heidari, D., Mazaheri, D., & Davvaz, B. (2019). Chemical Salt Reactions as Algebraic Hyperstructures. *Iranian Journal of Mathematics Chemistry, 10*(2), 93–102.

Hirschfeld, J. W. P. (1979). *Projective Geometries over Finite Fields*. Clarendon Press.

Horlemann, A. L. (2022). *The Density of Optimal Error-Correcting Codes in Various Metric Spaces*. Paper presented at the Sixth Irsee Conference.

Hur, K., Kim, J. H., & Ryou, J. H. (2004). Intuitionistic fuzzy topological spaces. *The Pure and Applied Mathematics, 11*, 243–265.

Ibrahim, M. A., & Agboola, A. A. A. (2020). Introduction to NeutroHyperGroups. *Neutrosophic Sets and System, 38*, 15–32.

Ibrahim, M. A., & Agboola, A. A. A. (2020a). NeutroVectorSpaces I. *Neutrosophic Sets and Systems, 36*, 328–351.

Ibrahim, M. A., & Agboola, A. A. A. (2020b). Introduction to NeutroHyperGroups. *Neutrosophic Sets and Systems, 38*(1), 2.

Ilić, J. (2022). *Secret sharing schemes arising from finite geometry* [Unpublished Doctoral thesis]. Univerza na Primorskem, Koper.

Iseri, H. (2002). Partially Paradoxist Smarandache Geometries. *Smarandache Notions Journal, 13*(1-2-3), 5-12.

Jiménez, D. S., Mayorga, J. A. V., Ubilla, M. E. R., & Hernández, N. B. (2021). NeutroAlgebra for the evaluation of barriers to migrants' access in Primary Health Care in Chile based on PROSPECTOR function. *Neutrosophic Sets and Systems*, *39*, 1–9.

Kapoor, P., & Singh, P. (2020). IT act crime pattern analysis using regression and correlation matrix. *Proc. 8th Int. Conf. Reliabil., Infocom Technologies and Optimization*, 1102–1106.

Kärteszi, F. (1976). *Introduction to finite geometries*. American Elsevier Publishing Company, Inc.

Kiss, G., & Szonyi, T. (2020). *Finite Geometries*. CRC Press Taylor & Francis Group.

Klein, A., & Storme, L. (2011). Applications of finite geometry in coding theory and cryptography. In *Information Security, Coding Theory and Related Combinatorics* (pp. 38–58). IOS Press.

Klir, G. J., & Folger, T. A. (1988). *Fuzzy sets, Uncertainty and Information*. Prentice-Hall.

Koestler, A. (1979). *The Ghost in the Machine*. Hutchinson.

Korchmáros, G., & Sonnino, A. (2012). Finite Bolyai-Lobachevskii Planes. *Acta Mathematica Hungarica*, *134*(4), 405–415. doi:10.100710474-011-0181-4

Korchmáros, G., & Sonnino, A. (2014). Old and Recent Results on Finite Bolyai-Lobachevsky Planes. *Mathematica*, *56*(79), 59–73.

Kosko, B. (1993). *Fuzzy thinking: The new science of fuzzy logic*. Hyperion.

Kou, Y., Lin, S., & Fossorier, M. P. C. (2001). Low-Density Parity-Check Codes Based on Finite Geometries: A Rediscovery and New Results. *IEEE Transactions on Information Theory*, *47*(7), 2711–2736. doi:10.1109/18.959255

Kuciuk, L., & Antholy, M. (2003). An introduction to the Smarandache geometries. *Infinite Study*.

Lam, C. W. H. (1991). The Search for a Finite Projective Plane of Order 10. *The American Mathematical Monthly*, *98*(4), 305–318. doi:10.1080/00029890.1991.12000759

Lam, C. W. H., Thiel, L., & Swiercz, S. (1989). The Non-Existence of Finite Projective Planes Of Order 10. *Canadian Journal of Mathematics*, *41*(6), 1117–1123. doi:10.4153/CJM-1989-049-4

Lashin, E. F., Kozae, A. M., Khadra, A. A., & Medhat, T. (2005). Rough set theory for topological spaces. *International Journal of Approximate Reasoning*, *40*(1-2), 35–43. doi:10.1016/j.ijar.2004.11.007

Lee, J. G., & Hur, K. (2020). Hesitant fuzzy topological spaces. *Mathematics*, *8*(2), 188. doi:10.3390/math8020188

Lobachevsky, N. (2010). Pangeometry, translator and editor: A. Papadopoulos. Heritage of European Mathematics Series. European Mathematical Society, 4.

Lutz, M. (2011). Programming Python (4th ed.). O'Reilly.

Maji, P. K., Biswas, R., & Roy, A. R. (2003). Soft Set Theory. *Computers & Mathematics with Applications (Oxford, England)*, *45*(4-5), 555–562. doi:10.1016/S0898-1221(03)00016-6

Mao, L. (2006). *Smarandache Geometries & Map Theory with Applications (I)*. Chinese Branch Xiquan House.

Mao, L. F. (2007). An introduction to Smarandache multi-spaces and mathematical combinatorics. *Scientia Magna*, *3*, 54–80.

Martelli, B. (2016). *An introduction to geometric topology*. arXiv preprint arXiv:1610.02592.

Martelli, F. (2021). Topology and complexity of the hydrogen bond network in classical models of water. *Journal of Molecular Liquids*, *329*, 115530. doi:10.1016/j.molliq.2021.115530

Marty, F. (1934). Sur une generalization de la notion de groupe. *Proceedings of the 8th Congress on Mathmatics, Scandinaves*, 45–49.

Mohanasundari, M., & Mohana, K. (2020). Quadripartitioned Single valued Neutrosophic Dombi Weighted Aggregation Operators for Multiple Attribute Decision Making. *Neutrosophic Sets and Systems*, *32*, 107–122.

Molodtsov, D. (1999). Soft set theory – first results. *Computers & Mathematics with Applications (Oxford, England)*, *37*(4-5), 19–31. doi:10.1016/S0898-1221(99)00056-5

Mondal, T. K., & Samanta, S. K. (2001). Topology of interval-valued intuitionistic fuzzy sets. *Fuzzy Sets and Systems*, *119*(3), 483–494. doi:10.1016/S0165-0114(98)00436-9

Morandi, G. (1992). *The role of topology in classical and quantum physics* (Vol. 7). Springer Science & Business Media. doi:10.1007/978-3-540-46688-8

Munkres, J. R. (2018). *Elements of algebraic topology*. CRC press. doi:10.1201/9780429493911

Mwchahary, D. D., & Basumatary, B. (2020). A note on neutrosophic bitopological space. *Neutrosophic Sets and Systems*, *33*, 134–144.

Neri, A., Santonastaso, P., & Zullo, F. (2023). The geometry of one-weight codes in the sum rank metric. *Journal of Combinatorial Theory Series A*, *194*, 105703. doi:10.1016/j.jcta.2022.105703

Nezhad, A. D., Nejad, S. M. M., Nadjafikhah, M., & Davvaz, B. (2012). A Physical Example of Algebraic Hyperstructures: Leptons. *Indian Journal of Physics and Proceedings of the Indian Association for the Cultivation of Science*, *86*(11), 1027–1032. doi:10.100712648-012-0151-x

Nicolis, J. S. (1991). *Chaos and Information Processing*. World Scientific Publishers. doi:10.1142/0984

Nodelman, (Eds.). (2022). https://plato.stanford.edu/archives/fall2022/entries/moral-theory

O'Keefe, C. M. (1993). Applications of finite geometries to information security. *Australasian Journal of Combinatorics*, *7*, 195–212.

Olgun, S., & Gunaltili, I. (2007). On Finite Homogeneous Bolyai-Lobachevsky (B-L) n– Spaces, n ≥ 2. *International Mathematical Forum, 2*(2), 69 - 73.

Olgun, M., Ünver, M., & Yardımcı, Ş. (2019). Pythagorean fuzzy topological spaces. *Complex & Intelligent Systems, 5*(2), 177–183. doi:10.100740747-019-0095-2

Pandey, L. K., Ojha, K. K., Singh, P. K., Singh, C. S., Dwivedi, S., & Bergey, E. A. (2016). Diatoms image database of India (DIDI): A research tool. *Environmental Technology & Innovation, 5*, 148–160. doi:10.1016/j.eti.2016.02.001

Papaioannou, Y. (2013). *Fuzzy Logic in Science, Law and Ethics, 2013.* Available online: https://blogs.elpais.com/atomium-culture/2013/01/fuzzy-logic-in-science-law-and-ethics.html

Pawlak, Z. (1991). *Rough Sets: Aspects of Reasoning about Data.* Kluer Academic Publishers. doi:10.1007/978-94-011-3534-4

Polster, B. (1998). *A Geometrical Picture Book.* Springer. doi:10.1007/978-1-4419-8526-2

Popov, M. R. (1996). The Smarandache non-geometry. *Abstracts of Papers Presented to the American Mathematical Society Meetings, 17*, 595.

Raber, N. C. (1972). On Ostrom's Finite Hyperbolic Planes [Unpublished Ph.D.]. The Ohio State University.

Rabounski, D. (2010). Smarandache Spaces as a New Extension of the Basic Space-Time of General Relativity. *Progress in Physics, 2*, L1–L2.

Razis, D. V. (1989). Modern Cassandras and our survival as a species - a new role for medicine? *Journal of the Royal Society of Medicine, 82*(10), 575–576. doi:10.1177/014107688908201001 PMID:2810296

Razis, D. V. (1990). Ethics and the Universality of Morality. *Quality Assurance in Health Care, 2*(3-4), 369–373. doi:10.1093/intqhc/2.3-4.369 PMID:1983258

Rezaei, A. & Smarandache, F. (2020a). On Neutro-BE-algebras and Anti-BE-algebras, *International Journal of Neutrosophic Science, 4*(1), 8-15.

Rezaei, A., &Smarandache, F. (2020). On neutro-BE-algebras and anti-BE-algebras.*International Journal of Neutrosophic Science, 4*, 8-15. doi:10.54216/IJNS.040101

Rezaei, A., Borumand Saeid, A., & Smarandache, F. (2015). Neutrosophic filters in BE-algebras. *Ratio Mathematica, 29*, 65–79.

Rezaei, A., & Smarandache, F. (2020b). The neutrosophic triplet of BI-algebras. *Neutrosophic Sets and Systems, 33*, 313–321.

Rezaei, A., Smarandache, F., & Mirvakili, S. (2021). Applications of (neutro/anti) sophications to semihypergroups. *Journal of Mathematics, 2021*, 2021. doi:10.1155/2021/6649349

Ross, W. D. (1930). The Right and the Good. Oxford University Press.

Şahin, M., Ecemiş, O., Uluçay, V., & Kargın, A. (2017). Some new generalized aggregation operators based on centroid single valued triangular neutrosophic numbers and their applications in multi-attribute decision making. *Asian Journal of Mathematics and Computer Research, 16*(2), 63–84.

Şahin, M., Kargın, A., & Altun, A. (2021). Neutro-metric spaces. *NeutroAlgebra Theory, I*, 71.

Şahin, M., Kargın, A., & Smarandache, F. (2021). Neutro–G modules and anti–G modules. *NeutroAlgebra Theory, I*, 50.

Şahin, M., Kargın, A., & Uz, M. S. (2021). Neutro-Lie Algebras and Anti-Lie Algebras. *NeutroAlgebra Theory, I*, 101.

Şahin, M., Kargın, A., & Yücel, M. (2021). Neutro-Topological space and Anti-Topological space. *NeutroAlgebra Theory, I*, 16.

Şahin, M., Kargın, A., & Yücel, M. (2021). Neutro-topological space and anti-topological space. *NeutroAlgebra Theory, I*, 16–30.

Salama, A. A., Smarandache, F., & Alblowi, S. A. (2014). New neutrosophic crisp topological concepts. *Neutrosophic Sets and Systems, 4*, 50-54.

Salama, A. A., & Alblowi, S. A. (2012). Neutrosophic set and neutrosophic topological spaces. *IOSR Journal of Mathematics, 3*(4), 31–35. doi:10.9790/5728-0343135

Salazar, L. (1984). *Fundamentos de la geometría euclidiana*. Seccional Manizales.

Sandel, M. J. (2011). *Justice. What is the Right Thing to do?* (A. Kioupkiolis, Trans.). Polis Editions.

Sandler, R. (1963). Finite Homogeneous Bolyai-Lobachevsky Planes. *The American Mathematical Monthly, 70*(8), 853–855. doi:10.2307/2312671

Sarkar, M. (1981). On L-fuzzy topological spaces. *Journal of Mathematical Analysis and Applications, 84*(2), 431–442. doi:10.1016/0022-247X(81)90179-7

Shipman, J. W. (2013). *Tkinter 8.5 reference: a GUI for Python*. New Mexico Tech. Computer Center.

Sierpinski, W. (2020). *General topology*. University of Toronto press.

Singh, P. (2021). AntiGeometry and NeutroGeometry characterization of Non-Euclidean data set. *J. Neutrosophic Fuzzy Syst, 1*(2), 24–33. doi:10.54216/JNFS.010102

Singh, P. K. (2021). Data with Turiyam set for fourth dimension quantum information Processing. *Journal of Neutrosophic and Fuzzy Systems, 1*, 9–23. doi:10.54216/JNFS.010101

Singh, P. K. (2021). Fourth dimension data representation and its analysis using Turiyam Context. *Journal of Computer and Communications, 9*(6), 222–229. doi:10.4236/jcc.2021.96014

Singh, P. K. (2021). Turiyam set a fourth dimension data representation. *Zeitschrift für Angewandte Mathematik und Physik, 9*(7), 1821–1828. doi:10.4236/jamp.2021.97116

Singh, P. K. (2022). Data with Non-Euclidean Geometry and Its Characterization. *Journal of Artificial Intelligence and Technology*, *2*(1), 3–8.

Singh, P. K. (2022). Data with non-Euclidean geometry and its characterization. *Journal of Artificial Intelligence and Technology*, 2, 3–8. doi:10.37965/jait.2021.12001

Singh, P. K. (2022). NeutroAlgebra and neutrogeometry for dealing heteroclinic patterns. In *Theory and Applications of NeutroAlgebras as Generalizations of Classical Algebras* (pp. 90–102). IGI Global Publishers. doi:10.4018/978-1-6684-3495-6.ch006

Singh, P. K. (2022). Non-Euclidean, anti-geometry and neutrogeometry characterization. *International Journal of Neutrosophic Sciences*, *18*(3), 8–19. doi:10.54216/IJNS.180301

Singh, P. K., Ahmad, K. D., Bal, M., & Aswad, M. (2021). On the symbolic Turiyam rings. *Journal of Neutrosophic and Fuzzy Systems*, *1*, 80–88. doi:10.54216/JNFS.010204

Smarandache, F. (1969). *Mixed non euclidean geometries*. Infinite Study.

Smarandache, F. (1969). Mixed noneuclidean geometries. *Infinite Study*.

Smarandache, F. (1969). *Paradoxist geometry*. Academic Press.

Smarandache, F. (1998). Neutrosophy. Neutrosophic Probability, Set, and Logic. ProQuest Information & Learning.

Smarandache, F. (1998). Neutrosophy: Neutrosophic Probability, Set and Logic. Amer. Research Press.

Smarandache, F. (2002). *Neutrosophy, a new Branch of Philosophy*. Infinite Study.

Smarandache, F. (2009). *Degree of Negation of an Axiom*. Infinite Study.

Smarandache, F. (2013). N-valued refined neutrosophic logic and its applications to physics. *Progress in Physics, 4*, 143-146. http://fs.unm.edu/RefinedNeutrosophicSet.pdf

Smarandache, F. (2020b). NeutroAlgebra is a Generalization of Partial Algebra. *International Journal of Neutrosophic Science, 2*(1), 8-17.

Smarandache, F. (2021). NeutroGeometry & AntiGeometry are alternatives and generalizations of the Non-Euclidean Geometries. *Neutrosophic Sets and Systems, 46*(2021), 456-476.

Smarandache, F., Rezaei, A., Agboola, A. A. A., Jun, Y. B., Borzooei, R. A., Davvaz, B., ... Mirvakili, S. (2021). On Neutro Quadruple Groups. *Infinite Study*.

Smarandache, F. (1998). *Neutrosophy / Neutrosophic Probability*. Set, and Logic.

Smarandache, F. (1998). *Neutrosophy: Neutrosophic probability, set, and logic: analytic synthesis & synthetic analysis*. American Research Press.

Smarandache, F. (2019). Introduction to Neutroalgebraic structures and antialgebraic structures, in advances of standard and nonstandard neutrosophic theories. *Pons Publishing House Brussels, Belgium*, *6*, 240–265.

Smarandache, F. (2019). Introduction to NeutroAlgebraic Structures and AntiAlgebraic Structures. In *Advances of Standard and Nonstandard Neutrosophic Theories* (pp. 240–265). Pons Publishing House.

Smarandache, F. (2019). *Introduction to NeutroAlgebraic Structures and AntiAlgebraic Structures. In Advances of Standard and Nonstandard Neutrosophic Theories*. Pons Publishing House Brussels.

Smarandache, F. (2020). Introduction to neutro algebraic structures and anti algebraic structures (revisited). *Neutrosophic Sets and Systems*, *31*, 1–16. doi:10.5281/zenodo.3638232

Smarandache, F. (2020). Introduction to NeutroAlgebraic structures and AntiAlgebraic structures (revisited). *Neutrosophic Sets and System*, *31*, 1–16.

Smarandache, F. (2020). NeutroAlgebra is a generalization of partial algebra. *International Journal of Neutrosophic Science*, *2*, 8–17. doi:10.54216/IJNS.020103

Smarandache, F. (2020a). Introduction to NeutroAlgebraic Structures and AntiAlgebraic Structures (revisited). *Neutrosophic Sets and Systems*, *31*, 1–16.

Smarandache, F. (2020b). Generalizations and Alternatives of Classical Algebraic Structures to NeutroAlgebraic Structures and AntiAlgebraic Structures. *Journal of Fuzzy Extension and Applications*, *1*(2), 85–87.

Smarandache, F. (2020d). Structure, NeutroStructure, and AntiStructure in science. *International Journal of Neutrosophic Science*, *13*(1), 28–33.

Smarandache, F. (2021). Indeterminacy in Neutrosophic Theories and their Applications. *International Journal of Neutrosophic Science*, *15*(2), 89–95. doi:10.54216/IJNS.150203

Smarandache, F. (2021). Neutro geometry and anti geometry are alternatives and generalizations of the non-euclidean geometries (revisited). *Neutrosophic Sets and Systems*, *46*, 456–477.

Smarandache, F. (2021). NeutroGeometry & AntiGeometry are alternatives and generalizations of the Non-Euclidian geometries. *Neutrosophic Sets and Systems*, *46*, 456–477.

Smarandache, F., & Chang-Wei, H. (2012). *Vacuum*. Space-Time, Matter and the Models of Smarandache Geometry.

Smarandache, F., & Hamidi, M. (2020). Neutro-BCK-algebra. *International Journal of Neutrosophic Science*, *8*, 110–117.

Smarandache, F., Rezaei, A., & Kim, H. S. (2020). A new trend to extensions of CI-algebras. *International Journal of Neutrosophic Science*, *5*(1), 8. doi:10.54216/IJNS.050101

Šostak, A. P. (1996). Basic structures of fuzzy topology. *Journal of Mathematical Sciences*, *78*(6), 662–701. doi:10.1007/BF02363065

Stillwell, J. (1993). *Classical topology and combinatorial group theory* (Vol. 72). Springer Science & Business Media. doi:10.1007/978-1-4612-4372-4

Sullivan, R. J. (1994). *An Introduction to Kant's Ethics*. Cambridge University Press. doi:10.1017/CBO9780511606151

Sumathi, I. R., & Arockiarani, I. (2015). Fuzzy neutrosophic groups. *Advanced in Fuzzy Mathematics*, *10*(2), 117–122.

Sumathi, I. R., & Arockiarani, I. (2016). Topological group structure of neutrosophic set. *Journal of Advanced Studies in Topology*, *7*(1), 12–20. doi:10.20454/jast.2016.1003

Szabó, Z. (2022). *Finite Geometric Methods in Quantum Theory* [Unpublished Diploma Thesis]. Budapest University of Technology and Economics.

The-GAP-Group. (2022). *GAP--Groups, Algorithms, and Programming, Version 4.12.2*. https://www.gap-system.org

Thomson, J. J. (1976). Killing, Letting Die, and the Trolley Problem. *The Monist*, *59*(2), 204–217. doi:10.5840/monist197659224 PMID:11662247

Timmer, E. (1994). Medical ethics and evolution. *Journal of the Royal Society of Medicine*, *87*(5), 250–252. doi:10.1177/014107689408700502 PMID:8207716

Todd, R. (n.d.). Manifold. *MathWorld*. https://mathworld.wolfram.com/Manifold.html

Uluçay, V., Şahin, M., & Hassan, N. (2018). Generalized neutrosophic soft expert set for multiple-criteria decision-making. *Symmetry*, *10*(10), 437. doi:10.3390ym10100437

Voskoglou, M. G. (2019). Generalizations of Fuzzy Sets and Related Theories. In M. Voskoglou (Ed.), *An Essential Guide to Fuzzy Systems* (pp. 345–352). Nova Science Publishers.

Voskoglou, M. G., & Athanassopoulos, E. (2020). Statistical Thinking in Problem Solving. *American Journal of Educational Research*, *8*(10), 754–761.

Vougiouklis, T. (1994). *Hyperstructures and Their Representations*. Hadronic Press, Inc.

Vougiouklis, T. (1996). -groups defined on same set. *Discrete Mathematics*, *155*(1-3), 259–265. doi:10.1016/0012-365X(94)00390-5

Wang, H., Smarandanche, F., Zhang, Y., & Sunderraman, R. (2010). Single Valued Neutrosophic Sets. *Review of the Air Force Academy (Brasov)*, *1*(16), 10–14.

Weisstein, E. W. (n.d.). Combinatorial Topology. *MathWorld*. https://mathworld.wolfram.com/CombinatorialTopology.html

Zadeh, L. A. (1965). Fuzzy Sets. *Information and Control*, *8*(3), 338–353. doi:10.1016/S0019-9958(65)90241-X

Zadeh, L. A. (1973). Outline of a new approach to the analysis of complex systems and decision processes. *IEEE Transactions on Systems, Man, and Cybernetics*, *3*(1), 28–44. doi:10.1109/TSMC.1973.5408575

Zadeh, L. A. (1988). Fuzzy logic. *Computer*, *21*(4), 83–93. doi:10.1109/2.53

Zangalho-Raposo, A. P. (2014). *Geometrias Finitas* [Unpublished Master Thesis]. University of Évora, Évora.

Zangalho-Raposo, A. P. (2014). *Geometrias Finitas* [Unpublished Master Thesis]. University of Évora.

Zhang, X., Yu, P., Smarandache, F., & Park, C. (2020). Redefined neutrosophic filters in BE-algebras. *International Journal of Pure and Applied Mathematics*, *43*, 158–176.

Zumbragel, J. (2016). *Designs and Codes in Affine Geometry.* arXiv e-prints arXiv -1605.03789.

Related References

To continue our tradition of advancing academic research, we have compiled a list of recommended IGI Global readings. These references will provide additional information and guidance to further enrich your knowledge and assist you with your own research and future publications.

Abbasnejad, B., Moeinzadeh, S., Ahankoob, A., & Wong, P. S. (2021). The Role of Collaboration in the Implementation of BIM-Enabled Projects. In J. Underwood & M. Shelbourn (Eds.), *Handbook of Research on Driving Transformational Change in the Digital Built Environment* (pp. 27–62). IGI Global. https://doi.org/10.4018/978-1-7998-6600-8.ch002

Abdulrahman, K. O., Mahamood, R. M., & Akinlabi, E. T. (2022). Additive Manufacturing (AM): Processing Technique for Lightweight Alloys and Composite Material. In K. Kumar, B. Babu, & J. Davim (Ed.), *Handbook of Research on Advancements in the Processing, Characterization, and Application of Lightweight Materials* (pp. 27-48). IGI Global. https://doi.org/10.4018/978-1-7998-7864-3.ch002

Agrawal, R., Sharma, P., & Saxena, A. (2021). A Diamond Cut Leather Substrate Antenna for BAN (Body Area Network) Application. In V. Singh, V. Dubey, A. Saxena, R. Tiwari, & H. Sharma (Eds.), *Emerging Materials and Advanced Designs for Wearable Antennas* (pp. 54–59). IGI Global. https://doi.org/10.4018/978-1-7998-7611-3.ch004

Ahmad, F., Al-Ammar, E. A., & Alsaidan, I. (2022). Battery Swapping Station: A Potential Solution to Address the Limitations of EV Charging Infrastructure. In M. Alam, R. Pillai, & N. Murugesan (Eds.), *Developing Charging Infrastructure and Technologies for Electric Vehicles* (pp. 195–207). IGI Global. doi:10.4018/978-1-7998-6858-3.ch010

Aikhuele, D. (2018). A Study of Product Development Engineering and Design Reliability Concerns. *International Journal of Applied Industrial Engineering*, *5*(1), 79–89. doi:10.4018/IJAIE.2018010105

Al-Khatri, H., & Al-Atrash, F. (2021). Occupants' Habits and Natural Ventilation in a Hot Arid Climate. In R. González-Lezcano (Ed.), *Advancements in Sustainable Architecture and Energy Efficiency* (pp. 146–168). IGI Global. https://doi.org/10.4018/978-1-7998-7023-4.ch007

Al-Shebeeb, O. A., Rangaswamy, S., Gopalakrishan, B., & Devaru, D. G. (2017). Evaluation and Indexing of Process Plans Based on Electrical Demand and Energy Consumption. *International Journal of Manufacturing, Materials, and Mechanical Engineering*, *7*(3), 1–19. doi:10.4018/IJMMME.2017070101

Amuda, M. O., Lawal, T. F., & Akinlabi, E. T. (2017). Research Progress on Rheological Behavior of AA7075 Aluminum Alloy During Hot Deformation. *International Journal of Materials Forming and Machining Processes*, *4*(1), 53–96. doi:10.4018/IJMFMP.2017010104

Amuda, M. O., Lawal, T. F., & Mridha, S. (2021). Microstructure and Mechanical Properties of Silicon Carbide-Treated Ferritic Stainless Steel Welds. In L. Burstein (Ed.), *Handbook of Research on Advancements in Manufacturing, Materials, and Mechanical Engineering* (pp. 395–411). IGI Global. https://doi.org/10.4018/978-1-7998-4939-1.ch019

Anikeev, V., Gasem, K. A., & Fan, M. (2021). Application of Supercritical Technologies in Clean Energy Production: A Review. In L. Chen (Ed.), *Handbook of Research on Advancements in Supercritical Fluids Applications for Sustainable Energy Systems* (pp. 792–821). IGI Global. https://doi.org/10.4018/978-1-7998-5796-9.ch022

Arafat, M. Y., Saleem, I., & Devi, T. P. (2022). Drivers of EV Charging Infrastructure Entrepreneurship in India. In M. Alam, R. Pillai, & N. Murugesan (Eds.), *Developing Charging Infrastructure and Technologies for Electric Vehicles* (pp. 208–219). IGI Global. https://doi.org/10.4018/978-1-7998-6858-3.ch011

Araujo, A., & Manninen, H. (2022). Contribution of Project-Based Learning on Social Skills Development: An Industrial Engineer Perspective. In A. Alves & N. van Hattum-Janssen (Eds.), *Training Engineering Students for Modern Technological Advancement* (pp. 119–145). IGI Global. https://doi.org/10.4018/978-1-7998-8816-1.ch006

Armutlu, H. (2018). Intelligent Biomedical Engineering Operations by Cloud Computing Technologies. In U. Kose, G. Guraksin, & O. Deperlioglu (Eds.), *Nature-Inspired Intelligent Techniques for Solving Biomedical Engineering Problems* (pp. 297–317). Hershey, PA: IGI Global. doi:10.4018/978-1-5225-4769-3.ch015

Atik, M., Sadek, M., & Shahrour, I. (2017). Single-Run Adaptive Pushover Procedure for Shear Wall Structures. In V. Plevris, G. Kremmyda, & Y. Fahjan (Eds.), *Performance-Based Seismic Design of Concrete Structures and Infrastructures* (pp. 59–83). Hershey, PA: IGI Global. doi:10.4018/978-1-5225-2089-4.ch003

Attia, H. (2021). Smart Power Microgrid Impact on Sustainable Building. In R. González-Lezcano (Ed.), *Advancements in Sustainable Architecture and Energy Efficiency* (pp. 169–194). IGI Global. https://doi.org/10.4018/978-1-7998-7023-4.ch008

Aydin, A., Akyol, E., Gungor, M., Kaya, A., & Tasdelen, S. (2018). Geophysical Surveys in Engineering Geology Investigations With Field Examples. In N. Ceryan (Ed.), *Handbook of Research on Trends and Digital Advances in Engineering Geology* (pp. 257–280). Hershey, PA: IGI Global. doi:10.4018/978-1-5225-2709-1.ch007

Ayoobkhan, M. U. D., Y., A., J., Easwaran, B., & R., T. (2021). Smart Connected Digital Products and IoT Platform With the Digital Twin. In P. Vasant, G. Weber, & W. Punurai (Ed.), Research Advancements in Smart Technology, Optimization, and Renewable Energy (pp. 330-350). IGI Global. https://doi.org/ doi:10.4018/978-1-7998-3970-5.ch016

Baeza Moyano, D., & González Lezcano, R. A. (2021). The Importance of Light in Our Lives: Towards New Lighting in Schools. In R. González-Lezcano (Ed.), *Advancements in Sustainable Architecture and Energy Efficiency* (pp. 239–256). IGI Global. https://doi.org/10.4018/978-1-7998-7023-4.ch011

Bagdadee, A. H. (2021). A Brief Assessment of the Energy Sector of Bangladesh. *International Journal of Energy Optimization and Engineering, 10*(1), 36–55. doi:10.4018/IJEOE.2021010103

Baklezos, A. T., & Hadjigeorgiou, N. G. (2021). Magnetic Sensors for Space Applications and Magnetic Cleanliness Considerations. In C. Nikolopoulos (Ed.), *Recent Trends on Electromagnetic Environmental Effects for Aeronautics and Space Applications* (pp. 147–185). IGI Global. https://doi.org/10.4018/978-1-7998-4879-0.ch006

Bas, T. G. (2017). Nutraceutical Industry with the Collaboration of Biotechnology and Nutrigenomics Engineering: The Significance of Intellectual Property in the Entrepreneurship and Scientific Research Ecosystems. In T. Bas & J. Zhao (Eds.), *Comparative Approaches to Biotechnology Development and Use in Developed and Emerging Nations* (pp. 1–17). Hershey, PA: IGI Global. doi:10.4018/978-1-5225-1040-6.ch001

Bazeer Ahamed, B., & Periakaruppan, S. (2021). Taxonomy of Influence Maximization Techniques in Unknown Social Networks. In P. Vasant, G. Weber, & W. Punurai (Eds.), *Research Advancements in Smart Technology, Optimization, and Renewable Energy* (pp. 351-363). IGI Global. https://doi.org/10.4018/978-1-7998-3970-5.ch017

Beale, R., & André, J. (2017). *Design Solutions and Innovations in Temporary Structures*. Hershey, PA: IGI Global. doi:10.4018/978-1-5225-2199-0

Behnam, B. (2017). Simulating Post-Earthquake Fire Loading in Conventional RC Structures. In P. Samui, S. Chakraborty, & D. Kim (Eds.), *Modeling and Simulation Techniques in Structural Engineering* (pp. 425–444). Hershey, PA: IGI Global. doi:10.4018/978-1-5225-0588-4.ch015

Ben Hamida, I., Salah, S. B., Msahli, F., & Mimouni, M. F. (2018). Distribution Network Reconfiguration Using SPEA2 for Power Loss Minimization and Reliability Improvement. *International Journal of Energy Optimization and Engineering*, *7*(1), 50–65. doi:10.4018/IJEOE.2018010103

Bentarzi, H. (2021). Fault Tree-Based Root Cause Analysis Used to Study Mal-Operation of a Protective Relay in a Smart Grid. In A. Recioui & H. Bentarzi (Eds.), *Optimizing and Measuring Smart Grid Operation and Control* (pp. 289–308). IGI Global. https://doi.org/10.4018/978-1-7998-4027-5.ch012

Beysens, D. A., Garrabos, Y., & Zappoli, B. (2021). Thermal Effects in Near-Critical Fluids: Piston Effect and Related Phenomena. In L. Chen (Ed.), *Handbook of Research on Advancements in Supercritical Fluids Applications for Sustainable Energy Systems* (pp. 1–31). IGI Global. https://doi.org/10.4018/978-1-7998-5796-9.ch001

Bhaskar, S. V., & Kudal, H. N. (2017). Effect of TiCN and AlCrN Coating on Tribological Behaviour of Plasma-nitrided AISI 4140 Steel. *International Journal of Surface Engineering and Interdisciplinary Materials Science*, *5*(2), 1–17. doi:10.4018/IJSEIMS.2017070101

Bhuyan, D. (2018). Designing of a Twin Tube Shock Absorber: A Study in Reverse Engineering. In K. Kumar & J. Davim (Eds.), *Design and Optimization of Mechanical Engineering Products* (pp. 83–104). Hershey, PA: IGI Global. doi:10.4018/978-1-5225-3401-3.ch005

Blumberg, G. (2021). Blockchains for Use in Construction and Engineering Projects. In J. Underwood & M. Shelbourn (Eds.), *Handbook of Research on Driving Transformational Change in the Digital Built Environment* (pp. 179–208). IGI Global. https://doi.org/10.4018/978-1-7998-6600-8.ch008

Bolboaca, A. M. (2021). Considerations Regarding the Use of Fuel Cells in Combined Heat and Power for Stationary Applications. In G. Badea, R. Felseghi, & I. Aşchilean (Eds.), *Hydrogen Fuel Cell Technology for Stationary Applications* (pp. 239–275). IGI Global. https://doi.org/10.4018/978-1-7998-4945-2.ch010

Burstein, L. (2021). Simulation Tool for Cable Design. In L. Burstein (Ed.), *Handbook of Research on Advancements in Manufacturing, Materials, and Mechanical Engineering* (pp. 54–74). IGI Global. https://doi.org/10.4018/978-1-7998-4939-1.ch003

Calderon, F. A., Giolo, E. G., Frau, C. D., Rengel, M. G., Rodriguez, H., Tornello, M., ... Gallucci, R. (2018). Seismic Microzonation and Site Effects Detection Through Microtremors Measures: A Review. In N. Ceryan (Ed.), *Handbook of Research on Trends and Digital Advances in Engineering Geology* (pp. 326–349). Hershey, PA: IGI Global. doi:10.4018/978-1-5225-2709-1.ch009

Ceryan, N., & Can, N. K. (2018). Prediction of The Uniaxial Compressive Strength of Rocks Materials. In N. Ceryan (Ed.), *Handbook of Research on Trends and Digital Advances in Engineering Geology* (pp. 31–96). Hershey, PA: IGI Global. doi:10.4018/978-1-5225-2709-1.ch002

Ceryan, S. (2018). Weathering Indices Used in Evaluation of the Weathering State of Rock Material. In N. Ceryan (Ed.), *Handbook of Research on Trends and Digital Advances in Engineering Geology* (pp. 132–186). Hershey, PA: IGI Global. doi:10.4018/978-1-5225-2709-1.ch004

Chen, H., Padilla, R. V., & Besarati, S. (2017). Supercritical Fluids and Their Applications in Power Generation. In L. Chen & Y. Iwamoto (Eds.), *Advanced Applications of Supercritical Fluids in Energy Systems* (pp. 369–402). Hershey, PA: IGI Global. doi:10.4018/978-1-5225-2047-4.ch012

Chen, H., Padilla, R. V., & Besarati, S. (2021). Supercritical Fluids and Their Applications in Power Generation. In L. Chen (Ed.), *Handbook of Research on Advancements in Supercritical Fluids Applications for Sustainable Energy Systems* (pp. 566–599). IGI Global. https://doi.org/10.4018/978-1-7998-5796-9.ch016

Chen, L. (2017). Principles, Experiments, and Numerical Studies of Supercritical Fluid Natural Circulation System. In L. Chen & Y. Iwamoto (Eds.), *Advanced Applications of Supercritical Fluids in Energy Systems* (pp. 136–187). Hershey, PA: IGI Global. doi:10.4018/978-1-5225-2047-4.ch005

Chen, L. (2021). Principles, Experiments, and Numerical Studies of Supercritical Fluid Natural Circulation System. In L. Chen (Ed.), *Handbook of Research on Advancements in Supercritical Fluids Applications for Sustainable Energy Systems* (pp. 219–269). IGI Global. https://doi.org/10.4018/978-1-7998-5796-9.ch007

Chiba, Y., Marif, Y., Henini, N., & Tlemcani, A. (2021). Modeling of Magnetic Refrigeration Device by Using Artificial Neural Networks Approach. *International Journal of Energy Optimization and Engineering*, *10*(4), 68–76. https://doi.org/10.4018/IJEOE.2021100105

Clementi, F., Di Sciascio, G., Di Sciascio, S., & Lenci, S. (2017). Influence of the Shear-Bending Interaction on the Global Capacity of Reinforced Concrete Frames: A Brief Overview of the New Perspectives. In V. Plevris, G. Kremmyda, & Y. Fahjan (Eds.), *Performance-Based Seismic Design of Concrete Structures and Infrastructures* (pp. 84–111). Hershey, PA: IGI Global. doi:10.4018/978-1-5225-2089-4.ch004

Codinhoto, R., Fialho, B. C., Pinti, L., & Fabricio, M. M. (2021). BIM and IoT for Facilities Management: Understanding Key Maintenance Issues. In J. Underwood & M. Shelbourn (Eds.), *Handbook of Research on Driving Transformational Change in the Digital Built Environment* (pp. 209–231). IGI Global. doi:10.4018/978-1-7998-6600-8.ch009

Cortés-Polo, D., Calle-Cancho, J., Carmona-Murillo, J., & González-Sánchez, J. (2017). Future Trends in Mobile-Fixed Integration for Next Generation Networks: Classification and Analysis. *International Journal of Vehicular Telematics and Infotainment Systems*, *1*(1), 33–53. doi:10.4018/IJVTIS.2017010103

Costa, H. G., Sheremetieff, F. H., & Araújo, E. A. (2022). Influence of Game-Based Methods in Developing Engineering Competences. In A. Alves & N. van Hattum-Janssen (Eds.), *Training Engineering Students for Modern Technological Advancement* (pp. 69–88). IGI Global. https://doi.org/10.4018/978-1-7998-8816-1.ch004

Cui, X., Zeng, S., Li, Z., Zheng, Q., Yu, X., & Han, B. (2018). Advanced Composites for Civil Engineering Infrastructures. In K. Kumar & J. Davim (Eds.), *Composites and Advanced Materials for Industrial Applications* (pp. 212–248). Hershey, PA: IGI Global. doi:10.4018/978-1-5225-5216-1.ch010

Dalgıç, S., & Kuşku, İ. (2018). Geological and Geotechnical Investigations in Tunneling. In N. Ceryan (Ed.), *Handbook of Research on Trends and Digital Advances in Engineering Geology* (pp. 482–529). Hershey, PA: IGI Global. doi:10.4018/978-1-5225-2709-1.ch014

Dang, C., & Hihara, E. (2021). Study on Cooling Heat Transfer of Supercritical Carbon Dioxide Applied to Transcritical Carbon Dioxide Heat Pump. In L. Chen (Ed.), *Handbook of Research on Advancements in Supercritical Fluids Applications for Sustainable Energy Systems* (pp. 451–493). IGI Global. https://doi.org/10.4018/978-1-7998-5796-9.ch013

Daus, Y., Kharchenko, V., & Yudaev, I. (2021). Research of Solar Energy Potential of Photovoltaic Installations on Enclosing Structures of Buildings. *International Journal of Energy Optimization and Engineering, 10*(4), 18–34. https://doi.org/10.4018/IJEOE.2021100102

Daus, Y., Kharchenko, V., & Yudaev, I. (2021). Optimizing Layout of Distributed Generation Sources of Power Supply System of Agricultural Object. *International Journal of Energy Optimization and Engineering, 10*(3), 70–84. https://doi.org/10.4018/IJEOE.2021070104

de la Varga, D., Soto, M., Arias, C. A., van Oirschot, D., Kilian, R., Pascual, A., & Álvarez, J. A. (2017). Constructed Wetlands for Industrial Wastewater Treatment and Removal of Nutrients. In Á. Val del Río, J. Campos Gómez, & A. Mosquera Corral (Eds.), *Technologies for the Treatment and Recovery of Nutrients from Industrial Wastewater* (pp. 202–230). Hershey, PA: IGI Global. doi:10.4018/978-1-5225-1037-6.ch008

Deb, S., Ammar, E. A., AlRajhi, H., Alsaidan, I., & Shariff, S. M. (2022). V2G Pilot Projects: Review and Lessons Learnt. In M. Alam, R. Pillai, & N. Murugesan (Eds.), *Developing Charging Infrastructure and Technologies for Electric Vehicles* (pp. 252–267). IGI Global. https://doi.org/10.4018/978-1-7998-6858-3.ch014

Dekhandji, F. Z., & Rais, M. C. (2021). A Comparative Study of Power Quality Monitoring Using Various Techniques. In A. Recioui & H. Bentarzi (Eds.), *Optimizing and Measuring Smart Grid Operation and Control* (pp. 259–288). IGI Global. https://doi.org/10.4018/978-1-7998-4027-5.ch011

Deperlioglu, O. (2018). Intelligent Techniques Inspired by Nature and Used in Biomedical Engineering. In U. Kose, G. Guraksin, & O. Deperlioglu (Eds.), *Nature-Inspired Intelligent Techniques for Solving Biomedical Engineering Problems* (pp. 51–77). Hershey, PA: IGI Global. doi:10.4018/978-1-5225-4769-3.ch003

Dhurpate, P. R., & Tang, H. (2021). Quantitative Analysis of the Impact of Inter-Line Conveyor Capacity for Throughput of Manufacturing Systems. *International Journal of Manufacturing, Materials, and Mechanical Engineering, 11*(1), 1–17. https://doi.org/10.4018/IJMMME.2021010101

Dinkar, S., & Deep, K. (2021). A Survey of Recent Variants and Applications of Antlion Optimizer. *International Journal of Energy Optimization and Engineering, 10*(2), 48–73. doi:10.4018/IJEOE.2021040103

Dixit, A. (2018). Application of Silica-Gel-Reinforced Aluminium Composite on the Piston of Internal Combustion Engine: Comparative Study of Silica-Gel-Reinforced Aluminium Composite Piston With Aluminium Alloy Piston. In K. Kumar & J. Davim (Eds.), *Composites and Advanced Materials for Industrial Applications* (pp. 63–98). Hershey, PA: IGI Global. doi:10.4018/978-1-5225-5216-1.ch004

Drabecki, M. P., & Kułak, K. B. (2021). Global Pandemics on European Electrical Energy Markets: Lessons Learned From the COVID-19 Outbreak. *International Journal of Energy Optimization and Engineering, 10*(3), 24–46. https://doi.org/10.4018/IJEOE.2021070102

Dutta, M. M. (2021). Nanomaterials for Food and Agriculture. In M. Bhat, I. Wani, & S. Ashraf (Eds.), *Applications of Nanomaterials in Agriculture, Food Science, and Medicine* (pp. 75–97). IGI Global. doi:10.4018/978-1-7998-5563-7.ch004

Dutta, M. M., & Goswami, M. (2021). Coating Materials: Nano-Materials. In S. Roy & G. Bose (Eds.), *Advanced Surface Coating Techniques for Modern Industrial Applications* (pp. 1–30). IGI Global. doi:10.4018/978-1-7998-4870-7.ch001

Elsayed, A. M., Dakkama, H. J., Mahmoud, S., Al-Dadah, R., & Kaialy, W. (2017). Sustainable Cooling Research Using Activated Carbon Adsorbents and Their Environmental Impact. In T. Kobayashi (Ed.), *Applied Environmental Materials Science for Sustainability* (pp. 186–221). Hershey, PA: IGI Global. doi:10.4018/978-1-5225-1971-3.ch009

Ercanoglu, M., & Sonmez, H. (2018). General Trends and New Perspectives on Landslide Mapping and Assessment Methods. In N. Ceryan (Ed.), *Handbook of Research on Trends and Digital Advances in Engineering Geology* (pp. 350–379). Hershey, PA: IGI Global. doi:10.4018/978-1-5225-2709-1.ch010

Faroz, S. A., Pujari, N. N., Rastogi, R., & Ghosh, S. (2017). Risk Analysis of Structural Engineering Systems Using Bayesian Inference. In P. Samui, S. Chakraborty, & D. Kim (Eds.), *Modeling and Simulation Techniques in Structural Engineering* (pp. 390–424). Hershey, PA: IGI Global. doi:10.4018/978-1-5225-0588-4.ch014

Fekik, A., Hamida, M. L., Denoun, H., Azar, A. T., Kamal, N. A., Vaidyanathan, S., Bousbaine, A., & Benamrouche, N. (2022). Multilevel Inverter for Hybrid Fuel Cell/PV Energy Conversion System. In A. Fekik & N. Benamrouche (Eds.), *Modeling and Control of Static Converters for Hybrid Storage Systems* (pp. 233–270). IGI Global. https://doi.org/10.4018/978-1-7998-7447-8.ch009

Fekik, A., Hamida, M. L., Houassine, H., Azar, A. T., Kamal, N. A., Denoun, H., Vaidyanathan, S., & Sambas, A. (2022). Power Quality Improvement for Grid-Connected Photovoltaic Panels Using Direct Power Control. In A. Fekik & N. Benamrouche (Eds.), *Modeling and Control of Static Converters for Hybrid Storage Systems* (pp. 107–142). IGI Global. https://doi.org/10.4018/978-1-7998-7447-8.ch005

Fernando, P. R., Hamigah, T., Disne, S., Wickramasingha, G. G., & Sutharshan, A. (2018). The Evaluation of Engineering Properties of Low Cost Concrete Blocks by Partial Doping of Sand with Sawdust: Low Cost Sawdust Concrete Block. *International Journal of Strategic Engineering*, *1*(2), 26–42. doi:10.4018/IJoSE.2018070103

Ferro, G., Minciardi, R., Parodi, L., & Robba, M. (2022). Optimal Charging Management of Microgrid-Integrated Electric Vehicles. In M. Alam, R. Pillai, & N. Murugesan (Eds.), *Developing Charging Infrastructure and Technologies for Electric Vehicles* (pp. 133–155). IGI Global. https://doi.org/10.4018/978-1-7998-6858-3.ch007

Flumerfelt, S., & Green, C. (2022). Graduate Lean Leadership Education: A Case Study of a Program. In A. Alves & N. van Hattum-Janssen (Eds.), *Training Engineering Students for Modern Technological Advancement* (pp. 202–224). IGI Global. https://doi.org/10.4018/978-1-7998-8816-1.ch010

Galli, B. J. (2021). Implications of Economic Decision Making to the Project Manager. *International Journal of Strategic Engineering*, *4*(1), 19–32. https://doi.org/10.4018/IJoSE.2021010102

Gento, A. M., Pimentel, C., & Pascual, J. A. (2022). Teaching Circular Economy and Lean Management in a Learning Factory. In A. Alves & N. van Hattum-Janssen (Eds.), *Training Engineering Students for Modern Technological Advancement* (pp. 183–201). IGI Global. https://doi.org/10.4018/978-1-7998-8816-1.ch009

Ghosh, S., Mitra, S., Ghosh, S., & Chakraborty, S. (2017). Seismic Reliability Analysis in the Framework of Metamodelling Based Monte Carlo Simulation. In P. Samui, S. Chakraborty, & D. Kim (Eds.), *Modeling and Simulation Techniques in Structural Engineering* (pp. 192–208). Hershey, PA: IGI Global. doi:10.4018/978-1-5225-0588-4.ch006

Gil, M., & Otero, B. (2017). Learning Engineering Skills through Creativity and Collaboration: A Game-Based Proposal. In R. Alexandre Peixoto de Queirós & M. Pinto (Eds.), *Gamification-Based E-Learning Strategies for Computer Programming Education* (pp. 14–29). Hershey, PA: IGI Global. doi:10.4018/978-1-5225-1034-5.ch002

Gill, J., Ayre, M., & Mills, J. (2017). Revisioning the Engineering Profession: How to Make It Happen! In M. Gray & K. Thomas (Eds.), *Strategies for Increasing Diversity in Engineering Majors and Careers* (pp. 156–175). Hershey, PA: IGI Global. doi:10.4018/978-1-5225-2212-6.ch008

Godzhaev, Z., Senkevich, S., Kuzmin, V., & Melikov, I. (2021). Use of the Neural Network Controller of Sprung Mass to Reduce Vibrations From Road Irregularities. In P. Vasant, G. Weber, & W. Punurai (Ed.), *Research Advancements in Smart Technology, Optimization, and Renewable Energy* (pp. 69-87). IGI Global. https://doi.org/10.4018/978-1-7998-3970-5.ch005

Gomes de Gusmão, C. M. (2022). Digital Competencies and Transformation in Higher Education: Upskilling With Extension Actions. In A. Alves & N. van Hattum-Janssen (Eds.), *Training Engineering Students for Modern Technological Advancement* (pp. 313–328). IGI Global. https://doi.org/10.4018/978-1-7998-8816-1.ch015A

Goyal, N., Ram, M., & Kumar, P. (2017). Welding Process under Fault Coverage Approach for Reliability and MTTF. In M. Ram & J. Davim (Eds.), *Mathematical Concepts and Applications in Mechanical Engineering and Mechatronics* (pp. 222–245). Hershey, PA: IGI Global. doi:10.4018/978-1-5225-1639-2.ch011

Gray, M., & Lundy, C. (2017). Engineering Study Abroad: High Impact Strategy for Increasing Access. In M. Gray & K. Thomas (Eds.), *Strategies for Increasing Diversity in Engineering Majors and Careers* (pp. 42–59). Hershey, PA: IGI Global. doi:10.4018/978-1-5225-2212-6.ch003

Güler, O., & Varol, T. (2021). Fabrication of Functionally Graded Metal and Ceramic Powders Synthesized by Electroless Deposition. In S. Roy & G. Bose (Eds.), *Advanced Surface Coating Techniques for Modern Industrial Applications* (pp. 150–187). IGI Global. https://doi.org/10.4018/978-1-7998-4870-7.ch007

Guraksin, G. E. (2018). Internet of Things and Nature-Inspired Intelligent Techniques for the Future of Biomedical Engineering. In U. Kose, G. Guraksin, & O. Deperlioglu (Eds.), *Nature-Inspired Intelligent Techniques for Solving Biomedical Engineering Problems* (pp. 263–282). Hershey, PA: IGI Global. doi:10.4018/978-1-5225-4769-3.ch013

Hamida, M. L., Fekik, A., Denoun, H., Ardjal, A., & Bokhtache, A. A. (2022). Flying Capacitor Inverter Integration in a Renewable Energy System. In A. Fekik & N. Benamrouche (Eds.), *Modeling and Control of Static Converters for Hybrid Storage Systems* (pp. 287–306). IGI Global. https://doi.org/10.4018/978-1-7998-7447-8.ch011

Hasegawa, N., & Takahashi, Y. (2021). Control of Soap Bubble Ejection Robot Using Facial Expressions. *International Journal of Manufacturing, Materials, and Mechanical Engineering*, *11*(2), 1–16. https://doi.org/10.4018/IJMMME.2021040101

Hejazi, T., & Akbari, L. (2017). A Multiresponse Optimization Model for Statistical Design of Processes with Discrete Variables. In M. Ram & J. Davim (Eds.), *Mathematical Concepts and Applications in Mechanical Engineering and Mechatronics* (pp. 17–37). Hershey, PA: IGI Global. doi:10.4018/978-1-5225-1639-2.ch002

Hejazi, T., & Hejazi, A. (2017). Monte Carlo Simulation for Reliability-Based Design of Automotive Complex Subsystems. In M. Ram & J. Davim (Eds.), *Mathematical Concepts and Applications in Mechanical Engineering and Mechatronics* (pp. 177–200). Hershey, PA: IGI Global. doi:10.4018/978-1-5225-1639-2.ch009

Hejazi, T., & Poursabbagh, H. (2017). Reliability Analysis of Engineering Systems: An Accelerated Life Testing for Boiler Tubes. In M. Ram & J. Davim (Eds.), *Mathematical Concepts and Applications in Mechanical Engineering and Mechatronics* (pp. 154–176). Hershey, PA: IGI Global. doi:10.4018/978-1-5225-1639-2.ch008

Henao, J., Poblano-Salas, C. A., Vargas, F., Giraldo-Betancur, A. L., Corona-Castuera, J., & Sotelo-Mazón, O. (2021). Principles and Applications of Thermal Spray Coatings. In S. Roy & G. Bose (Eds.), *Advanced Surface Coating Techniques for Modern Industrial Applications* (pp. 31–70). IGI Global. https://doi.org/10.4018/978-1-7998-4870-7.ch002

Henao, J., & Sotelo, O. (2018). Surface Engineering at High Temperature: Thermal Cycling and Corrosion Resistance. In A. Pakseresht (Ed.), *Production, Properties, and Applications of High Temperature Coatings* (pp. 131–159). Hershey, PA: IGI Global. doi:10.4018/978-1-5225-4194-3.ch006

Hrnčič, M. K., Cör, D., & Knez, Ž. (2021). Supercritical Fluids as a Tool for Green Energy and Chemicals. In L. Chen (Ed.), *Handbook of Research on Advancements in Supercritical Fluids Applications for Sustainable Energy Systems* (pp. 761–791). IGI Global. doi:10.4018/978-1-7998-5796-9.ch021

Ibrahim, O., Erdem, S., & Gurbuz, E. (2021). Studying Physical and Chemical Properties of Graphene Oxide and Reduced Graphene Oxide and Their Applications in Sustainable Building Materials. In R. González-Lezcano (Ed.), *Advancements in Sustainable Architecture and Energy Efficiency* (pp. 221–238). IGI Global. https://doi.org/10.4018/978-1-7998-7023-4.ch010

Ihianle, I. K., Islam, S., Naeem, U., & Ebenuwa, S. H. (2021). Exploiting Patterns of Object Use for Human Activity Recognition. In A. Nwajana & I. Ihianle (Eds.), *Handbook of Research on 5G Networks and Advancements in Computing, Electronics, and Electrical Engineering* (pp. 382–401). IGI Global. https://doi.org/10.4018/978-1-7998-6992-4.ch015

Ijemaru, G. K., Ngharamike, E. T., Oleka, E. U., & Nwajana, A. O. (2021). An Energy-Efficient Model for Opportunistic Data Collection in IoV-Enabled SC Waste Management. In A. Nwajana & I. Ihianle (Eds.), *Handbook of Research on 5G Networks and Advancements in Computing, Electronics, and Electrical Engineering* (pp. 1–19). IGI Global. https://doi.org/10.4018/978-1-7998-6992-4.ch001

Ilori, O. O., Adetan, D. A., & Umoru, L. E. (2017). Effect of Cutting Parameters on the Surface Residual Stress of Face-Milled Pearlitic Ductile Iron. *International Journal of Materials Forming and Machining Processes*, *4*(1), 38–52. doi:10.4018/IJMFMP.2017010103

Imam, M. H., Tasadduq, I. A., Ahmad, A., Aldosari, F., & Khan, H. (2017). Automated Generation of Course Improvement Plans Using Expert System. *International Journal of Quality Assurance in Engineering and Technology Education*, *6*(1), 1–12. doi:10.4018/IJQAETE.2017010101

Injeti, S. K., & Kumar, T. V. (2018). A WDO Framework for Optimal Deployment of DGs and DSCs in a Radial Distribution System Under Daily Load Pattern to Improve Techno-Economic Benefits. *International Journal of Energy Optimization and Engineering*, *7*(2), 1–38. doi:10.4018/IJEOE.2018040101

Ishii, N., Anami, K., & Knisely, C. W. (2018). *Dynamic Stability of Hydraulic Gates and Engineering for Flood Prevention*. Hershey, PA: IGI Global. doi:10.4018/978-1-5225-3079-4

Iwamoto, Y., & Yamaguchi, H. (2021). Application of Supercritical Carbon Dioxide for Solar Water Heater. In L. Chen (Ed.), *Handbook of Research on Advancements in Supercritical Fluids Applications for Sustainable Energy Systems* (pp. 370–387). IGI Global. https://doi.org/10.4018/978-1-7998-5796-9.ch010

Jayapalan, S. (2018). A Review of Chemical Treatments on Natural Fibers-Based Hybrid Composites for Engineering Applications. In K. Kumar & J. Davim (Eds.), *Composites and Advanced Materials for Industrial Applications* (pp. 16–37). Hershey, PA: IGI Global. doi:10.4018/978-1-5225-5216-1.ch002

Kapetanakis, T. N., Vardiambasis, I. O., Ioannidou, M. P., & Konstantaras, A. I. (2021). Modeling Antenna Radiation Using Artificial Intelligence Techniques: The Case of a Circular Loop Antenna. In C. Nikolopoulos (Ed.), *Recent Trends on Electromagnetic Environmental Effects for Aeronautics and Space Applications* (pp. 186–225). IGI Global. https://doi.org/10.4018/978-1-7998-4879-0.ch007

Karkalos, N. E., Markopoulos, A. P., & Dossis, M. F. (2017). Optimal Model Parameters of Inverse Kinematics Solution of a 3R Robotic Manipulator Using ANN Models. *International Journal of Manufacturing, Materials, and Mechanical Engineering*, *7*(3), 20–40. doi:10.4018/IJMMME.2017070102

Kelly, M., Costello, M., Nicholson, G., & O'Connor, J. (2021). The Evolving Integration of BIM Into Built Environment Programmes in a Higher Education Institute. In J. Underwood & M. Shelbourn (Eds.), *Handbook of Research on Driving Transformational Change in the Digital Built Environment* (pp. 294–326). IGI Global. https://doi.org/10.4018/978-1-7998-6600-8.ch012

Kesimal, A., Karaman, K., Cihangir, F., & Ercikdi, B. (2018). Excavatability Assessment of Rock Masses for Geotechnical Studies. In N. Ceryan (Ed.), *Handbook of Research on Trends and Digital Advances in Engineering Geology* (pp. 231–256). Hershey, PA: IGI Global. doi:10.4018/978-1-5225-2709-1.ch006

Knoflacher, H. (2017). The Role of Engineers and Their Tools in the Transport Sector after Paradigm Change: From Assumptions and Extrapolations to Science. In H. Knoflacher & E. Ocalir-Akunal (Eds.), *Engineering Tools and Solutions for Sustainable Transportation Planning* (pp. 1–29). Hershey, PA: IGI Global. doi:10.4018/978-1-5225-2116-7.ch001

Kose, U. (2018). Towards an Intelligent Biomedical Engineering With Nature-Inspired Artificial Intelligence Techniques. In U. Kose, G. Guraksin, & O. Deperlioglu (Eds.), *Nature-Inspired Intelligent Techniques for Solving Biomedical Engineering Problems* (pp. 1–26). Hershey, PA: IGI Global. doi:10.4018/978-1-5225-4769-3.ch001

Kostić, S. (2018). A Review on Enhanced Stability Analyses of Soil Slopes Using Statistical Design. In N. Ceryan (Ed.), *Handbook of Research on Trends and Digital Advances in Engineering Geology* (pp. 446–481). Hershey, PA: IGI Global. doi:10.4018/978-1-5225-2709-1.ch013

Kumar, A., Patil, P. P., & Prajapati, Y. K. (2018). *Advanced Numerical Simulations in Mechanical Engineering*. Hershey, PA: IGI Global. doi:10.4018/978-1-5225-3722-9

Kumar, G. R., Rajyalakshmi, G., & Manupati, V. K. (2017). Surface Micro Patterning of Aluminium Reinforced Composite through Laser Peening. *International Journal of Manufacturing, Materials, and Mechanical Engineering*, *7*(4), 15–27. doi:10.4018/IJMMME.2017100102

Kumar, N., Basu, D. N., & Chen, L. (2021). Effect of Flow Acceleration and Buoyancy on Thermalhydraulics of sCO2 in Mini/Micro-Channel. In L. Chen (Ed.), *Handbook of Research on Advancements in Supercritical Fluids Applications for Sustainable Energy Systems* (pp. 161–182). IGI Global. doi:10.4018/978-1-7998-5796-9.ch005

Kumari, N., & Kumar, K. (2018). Fabrication of Orthotic Calipers With Epoxy-Based Green Composite. In K. Kumar & J. Davim (Eds.), *Composites and Advanced Materials for Industrial Applications* (pp. 157–176). Hershey, PA: IGI Global. doi:10.4018/978-1-5225-5216-1.ch008

Kuppusamy, R. R. (2018). Development of Aerospace Composite Structures Through Vacuum-Enhanced Resin Transfer Moulding Technology (VERTMTy): Vacuum-Enhanced Resin Transfer Moulding. In K. Kumar & J. Davim (Eds.), *Composites and Advanced Materials for Industrial Applications* (pp. 99–111). Hershey, PA: IGI Global. doi:10.4018/978-1-5225-5216-1.ch005

Kurganov, V. A., Zeigarnik, Y. A., & Maslakova, I. V. (2021). Normal and Deteriorated Heat Transfer Under Heating Turbulent Supercritical Pressure Coolants Flows in Round Tubes. In L. Chen (Ed.), *Handbook of Research on Advancements in Supercritical Fluids Applications for Sustainable Energy Systems* (pp. 494–532). IGI Global. https://doi.org/10.4018/978-1-7998-5796-9.ch014

Li, H., & Zhang, Y. (2021). Heat Transfer and Fluid Flow Modeling for Supercritical Fluids in Advanced Energy Systems. In L. Chen (Ed.), *Handbook of Research on Advancements in Supercritical Fluids Applications for Sustainable Energy Systems* (pp. 388–422). IGI Global. https://doi.org/10.4018/978-1-7998-5796-9.ch011

Loy, J., Howell, S., & Cooper, R. (2017). Engineering Teams: Supporting Diversity in Engineering Education. In M. Gray & K. Thomas (Eds.), *Strategies for Increasing Diversity in Engineering Majors and Careers* (pp. 106–129). Hershey, PA: IGI Global. doi:10.4018/978-1-5225-2212-6.ch006

Macher, G., Armengaud, E., Kreiner, C., Brenner, E., Schmittner, C., Ma, Z., ... Krammer, M. (2018). Integration of Security in the Development Lifecycle of Dependable Automotive CPS. In N. Druml, A. Genser, A. Krieg, M. Menghin, & A. Hoeller (Eds.), *Solutions for Cyber-Physical Systems Ubiquity* (pp. 383–423). Hershey, PA: IGI Global. doi:10.4018/978-1-5225-2845-6.ch015

Madhu, M. N., Singh, J. G., Mohan, V., & Ongsakul, W. (2021). Transmission Risk Optimization in Interconnected Systems: Risk-Adjusted Available Transfer Capability. In P. Vasant, G. Weber, & W. Punurai (Ed.), *Research Advancements in Smart Technology, Optimization, and Renewable Energy* (pp. 183-199). IGI Global. https://doi.org/10.4018/978-1-7998-3970-5.ch010

Mahendramani, G., & Lakshmana Swamy, N. (2018). Effect of Weld Groove Area on Distortion of Butt Welded Joints in Submerged Arc Welding. *International Journal of Manufacturing, Materials, and Mechanical Engineering*, *8*(2), 33–44. doi:10.4018/IJMMME.2018040103

Makropoulos, G., Koumaras, H., Setaki, F., Filis, K., Lutz, T., Montowtt, P., Tomaszewski, L., Dybiec, P., & Järvet, T. (2021). 5G and Unmanned Aerial Vehicles (UAVs) Use Cases: Analysis of the Ecosystem, Architecture, and Applications. In A. Nwajana & I. Ihianle (Eds.), *Handbook of Research on 5G Networks and Advancements in Computing, Electronics, and Electrical Engineering* (pp. 36–69). IGI Global. https://doi.org/10.4018/978-1-7998-6992-4.ch003

Meric, E. M., Erdem, S., & Gurbuz, E. (2021). Application of Phase Change Materials in Construction Materials for Thermal Energy Storage Systems in Buildings. In R. González-Lezcano (Ed.), *Advancements in Sustainable Architecture and Energy Efficiency* (pp. 1–20). IGI Global. https://doi.org/10.4018/978-1-7998-7023-4.ch001

Mihret, E. T., & Yitayih, K. A. (2021). Operation of VANET Communications: The Convergence of UAV System With LTE/4G and WAVE Technologies. *International Journal of Smart Vehicles and Smart Transportation*, *4*(1), 29–51. https://doi.org/10.4018/IJSVST.2021010103

Mir, M. A., Bhat, B. A., Sheikh, B. A., Rather, G. A., Mehraj, S., & Mir, W. R. (2021). Nanomedicine in Human Health Therapeutics and Drug Delivery: Nanobiotechnology and Nanobiomedicine. In M. Bhat, I. Wani, & S. Ashraf (Eds.), *Applications of Nanomaterials in Agriculture, Food Science, and Medicine* (pp. 229–251). IGI Global. doi:10.4018/978-1-7998-5563-7.ch013

Mohammadzadeh, S., & Kim, Y. (2017). Nonlinear System Identification of Smart Buildings. In P. Samui, S. Chakraborty, & D. Kim (Eds.), *Modeling and Simulation Techniques in Structural Engineering* (pp. 328–347). Hershey, PA: IGI Global. doi:10.4018/978-1-5225-0588-4.ch011

Molina, G. J., Aktaruzzaman, F., Soloiu, V., & Rahman, M. (2017). Design and Testing of a Jet-Impingement Instrument to Study Surface-Modification Effects by Nanofluids. *International Journal of Surface Engineering and Interdisciplinary Materials Science*, *5*(2), 43–61. doi:10.4018/IJSEIMS.2017070104

Moreno-Rangel, A., & Carrillo, G. (2021). Energy-Efficient Homes: A Heaven for Respiratory Illnesses. In R. González-Lezcano (Ed.), *Advancements in Sustainable Architecture and Energy Efficiency* (pp. 49–71). IGI Global. https://doi.org/10.4018/978-1-7998-7023-4.ch003

Msomi, V., & Jantjies, B. T. (2021). Correlative Analysis Between Tensile Properties and Tool Rotational Speeds of Friction Stir Welded Similar Aluminium Alloy Joints. *International Journal of Surface Engineering and Interdisciplinary Materials Science*, *9*(2), 58–78. https://doi.org/10.4018/IJSEIMS.2021070104

Muigai, M. N., Mwema, F. M., Akinlabi, E. T., & Obiko, J. O. (2021). Surface Engineering of Materials Through Weld-Based Technologies: An Overview. In S. Roy & G. Bose (Eds.), *Advanced Surface Coating Techniques for Modern Industrial Applications* (pp. 247–260). IGI Global. doi:10.4018/978-1-7998-4870-7.ch011

Mukherjee, A., Saeed, R. A., Dutta, S., & Naskar, M. K. (2017). Fault Tracking Framework for Software-Defined Networking (SDN). In C. Singhal & S. De (Eds.), *Resource Allocation in Next-Generation Broadband Wireless Access Networks* (pp. 247–272). Hershey, PA: IGI Global. doi:10.4018/978-1-5225-2023-8.ch011

Mukhopadhyay, A., Barman, T. K., & Sahoo, P. (2018). Electroless Nickel Coatings for High Temperature Applications. In K. Kumar & J. Davim (Eds.), *Composites and Advanced Materials for Industrial Applications* (pp. 297–331). Hershey, PA: IGI Global. doi:10.4018/978-1-5225-5216-1.ch013

Mwema, F. M., & Wambua, J. M. (2022). Machining of Poly Methyl Methacrylate (PMMA) and Other Olymeric Materials: A Review. In K. Kumar, B. Babu, & J. Davim (Eds.), *Handbook of Research on Advancements in the Processing, Characterization, and Application of Lightweight Materials* (pp. 363–379). IGI Global. https://doi.org/10.4018/978-1-7998-7864-3.ch016

Mykhailyshyn, R., Savkiv, V., Boyko, I., Prada, E., & Virgala, I. (2021). Substantiation of Parameters of Friction Elements of Bernoulli Grippers With a Cylindrical Nozzle. *International Journal of Manufacturing, Materials, and Mechanical Engineering, 11*(2), 17–39. https://doi.org/10.4018/IJMMME.2021040102

Náprstek, J., & Fischer, C. (2017). Dynamic Stability and Post-Critical Processes of Slender Auto-Parametric Systems. In V. Plevris, G. Kremmyda, & Y. Fahjan (Eds.), *Performance-Based Seismic Design of Concrete Structures and Infrastructures* (pp. 128–171). Hershey, PA: IGI Global. doi:10.4018/978-1-5225-2089-4.ch006

Nautiyal, L., Shivach, P., & Ram, M. (2018). Optimal Designs by Means of Genetic Algorithms. In M. Ram & J. Davim (Eds.), *Soft Computing Techniques and Applications in Mechanical Engineering* (pp. 151–161). Hershey, PA: IGI Global. doi:10.4018/978-1-5225-3035-0.ch007

Nazir, R. (2017). Advanced Nanomaterials for Water Engineering and Treatment: Nano-Metal Oxides and Their Nanocomposites. In T. Saleh (Ed.), *Advanced Nanomaterials for Water Engineering, Treatment, and Hydraulics* (pp. 84–126). Hershey, PA: IGI Global. doi:10.4018/978-1-5225-2136-5.ch005

Nikolopoulos, C. D. (2021). Recent Advances on Measuring and Modeling ELF-Radiated Emissions for Space Applications. In C. Nikolopoulos (Ed.), *Recent Trends on Electromagnetic Environmental Effects for Aeronautics and Space Applications* (pp. 1–38). IGI Global. https://doi.org/10.4018/978-1-7998-4879-0.ch001

Nogueira, A. F., Ribeiro, J. C., Fernández de Vega, F., & Zenha-Rela, M. A. (2018). Evolutionary Approaches to Test Data Generation for Object-Oriented Software: Overview of Techniques and Tools. In M. Khosrow-Pour, D.B.A. (Ed.), Incorporating Nature-Inspired Paradigms in Computational Applications (pp. 162-194). Hershey, PA: IGI Global. https://doi.org/ doi:10.4018/978-1-5225-5020-4.ch006

Nwajana, A. O., Obi, E. R., Ijemaru, G. K., Oleka, E. U., & Anthony, D. C. (2021). Fundamentals of RF/Microwave Bandpass Filter Design. In A. Nwajana & I. Ihianle (Eds.), *Handbook of Research on 5G Networks and Advancements in Computing, Electronics, and Electrical Engineering* (pp. 149–164). IGI Global. https://doi.org/10.4018/978-1-7998-6992-4.ch005

Ogbodo, E. A. (2021). Comparative Study of Transmission Line Junction vs. Asynchronously Coupled Junction Diplexers. In A. Nwajana & I. Ihianle (Eds.), *Handbook of Research on 5G Networks and Advancements in Computing, Electronics, and Electrical Engineering* (pp. 326–336). IGI Global. https://doi.org/10.4018/978-1-7998-6992-4.ch013

Orosa, J. A., Vergara, D., Fraguela, F., & Masdías-Bonome, A. (2021). Statistical Understanding and Optimization of Building Energy Consumption and Climate Change Consequences. In R. González-Lezcano (Ed.), *Advancements in Sustainable Architecture and Energy Efficiency* (pp. 195–220). IGI Global. https://doi.org/10.4018/978-1-7998-7023-4.ch009

Osho, M. B. (2018). Industrial Enzyme Technology: Potential Applications. In S. Bharati & P. Chaurasia (Eds.), *Research Advancements in Pharmaceutical, Nutritional, and Industrial Enzymology* (pp. 375–394). Hershey, PA: IGI Global. doi:10.4018/978-1-5225-5237-6.ch017

Ouadi, A., & Zitouni, A. (2021). Phasor Measurement Improvement Using Digital Filter in a Smart Grid. In A. Recioui & H. Bentarzi (Eds.), *Optimizing and Measuring Smart Grid Operation and Control* (pp. 100–117). IGI Global. https://doi.org/10.4018/978-1-7998-4027-5.ch005

Padmaja, P., & Marutheswar, G. (2017). Certain Investigation on Secured Data Transmission in Wireless Sensor Networks. *International Journal of Mobile Computing and Multimedia Communications*, *8*(1), 48–61. doi:10.4018/IJMCMC.2017010104

Palmer, S., & Hall, W. (2017). An Evaluation of Group Work in First-Year Engineering Design Education. In R. Tucker (Ed.), *Collaboration and Student Engagement in Design Education* (pp. 145–168). Hershey, PA: IGI Global. doi:10.4018/978-1-5225-0726-0.ch007

Panchenko, V. (2021). Prospects for Energy Supply of the Arctic Zone Objects of Russia Using Frost-Resistant Solar Modules. In P. Vasant, G. Weber, & W. Punurai (Eds.), *Research Advancements in Smart Technology, Optimization, and Renewable Energy* (pp. 149-169). IGI Global. https://doi.org/10.4018/978-1-7998-3970-5.ch008

Panchenko, V. (2021). Photovoltaic Thermal Module With Paraboloid Type Solar Concentrators. *International Journal of Energy Optimization and Engineering*, *10*(2), 1–23. https://doi.org/10.4018/IJEOE.2021040101

Pandey, K., & Datta, S. (2021). Dry Machining of Inconel 825 Superalloys: Performance of Tool Inserts (Carbide, Cermet, and SiAlON). *International Journal of Manufacturing, Materials, and Mechanical Engineering*, *11*(4), 26–39. doi:10.4018/IJMMME.2021100102

Panneer, R. (2017). Effect of Composition of Fibers on Properties of Hybrid Composites. *International Journal of Manufacturing, Materials, and Mechanical Engineering*, *7*(4), 28–43. doi:10.4018/IJMMME.2017100103

Pany, C. (2021). Estimation of Correct Long-Seam Mismatch Using FEA to Compare the Measured Strain in a Non-Destructive Testing of a Pressurant Tank: A Reverse Problem. *International Journal of Smart Vehicles and Smart Transportation*, *4*(1), 16–28. doi:10.4018/IJSVST.2021010102

Paul, S., & Roy, P. (2018). Optimal Design of Power System Stabilizer Using a Novel Evolutionary Algorithm. *International Journal of Energy Optimization and Engineering*, *7*(3), 24–46. doi:10.4018/IJEOE.2018070102

Paul, S., & Roy, P. K. (2021). Oppositional Differential Search Algorithm for the Optimal Tuning of Both Single Input and Dual Input Power System Stabilizer. In P. Vasant, G. Weber, & W. Punurai (Eds.), *Research Advancements in Smart Technology, Optimization, and Renewable Energy* (pp. 256-282). IGI Global. https://doi.org/10.4018/978-1-7998-3970-5.ch013

Pavaloiu, A. (2018). Artificial Intelligence Ethics in Biomedical-Engineering-Oriented Problems. In U. Kose, G. Guraksin, & O. Deperlioglu (Eds.), *Nature-Inspired Intelligent Techniques for Solving Biomedical Engineering Problems* (pp. 219–231). Hershey, PA: IGI Global. doi:10.4018/978-1-5225-4769-3.ch010

Pioro, I., Mahdi, M., & Popov, R. (2017). Application of Supercritical Pressures in Power Engineering. In L. Chen & Y. Iwamoto (Eds.), *Advanced Applications of Supercritical Fluids in Energy Systems* (pp. 404–457). Hershey, PA: IGI Global. doi:10.4018/978-1-5225-2047-4.ch013

Plaksina, T., & Gildin, E. (2017). Rigorous Integrated Evolutionary Workflow for Optimal Exploitation of Unconventional Gas Assets. *International Journal of Energy Optimization and Engineering*, *6*(1), 101–122. doi:10.4018/IJEOE.2017010106

Popat, J., Kakadiya, H., Tak, L., Singh, N. K., Majeed, M. A., & Mahajan, V. (2021). Reliability of Smart Grid Including Cyber Impact: A Case Study. In R. Singh, A. Singh, A. Dwivedi, & P. Nagabhushan (Eds.), *Computational Methodologies for Electrical and Electronics Engineers* (pp. 163–174). IGI Global. https://doi.org/10.4018/978-1-7998-3327-7.ch013

Quiza, R., La Fé-Perdomo, I., Rivas, M., & Ramtahalsing, V. (2021). Triple Bottom Line-Focused Optimization of Oblique Turning Processes Based on Hybrid Modeling: A Study Case on AISI 1045 Steel Turning. In L. Burstein (Ed.), *Handbook of Research on Advancements in Manufacturing, Materials, and Mechanical Engineering* (pp. 215–241). IGI Global. https://doi.org/10.4018/978-1-7998-4939-1.ch010

Rahmani, M. K. (2022). Blockchain Technology: Principles and Algorithms. In S. Khan, M. Syed, R. Hammad, & A. Bushager (Eds.), *Blockchain Technology and Computational Excellence for Society 5.0* (pp. 16–27). IGI Global. https://doi.org/10.4018/978-1-7998-8382-1.ch002

Ramdani, N., & Azibi, M. (2018). Polymer Composite Materials for Microelectronics Packaging Applications: Composites for Microelectronics Packaging. In K. Kumar & J. Davim (Eds.), *Composites and Advanced Materials for Industrial Applications* (pp. 177–211). Hershey, PA: IGI Global. doi:10.4018/978-1-5225-5216-1.ch009

Ramesh, M., Garg, R., & Subrahmanyam, G. V. (2017). Investigation of Influence of Quenching and Annealing on the Plane Fracture Toughness and Brittle to Ductile Transition Temperature of the Zinc Coated Structural Steel Materials. *International Journal of Surface Engineering and Interdisciplinary Materials Science*, *5*(2), 33–42. doi:10.4018/IJSEIMS.2017070103

Robinson, J., & Beneroso, D. (2022). Project-Based Learning in Chemical Engineering: Curriculum and Assessment, Culture and Learning Spaces. In A. Alves & N. van Hattum-Janssen (Eds.), *Training Engineering Students for Modern Technological Advancement* (pp. 1–19). IGI Global. https://doi.org/10.4018/978-1-7998-8816-1.ch001

Rondon, B. (2021). Experimental Characterization of Admittance Meter With Crude Oil Emulsions. *International Journal of Electronics, Communications, and Measurement Engineering*, *10*(2), 51–59. https://doi.org/10.4018/IJECME.2021070104

Rudolf, S., Biryuk, V. V., & Volov, V. (2018). Vortex Effect, Vortex Power: Technology of Vortex Power Engineering. In V. Kharchenko & P. Vasant (Eds.), *Handbook of Research on Renewable Energy and Electric Resources for Sustainable Rural Development* (pp. 500–533). Hershey, PA: IGI Global. doi:10.4018/978-1-5225-3867-7.ch021

Sah, A., Bhadula, S. J., Dumka, A., & Rawat, S. (2018). A Software Engineering Perspective for Development of Enterprise Applications. In A. Elçi (Ed.), *Handbook of Research on Contemporary Perspectives on Web-Based Systems* (pp. 1–23). Hershey, PA: IGI Global. doi:10.4018/978-1-5225-5384-7.ch001

Sahli, Y., Zitouni, B., & Hocine, B. M. (2021). Three-Dimensional Numerical Study of Overheating of Two Intermediate Temperature P-AS-SOFC Geometrical Configurations. In G. Badea, R. Felseghi, & I. Aşchilean (Eds.), *Hydrogen Fuel Cell Technology for Stationary Applications* (pp. 186–222). IGI Global. https://doi.org/10.4018/978-1-7998-4945-2.ch008

Sahoo, P., & Roy, S. (2017). Tribological Behavior of Electroless Ni-P, Ni-P-W and Ni-P-Cu Coatings: A Comparison. *International Journal of Surface Engineering and Interdisciplinary Materials Science*, *5*(1), 1–15. doi:10.4018/IJSEIMS.2017010101

Sahoo, S. (2018). Laminated Composite Hypar Shells as Roofing Units: Static and Dynamic Behavior. In K. Kumar & J. Davim (Eds.), *Composites and Advanced Materials for Industrial Applications* (pp. 249–269). Hershey, PA: IGI Global. doi:10.4018/978-1-5225-5216-1.ch011

Sahu, H., & Hungyo, M. (2018). Introduction to SDN and NFV. In A. Dumka (Ed.), *Innovations in Software-Defined Networking and Network Functions Virtualization* (pp. 1–25). Hershey, PA: IGI Global. doi:10.4018/978-1-5225-3640-6.ch001

Salem, A. M., & Shmelova, T. (2018). Intelligent Expert Decision Support Systems: Methodologies, Applications, and Challenges. In T. Shmelova, Y. Sikirda, N. Rizun, A. Salem, & Y. Kovalyov (Eds.), *Socio-Technical Decision Support in Air Navigation Systems: Emerging Research and Opportunities* (pp. 215–242). Hershey, PA: IGI Global. doi:10.4018/978-1-5225-3108-1.ch007

Samal, M. (2017). FE Analysis and Experimental Investigation of Cracked and Un-Cracked Thin-Walled Tubular Components to Evaluate Mechanical and Fracture Properties. In P. Samui, S. Chakraborty, & D. Kim (Eds.), *Modeling and Simulation Techniques in Structural Engineering* (pp. 266–293). Hershey, PA: IGI Global. doi:10.4018/978-1-5225-0588-4.ch009

Samal, M., & Balakrishnan, K. (2017). Experiments on a Ring Tension Setup and FE Analysis to Evaluate Transverse Mechanical Properties of Tubular Components. In P. Samui, S. Chakraborty, & D. Kim (Eds.), *Modeling and Simulation Techniques in Structural Engineering* (pp. 91–115). Hershey, PA: IGI Global. doi:10.4018/978-1-5225-0588-4.ch004

Samarasinghe, D. A., & Wood, E. (2021). Innovative Digital Technologies. In J. Underwood & M. Shelbourn (Eds.), *Handbook of Research on Driving Transformational Change in the Digital Built Environment* (pp. 142–163). IGI Global. https://doi.org/10.4018/978-1-7998-6600-8.ch006

Sawant, S. (2018). Deep Learning and Biomedical Engineering. In U. Kose, G. Guraksin, & O. Deperlioglu (Eds.), *Nature-Inspired Intelligent Techniques for Solving Biomedical Engineering Problems* (pp. 283–296). Hershey, PA: IGI Global. doi:10.4018/978-1-5225-4769-3.ch014

Schulenberg, T. (2021). Energy Conversion Using the Supercritical Steam Cycle. In L. Chen (Ed.), *Handbook of Research on Advancements in Supercritical Fluids Applications for Sustainable Energy Systems* (pp. 659–681). IGI Global. doi:10.4018/978-1-7998-5796-9.ch018

Sezgin, H., & Berkalp, O. B. (2018). Textile-Reinforced Composites for the Automotive Industry. In K. Kumar & J. Davim (Eds.), *Composites and Advanced Materials for Industrial Applications* (pp. 129–156). Hershey, PA: IGI Global. doi:10.4018/978-1-5225-5216-1.ch007

Shaaban, A. A., & Shehata, O. M. (2021). Combining Response Surface Method and Metaheuristic Algorithms for Optimizing SPIF Process. *International Journal of Manufacturing, Materials, and Mechanical Engineering, 11*(4), 1–25. https://doi.org/10.4018/IJMMME.2021100101

Shafaati Shemami, M., & Sefid, M. (2022). Implementation and Demonstration of Electric Vehicle-to-Home (V2H) Application: A Case Study. In M. Alam, R. Pillai, & N. Murugesan (Eds.), *Developing Charging Infrastructure and Technologies for Electric Vehicles* (pp. 268–293). IGI Global. https://doi.org/10.4018/978-1-7998-6858-3.ch015

Shah, M. Z., Gazder, U., Bhatti, M. S., & Hussain, M. (2018). Comparative Performance Evaluation of Effects of Modifier in Asphaltic Concrete Mix. *International Journal of Strategic Engineering, 1*(2), 13–25. doi:10.4018/IJoSE.2018070102

Sharma, N., & Kumar, K. (2018). Fabrication of Porous NiTi Alloy Using Organic Binders. In K. Kumar & J. Davim (Eds.), *Composites and Advanced Materials for Industrial Applications* (pp. 38–62). Hershey, PA: IGI Global. doi:10.4018/978-1-5225-5216-1.ch003

Shivach, P., Nautiyal, L., & Ram, M. (2018). Applying Multi-Objective Optimization Algorithms to Mechanical Engineering. In M. Ram & J. Davim (Eds.), *Soft Computing Techniques and Applications in Mechanical Engineering* (pp. 287–301). Hershey, PA: IGI Global. doi:10.4018/978-1-5225-3035-0.ch014

Shmelova, T. (2018). Stochastic Methods for Estimation and Problem Solving in Engineering: Stochastic Methods of Decision Making in Aviation. In S. Kadry (Ed.), *Stochastic Methods for Estimation and Problem Solving in Engineering* (pp. 139–160). Hershey, PA: IGI Global. doi:10.4018/978-1-5225-5045-7.ch006

Siero González, L. R., & Romo Vázquez, A. (2017). Didactic Sequences Teaching Mathematics for Engineers With Focus on Differential Equations. In M. Ramírez-Montoya (Ed.), *Handbook of Research on Driving STEM Learning With Educational Technologies* (pp. 129–151). Hershey, PA: IGI Global. doi:10.4018/978-1-5225-2026-9.ch007

Sim, M. S., You, K. Y., Esa, F., & Chan, Y. L. (2021). Nanostructured Electromagnetic Metamaterials for Sensing Applications. In M. Bhat, I. Wani, & S. Ashraf (Eds.), *Applications of Nanomaterials in Agriculture, Food Science, and Medicine* (pp. 141–164). IGI Global. https://doi.org/10.4018/978-1-7998-5563-7.ch009

Singh, R., & Dutta, S. (2018). Visible Light Active Nanocomposites for Photocatalytic Applications. In K. Kumar & J. Davim (Eds.), *Composites and Advanced Materials for Industrial Applications* (pp. 270–296). Hershey, PA: IGI Global. doi:10.4018/978-1-5225-5216-1.ch012

Skripov, P. V., Yampol'skiy, A. D., & Rutin, S. B. (2021). High-Power Heat Transfer in Supercritical Fluids: Microscale Times and Sizes. In L. Chen (Ed.), *Handbook of Research on Advancements in Supercritical Fluids Applications for Sustainable Energy Systems* (pp. 424–450). IGI Global. https://doi.org/10.4018/978-1-7998-5796-9.ch012

Sözbilir, H., Özkaymak, Ç., Uzel, B., & Sümer, Ö. (2018). Criteria for Surface Rupture Microzonation of Active Faults for Earthquake Hazards in Urban Areas. In N. Ceryan (Ed.), *Handbook of Research on Trends and Digital Advances in Engineering Geology* (pp. 187–230). Hershey, PA: IGI Global. doi:10.4018/978-1-5225-2709-1.ch005

Stanciu, I. (2018). Stochastic Methods in Microsystems Engineering. In S. Kadry (Ed.), *Stochastic Methods for Estimation and Problem Solving in Engineering* (pp. 161–176). Hershey, PA: IGI Global. doi:10.4018/978-1-5225-5045-7.ch007

Strebkov, D., Nekrasov, A., Trubnikov, V., & Nekrasov, A. (2018). Single-Wire Resonant Electric Power Systems for Renewable-Based Electric Grid. In V. Kharchenko & P. Vasant (Eds.), *Handbook of Research on Renewable Energy and Electric Resources for Sustainable Rural Development* (pp. 449–474). Hershey, PA: IGI Global. doi:10.4018/978-1-5225-3867-7.ch019

Sukhyy, K., Belyanovskaya, E., & Sukhyy, M. (2021). *Basic Principles for Substantiation of Working Pair Choice*. IGI Global. doi:10.4018/978-1-7998-4432-7.ch002

Suri, M. S., & Kaliyaperumal, D. (2022). Extension of Aspiration Level Model for Optimal Planning of Fast Charging Stations. In A. Fekik & N. Benamrouche (Eds.), *Modeling and Control of Static Converters for Hybrid Storage Systems* (pp. 91–106). IGI Global. https://doi.org/10.4018/978-1-7998-7447-8.ch004

Tallet, E., Gledson, B., Rogage, K., Thompson, A., & Wiggett, D. (2021). Digitally-Enabled Design Management. In J. Underwood & M. Shelbourn (Eds.), *Handbook of Research on Driving Transformational Change in the Digital Built Environment* (pp. 63–89). IGI Global. https://doi.org/10.4018/978-1-7998-6600-8.ch003

Terki, A., & Boubertakh, H. (2021). A New Hybrid Binary-Real Coded Cuckoo Search and Tabu Search Algorithm for Solving the Unit-Commitment Problem. *International Journal of Energy Optimization and Engineering*, *10*(2), 104–119. https://doi.org/10.4018/IJEOE.2021040105

Tüdeş, Ş., Kumlu, K. B., & Ceryan, S. (2018). Integration Between Urban Planning and Natural Hazards For Resilient City. In N. Ceryan (Ed.), *Handbook of Research on Trends and Digital Advances in Engineering Geology* (pp. 591–630). Hershey, PA: IGI Global. doi:10.4018/978-1-5225-2709-1.ch017

Ulamis, K. (2018). Soil Liquefaction Assessment by Anisotropic Cyclic Triaxial Test. In N. Ceryan (Ed.), *Handbook of Research on Trends and Digital Advances in Engineering Geology* (pp. 631–664). Hershey, PA: IGI Global. doi:10.4018/978-1-5225-2709-1.ch018

Valente, M., & Milani, G. (2017). Seismic Assessment and Retrofitting of an Under-Designed RC Frame Through a Displacement-Based Approach. In V. Plevris, G. Kremmyda, & Y. Fahjan (Eds.), *Performance-Based Seismic Design of Concrete Structures and Infrastructures* (pp. 36–58). Hershey, PA: IGI Global. doi:10.4018/978-1-5225-2089-4.ch002

Vargas-Bernal, R. (2021). Advances in Electromagnetic Environmental Shielding for Aeronautics and Space Applications. In C. Nikolopoulos (Ed.), *Recent Trends on Electromagnetic Environmental Effects for Aeronautics and Space Applications* (pp. 80–96). IGI Global. https://doi.org/10.4018/978-1-7998-4879-0.ch003

Vasant, P. (2018). A General Medical Diagnosis System Formed by Artificial Neural Networks and Swarm Intelligence Techniques. In U. Kose, G. Guraksin, & O. Deperlioglu (Eds.), *Nature-Inspired Intelligent Techniques for Solving Biomedical Engineering Problems* (pp. 130–145). Hershey, PA: IGI Global. doi:10.4018/978-1-5225-4769-3.ch006

Verner, C. M., & Sarwar, D. (2021). Avoiding Project Failure and Achieving Project Success in NHS IT System Projects in the United Kingdom. *International Journal of Strategic Engineering*, *4*(1), 33–54. https://doi.org/10.4018/IJoSE.2021010103

Verrollot, J., Tolonen, A., Harkonen, J., & Haapasalo, H. J. (2018). Challenges and Enablers for Rapid Product Development. *International Journal of Applied Industrial Engineering*, *5*(1), 25–49. doi:10.4018/IJAIE.2018010102

Wan, A. C., Zulu, S. L., & Khosrow-Shahi, F. (2021). Industry Views on BIM for Site Safety in Hong Kong. In J. Underwood & M. Shelbourn (Eds.), *Handbook of Research on Driving Transformational Change in the Digital Built Environment* (pp. 120–140). IGI Global. https://doi.org/10.4018/978-1-7998-6600-8.ch005

Yardimci, A. G., & Karpuz, C. (2018). Fuzzy Rock Mass Rating: Soft-Computing-Aided Preliminary Stability Analysis of Weak Rock Slopes. In N. Ceryan (Ed.), *Handbook of Research on Trends and Digital Advances in Engineering Geology* (pp. 97–131). Hershey, PA: IGI Global. doi:10.4018/978-1-5225-2709-1.ch003

You, K. Y. (2021). Development Electronic Design Automation for RF/Microwave Antenna Using MATLAB GUI. In A. Nwajana & I. Ihianle (Eds.), *Handbook of Research on 5G Networks and Advancements in Computing, Electronics, and Electrical Engineering* (pp. 70–148). IGI Global. https://doi.org/10.4018/978-1-7998-6992-4.ch004

Yousefi, Y., Gratton, P., & Sarwar, D. (2021). Investigating the Opportunities to Improve the Thermal Performance of a Case Study Building in London. *International Journal of Strategic Engineering*, *4*(1), 1–18. https://doi.org/10.4018/IJoSE.2021010101

Zindani, D., & Kumar, K. (2018). Industrial Applications of Polymer Composite Materials. In K. Kumar & J. Davim (Eds.), *Composites and Advanced Materials for Industrial Applications* (pp. 1–15). Hershey, PA: IGI Global. doi:10.4018/978-1-5225-5216-1.ch001

Zindani, D., Maity, S. R., & Bhowmik, S. (2018). A Decision-Making Approach for Material Selection of Polymeric Composite Bumper Beam. In K. Kumar & J. Davim (Eds.), *Composites and Advanced Materials for Industrial Applications* (pp. 112–128). Hershey, PA: IGI Global. doi:10.4018/978-1-5225-5216-1.ch006

About the Contributors

Florentin Smarandache has published many papers and books on mathematics and science. See UNM's website: http://fs.gallup.unm.edu/FlorentinSmarandache.htm.bf1e70e1-e209-4fbc-86bf-9b367ca6636a

* * *

Fakhry Asad Agusfrianto is an undergraduate student who loves mathematics, especially algebra. The research topics he studies are hyperstructures, ring theory, and module theory.

Evangelos Athanassopoulos is a Chemical Engineer and Topographer of the National Technical University of Athens, Greece, as well as an independent researcher in the area of Philosophy of Science. He used to work for ten years as manager of the service department of the Hellenic Labware, a commercial company of chemical and medical instruments. Since 1995 he has been involved in the chemical analysis of wines driving forward the technology of this sector. Among others, Evangelos constructed an innovative hydrostatic balance of high accuracy connected to the proper software for measuring the alcoholic degree, as well as a modern glass-made plate distillatory supplied with a sophisticated reflux system. He has also proposed corrections and improvements for the international rules of examining the quality of wine (OIV). Since 1995 he has published articles in local newspapers and in social media on a variety of scientific topics. Started from 2005, he is the author of three books on human intelligence and logic and of a number of papers published in reputed international journals . His research interests are currently focused on applications of Probability, Statistics and Fuzzy Logic to philosophy and morals.

Bhimraj Basumatary is an Assistant Professor, Department of Math. Sciences, Bodoland University.

Said Broumi completed his MSc and PhD in Computer Science from University of Hassan II of Casablanca, Morocco. Currently, he is serving as an Assistant Professor at Regional Center for the Professions of Education and Training, Casablanca-Settat, Morocco. He is permanent member at Laboratory of Information Processing, Faculty of Science Ben M'Sik, University Hassan II, Casablanca, Morocco .His research interests are in the field of graph theory, extended fuzzy graphs, decision analysis and neutrosophic theory. He has diversity in his research work. he has published several work on neutrosophic graph theory, soft set theory, multi-attribute decision analysis and some alternative theories of Neutrosophic Mathematics. He has published more than 200 research articles in international peer-reviewed ISI Indexed /Impact factor journals. Some of his papers have been published in high impact journals including, Complex & Intelligent Systems, Computational and Applied Mathematics, Symmetry, and Journal of Neutrosophic Set and Systems. His work has largely been cited and total citation is more than 5992 (Scholar Google, H-index 41). He has presented his research work at some national as well as international conferences. He has been acting as a reviewer for a number of International journals. He is also an Editor in chief of the following journals: 1. International Journal of Neutrosophic Set and Systems: 2. International Journal of Neutrosophic Science: https://www.scopus.com/authid/detail.uri?authorId=55919041000.5bfacc0c-cf57-4602-9bfe-5285f6e73fd6

Somen Debnath has received his Ph.D degree in Mathematics(2020)from Tripura University(A Central University),Suryamaninagar, India, under the supervision of Prof. Anjan Mukherjee, ex-pro vice chancellor and faculty of Tripura University. Currently he is working as a Post Graduate Teacher in Mathematics at Umakanta Academy. His research interest includes fuzzy set, soft set, game theory, topology, decision making, hypersoft set, etc.

Joachim Feuerstein was born in Germany. He obtained university degrees in German and French Literature, Philosophy, Theology and Musicology. He worked as a teacher and inspector of Pedagogy in schools of France and Germany, and later as an advisor of the German Minister of Foreign Affairs for the study of the German language as a foreign language in Greece, Cyprus and Algeria. He was also involved in theater staging and used to be a violinist playing in many musical bands. He has published several articles and books on topics related to Ethics, Political Philosophy, History, Pedagogy and Theology. He is currently retired living in Saarbrucken, Germany and in Achaia, Greece.

Erick González Caballero received his PhD degree (2013) in technical science from the Technological University of Havana in Cuba, and his BS degree (1998) in Mathematics from the University of Havana. After graduating, he works as a scientist or teacher at various universities and research centers in Cuba. In 2011 he continued his teaching and scientific activities at the Technological University of Havana. He was a Postdoctoral Visitor (2016) at the University of Oviedo, Spain. His research interests include, decision-making under uncertainty, rough sets, fuzziness, image processing, expert systems, cooperative game theory, among others. He has published in important journals like "Group Decision and Negotiation", "International Journal of Computational Intelligence Systems", and "Journal of Intelligent & Fuzzy Systems", also in book chapters of Atlantis Press and Springer.

Carlos Granados got his master's degree at Universidad del Atl\'{a}ntico in 2019, later in 2021 he started his doctorate in Mathematics at Universidad de Antioquia. He has published many papers associated to general topology, mathematical education, neutrosophic theory, decision making and sequences spaces.

Yudi Mahatma was born on October 20, 1976, studied at the ITB Mathematics Department in 1993 and earned a Bachelor of Science degree in 1998 with the topic of module homomorphism thesis. In 1999 he continued his studies at the ITB Mathematics Postgraduate Program and earned a Master of Science degree in 2002 with the topic of Category Equivalence thesis. In 2004 the author started his career as a non-permanent lecturer at the Mathematics Study Program, Faculty of Science and Technology, UIN Jakarta. In 2008 he was accepted at the Jakarta State University and was appointed as a Civil Servant in 2010. In 2014 the author continued his studies at the ITB Mathematics Doctoral Program with a scholarship from BPPDN Dikti. The author conducted research related to representation theory and earned his Doctoral degree on March 4 2020. During his doctoral studies the author has produced several scientific papers on the topic of structural algebra which were presented in seminars and published.

Prem Kumar Singh is a researchers, speaker and well known academician for his work on graph analytics, uncertainty measurement and knowledge processing tasks. Currently, he is working as an Associate Professor at Department of Computer Science and Engineering, GITAM University-Visakhapatnam (Demmed-to-be-University). Before Joining the GITAM, Dr. Prem Kumar Singh was associated with Amity School of Engineering and Technology, Amity University-Noida Sector 125 as a capacity of Assistant Professor after completion of a Post-Doc from University of Malaya under the grant UM.C/625/1/HIR/MOE/FCSIT/03 (). In this way Dr. Prem Kumar has more than 12 years of experience under difference capacity like teaching, research and administration as NAAC, WASC, SCOPUS, NIRF data compilation as well as other activities.

Michael G. Voskoglou (B.Sc., M.Sc., M.Phil., Ph.D. in Mathematics) is currently an Emeritus Professor of Mathematical Sciences at the Graduate Technological Educational Institute of Western Greece. He used to be a Visiting Researcher at the Bulgarian Academy of Sciences (1997-2000), a Visiting Professor at the Universities of Warsaw (2009), of Applied Sciences of Berlin (2010) and at the National Institute of Technology of Durgapur, India (2016). Prof. Voskoglou is the author of 16 authored and 3 edited books and of more than 500 papers published in reputed mathematical journals and proceedings of international conferences of about 30 countries in the five continents (2020 Google Scholar Citations: 2401, H-index 20). He is also the Editor in Chief of the "International Journal of Applications of Fuzzy Sets and Artificial Intelligence" (ISSN 2241-1240), reviewer of the AMS and member of the Editorial Board or referee in many mathematical journals. His research interests include Algebra, Fuzzy Sets, Markov Chains, Artificial Intelligence, Philosophy of Science and Mathematics Education.

Index

www.ingramcontent.com/pod-product-compliance
Lightning Source LLC
LaVergne TN
LVHW080855300926
849162LV00003B/424